Springer
Berlin
Heidelberg
New York
Barcelona
Budapest
Hong Kong
London
Milan
Paris
Singapore
Tokyo

D. L. Mills

Nonlinear Optics

Basic Concepts

Second, Enlarged Edition
With 45 Figures

Springer

Professor D. L. Mills
Department of Physics and Astronomy
University of California
Irvine, CA 92697, USA
e-mail: dlmills@uci.edu

ISBN 3-540-64182-3 Springer-Verlag Berlin Heidelberg New York

ISBN 3-540-54192-6 First Edition Springer-Verlag Berlin Heidelberg New York

Library of Congress Cataloging-in-Publication Data. Mills, D. L. Nonlinear optics: basic concepts / D. L. Mills. – 2nd, enl. ed. p. cm. Includes bibliographical references and index. ISBN 3-540-64182-3 (alk. paper) 1. Nonlinear optics. I. Title. QC446.2.M55 1998 535'.2–dc21 98-6385 CIP

Typesetting: Edwards Bros. Inc., Ann Arbor, MI 48106, USA. The LaTeX files of the new Chapters 8, 9 and Appendix C of this second edition have been reformatted and edited by Kurt Mattes, Heidelberg, using the MathTime fonts.
Cover design: *design & production* GmbH, Heidelberg

SPIN 10481606 54/3143 – 5 4 3 2 1 0 – Printed on acid-free paper

Preface to the Second Edition

Since the book was first published in 1991, the field of surface nonlinear optics has grown substantially to the point where an exposition of the principles of this field will prove useful to many. Thus, in this second edition, Chapter 8 addresses this area. Also, optical probes of magnetism of very thin films and multilayers are now widely used, and magneto-optic devices of increasing sophistication have appeared. Chapter 9 is thus devoted to magneto-optics, and associated nonlinear phenomena. The earlier chapter on "Chaos" appears as Chapter 10. The philosophy which underlies the first edition was also employed in the writing of the two new chapters.

Irvine, CA
March 1998

D. L. Mills

Preface to the First Edition

One intriguing aspect of physics is its dynamic and rapidly evolving nature; exciting new fields can become moribund within relatively few years, only to revive and grow again in a dramatic and explosive manner in response to new developments. This has been the case for the fields of optics and atomic physics. In the 1950s, and perhaps into the early 1960s, both fields appeared mature, fully developed, and perhaps even a bit dull as a consequence. The appearance of the laser has turned both of these fields into dynamic areas of research, within which fundamental and profound questions are being explored. The research of the past two or three decades has led also to very important applications and to new devices. The dye laser, which enables a very narrow line to be tuned over an appreciable spectral range, has led to a virtual revolution in the spectroscopy of atoms, molecules, and the condensed phases of matter.

A parallel development, readily detectable in the recent literature of theoretical physics, has been the substantial advance in our understanding of highly nonlinear phenomena. Numerous texts are devoted to exposition of the theoretical methods which may be used to extract useful information from the important equations encountered in the various fields of physics.

A survey of the contemporary literature of nonlinear optics shows that in this area one encounters a large fraction of the basic equations and principles of nonlinear physics. For example, analysis of self-induced transparency leads to the sine-Gordon equation, and that of soliton propagation in optical fibers to the nonlinear Schrödinger equation. Other examples can be found as well. As a consequence, one has in hand real data which illustrate basic properties of those solutions of these important equations with no counterpart in linear response theory or in perturbation theoretic analyses of nonlinear terms. Other concepts central to nonlinear optics, such as the role of phase matching in various wave mixing experiments, are of very considerable importance in other subfields of physics. The field of nonlinear optics is thus a superb laboratory within which the student may encounter and explore key notions of nonlinear physics of general importance, while at the same time learning the foundations of a most important and fundamental area of contemporary physics.

Furthermore, the above issues are not addressed in many introductory graduate courses in electromagnetic theory; the emphasis is usually placed on more classical topics developed in an earlier era. This volume has its origin in a course given at Irvine by the author, directed toward the student who has completed the first year electromagnetic theory sequence. It is intended as an extension of interest not only to students who wish to pursue thesis research in optics or laser spectroscopy but also to the general student whose ultimate research specialty may lie within a very different subfield. The focus is then on general principles, with many technical points that are important to the specialist played down or set aside. It is the author's experience that existing texts on nonlinear optics use specialized terms that are not defined fully and present introductions to important basic issues that are perhaps too concise to be grasped easily by the general reader. The purpose of this volume is to bridge the gap between the classic texts on electromagnetic theory, which omit systematic exploration of modern optics, and the (often excellent!) specialized texts full of discussion essential to those who are pursuing research in the area, but which are too detailed and too terse for the general student of physics.

The specialist may thus find important topics omitted from this volume, or discussions of a number of technical points a bit incomplete. The intention here is to provide an overview; the literature on nonlinear optics is sufficiently complete and accessible that those who wish to pursue particular aspects in more detail can proceed further without difficulty.

It is important also for the reader to acquire an understanding of the optical properties of various solid materials, to appreciate the reasons for the choice of samples for a given experiment, and the constraints that limit one's ability to explore various phenomena. Also, one must acquire a certain vocabulary in this arena to understand the literature. Chapter 2, which is quite lengthy, presents an overview of the optical properties of materials, since it is the experience of the author that many students have a very limited grasp of this area, though they may know the mathematical details associated with various models of matter.

It is hoped that this volume will broaden the horizons of graduate students in the physical sciences, by introducing them to the fascinating field of nonlinear optics, and at the same time provide them with an introduction to general aspects of the physics of nonlinear systems.

Irvine, CA *D.L. Mills*
March 1991

Contents

1. Introductory Remarks

Throughout most of this volume, we shall be describing the interaction of electromagnetic radiation with matter within a macroscopic framework; we shall always be concerned with electric fields whose spatial variation involves length scales very large compared to the size of the atoms and molecules that are the basic constituents of the material of interest. In this regard, we note that in the visible portion of the spectrum, the wavelength of light is roughly 5×10^{-5} cm, while a typical molecular bond length or crystal lattice constant is 3×10^{-8} cm. We may then proceed by studying the solutions of Maxwell's equations, which have the form

$$\nabla \cdot \boldsymbol{D} = 0, \quad \nabla \cdot \boldsymbol{B} = 0\,, \tag{1.1a}$$

$$\nabla \times \boldsymbol{E} = -\frac{1}{c}\frac{\partial \boldsymbol{B}}{\partial t}, \quad \nabla \times \boldsymbol{H} = \frac{1}{c}\frac{\partial \boldsymbol{D}}{\partial t}\,. \tag{1.1b}$$

We assume no external charge is present, and no external current is present as well, so we set $\rho = \boldsymbol{J} = 0$.

To proceed, we require relations between the various fields which enter Maxwell's equations. In the limit of interest, we have [1.1]

$$\boldsymbol{D} = \boldsymbol{E} + 4\pi\boldsymbol{P} \tag{1.2a}$$

and

$$\boldsymbol{B} = \boldsymbol{H} + 4\pi\boldsymbol{M}\,, \tag{1.2b}$$

where $\boldsymbol{P}$ and $\boldsymbol{M}$ are the electric dipole moment per unit volume, and magnetic moment per unit volume, respectively. Until we reach Chap. 9, we shall confine our attention entirely to nonmagnetic media, and set $\boldsymbol{M} \equiv 0$. Nonlinearities very similar to those explored through Chap. 8 have their origin in the magnetic degrees of freedom of appropriate materials. Their study constitutes the very important field of magnetooptics.

We then have

$$\nabla \cdot \boldsymbol{E} + 4\pi\nabla \cdot \boldsymbol{P} = 0 \tag{1.3a}$$

and with $\boldsymbol{B} = \boldsymbol{H}$, the two remaining Maxwell equations are easily combined to give

$$\nabla^2\boldsymbol{E} - \nabla(\nabla \cdot \boldsymbol{E}) - \frac{1}{c^2}\frac{\partial^2\boldsymbol{E}}{\partial t^2} - \frac{4\pi\partial^2\boldsymbol{P}}{c^2\partial t^2} = 0\,. \tag{1.3b}$$

To proceed, we require information on the relationship between $\boldsymbol{P}$ and $\boldsymbol{E}$. In principle, one requires a full microscopic theory of the response of a particular material to relate $\boldsymbol{E}$, the macroscopic electric field [1.1], to the dipole moment per unit volume.

It would appear that by our neglect of the term involving the current density $\boldsymbol{J}$ on the right-hand side of the $\nabla \times \boldsymbol{H}$ Maxwell equation, our attention will be confined exclusively to insulating materials, rather than conductors such as metals or doped semiconductors. We shall see shortly that this is not the case; the conduction currents stimulated by time varying electric fields can in fact be viewed as contributing to the dipole moment per unit volume, as we shall come to appreciate. It is the case, however, that many of the nonlinear optical phenomena to be explored here require that the various beams involved have a long path length in the medium of interest. Thus, our attention will be directed largely toward applications to insulating materials that are nominally transparent at the relevant wavelengths. The reader should keep in mind that one encounters very interesting nonlinear optical phenomena in metals and in doped semiconductors, as discussed in the text by *Shen* [1.2].

One may proceed by noting the following: The largest electric fields encountered in practice fall into the range of 10^6 V/cm; most forms of matter exhibit electrical breakdown for fields in excess of this value. An electron bound to an atom or molecule, or moving through a solid or dense liquid, experiences electric fields in the range of 10^9 V/cm. This follows by noting that, over distances the order of an angstrom, the change in electrostatic potential can be several electron volts. The laboratory fields of interest are then small compared to the electric fields experienced by the electrons in the atoms and molecules from which dense matter is constructed. In this circumstance, we can expand the dipole moment per unit volume in a Taylor series in powers of the macroscopic field $\boldsymbol{E}$. For the moment, we suppose the dipole moment per unit volume $\boldsymbol{P}(\boldsymbol{r}, t)$, depends on the electric field $\boldsymbol{E}$ at the same point $\boldsymbol{r}$, and at the same time. We shall see shortly that in real materials, this assumption is overly restrictive. There are important qualitative implications of a more realistic relation between $\boldsymbol{P}$ and $\boldsymbol{E}$. But this simple assertion will allow us to begin the discussion. Then the αth Cartesian component of the dipole moment per unit volume, $P_\alpha(\boldsymbol{r}, t)$, is a function of the three Cartesian components of the electric field, $E_\beta(\boldsymbol{r}, t)$. Here α and β range over x, y, and z. The Taylor series then takes the form

$$P_\alpha(\boldsymbol{r}, t) = P_\alpha^{(0)} + \sum_\beta \left(\frac{\partial P_\alpha}{\partial E_\beta}\right)_0 E_\beta + \frac{1}{2!}\sum_{\beta\gamma}\left(\frac{\partial^2 P_\alpha}{\partial E_\beta \partial E_\gamma}\right)_0 E_\beta E_\gamma + \frac{1}{3!}\sum_{\beta\gamma\delta}\left(\frac{\partial^3 P_\alpha}{\partial E_\beta \partial E_\gamma \partial E_\delta}\right)_0 E_\beta E_\gamma E_\delta + \cdots . \tag{1.4}$$

In most of the common materials we encounter, the first term, which is the electric dipole moment per unit volume in zero electric field, vanishes identically. The usual situation is a dielectric material, within which the dipole moment in zero field vanishes; any dipole moment present is then induced by the external field.

There is an important class of materials, known as ferroelectrics, which possess a spontaneous electric dipole moment in zero field [1.3]. These are electrical analogues of the better known ferromagnets, which possess a spontaneous magnetization per unit volume $\boldsymbol{M}$. In a ferromagnet, as the material is heated, the spontaneous magnetization $\boldsymbol{M}$ decreases in magnitude, to vanish at a certain temperature T_c, known as the Curie temperature. Ferroelectrics behave in a similar manner; the polarization $\boldsymbol{P}$ will decrease with increasing temperature, again to vanish above a critical temperature.

In the common ferroelectrics, the origin of the spontaneous dipole moment is in the shift of an ion from a high symmetry site to a low symmetry site as the temperature is lowered. For example, in the ferroelectric material $BaTiO_3$, in the high temperature phase for which $P_\alpha^{(0)} = 0$, the Ti^{2+} ion sits at the center of an oxygen octahedron. As the temperature is lowered below the critical temperature, the Ti^{2+} shifts off this high symmetry site, so the Ti–O complex acquires a net electric dipole moment. This is a collective phenomenon; all the Ti^{2+} ions shift in a coherent manner, so the crystal as a whole acquires an electric dipole moment. Similarly, in KH_2PO_4, it is the H that shifts.

Such materials are influenced strongly by dc or low frequency electric fields, particularly near their transition temperature, because the unstable ionic species respond strongly to an applied electric field. Ferroelectrics are thus useful in a variety of electro-optic devices.

In a ferroelectric, the dipole moment per unit volume, $\boldsymbol{P}^{(0)}$, in the absence of an electric field, is independent of time, but may vary with position in the sample. In general, the presence of a nonzero polarization leads to the presence of a static, macroscopic electric field, $\boldsymbol{E}^{(0)}(\boldsymbol{r})$. This static field obeys the two equations

$$\nabla \times \boldsymbol{E}^{(0)} = 0 \tag{1.5a}$$

and

$$\nabla \cdot \boldsymbol{E}^{(0)} = -4\pi \nabla \cdot \boldsymbol{P}^{(0)} . \tag{1.5b}$$

Such fields may be analyzed by the methods of electrostatics [1.4]. We may write, from (1.5 a), $\boldsymbol{E}^{(0)} = -\nabla \phi^{(0)}$, and (1.5 b) becomes Poisson's equation with effective charge density $\rho_p = -\nabla \cdot \boldsymbol{P}^{(0)}$.

Our interest will be in the study of various time-dependent phenomena, in response to externally applied electric fields, usually with their origin in incident laser radiation. Thus the first term in (1.4) and the static fields it may generate in a ferroelectric are of little interest. The second and subsequent terms describe the influence of a time-dependent, macroscopic electric field; in what follows, we consider only these contributions to the dipole moment per unit volume and their consequences, assuming the effect of $\boldsymbol{P}^{(0)}$ and the static electric fields it generates are accounted for as described above, if necessary.

We may then treat the response of dielectric and ferroelectric materials within the same framework. For each, we will henceforth write

$$P_\alpha(\boldsymbol{r}, t) = \sum_\beta \chi^{(1)}_{\alpha\beta} E_\beta(\boldsymbol{r}, t) + \sum_{\beta\gamma} \chi^{(2)}_{\alpha\beta\gamma} E_\beta(\boldsymbol{r}, t) E_\gamma(\boldsymbol{r}, t) + \sum_{\beta\gamma\delta} \chi^{(3)}_{\alpha\beta\gamma\delta} E_\beta(\boldsymbol{r}, t) E_\gamma(\boldsymbol{r}, t) E_\delta(\boldsymbol{r}, t) + \cdots , \tag{1.6}$$

where $\chi^{(2)}_{\alpha\beta\gamma}$, $\chi^{(3)}_{\alpha\beta\gamma\delta}$ are referred to as the second and third order susceptibilities, respectively, and $\chi^{(1)}_{\alpha\beta}$ is the susceptibility tensor of ordinary dieletric theory.

As the notation in (1.6) indicates, the various susceptibilities are tensor objects. Thus, $\chi^{(1)}_{\alpha\beta}$ is a second rank tensor. For an isotropic material, such as a gas or liquid, $\chi^{(1)}_{\alpha\beta}$ is diagonal: $\chi^{(1)}_{\alpha\beta} = \chi^{(1)}\delta_{\alpha\beta}$. It is also established easily that for a cubic crystal, $\chi^{(1)}_{\alpha\beta}$ is diagonal. Since both $\boldsymbol{P}$ and $\boldsymbol{E}$ are vectors, and thus are odd under inversion symmetry, $\chi^{(2)}_{\alpha\beta\gamma}$ must vanish in any material that is left invariant in form under inversion. This is the case for liquids, gases, and for a number of common crystals such as the alkali halides, and also for the semiconductors Si and Ge. Notice that the fourth rank tensor $\chi^{(3)}_{\alpha\beta\gamma\delta}$ has the same transformation properties as the elastic constants of elasticity theory. One may consult analyses of elastic constants, in any given case, to investigate which elements of this tensor are nonvanishing [1.5]. *Shen* [1.2] has given very useful compilations of the nonzero elements of $\chi^{(2)}_{\alpha\beta\gamma}$ and $\chi^{(3)}_{\alpha\beta\gamma\delta}$ for crystals of various symmetry.

We separate P_α in (1.6) into two pieces, one linear in the electric field, and one nonlinear:

$$P_\alpha(\boldsymbol{r}, t) = P^{(\mathrm{L})}_\alpha(\boldsymbol{r}, t) + P^{(\mathrm{NL})}_\alpha(\boldsymbol{r}, t) , \tag{1.7a}$$

where

$$P^{(\mathrm{L})}_\alpha(\boldsymbol{r}, t) = \sum_\beta \chi^{(1)}_{\alpha\beta} E_\beta(\boldsymbol{r}, t) \tag{1.7b}$$

and

$$P^{(\mathrm{NL})}_\alpha(\boldsymbol{r}, t) = \sum_{\beta\gamma} \chi^{(2)}_{\alpha\beta\gamma} E_\beta(\boldsymbol{r}, t) E_\gamma(\boldsymbol{r}, t) + \sum_{\beta\gamma\delta} \chi^{(3)}_{\alpha\beta\gamma\delta} E_\beta(\boldsymbol{r}, t) E_\gamma(\boldsymbol{r}, t) E_\delta(\boldsymbol{r}, t) \cdots . \tag{1.7c}$$

If one retains only $P^{(\mathrm{L})}_\alpha(\boldsymbol{r}, t)$ in the analysis, and combines this with the Maxwell equations (1.3) then one obtains a description of electromagnetic wave propagation in media, possibly crystalline in nature, and thus described by an electric susceptibility tensor $\chi_{\alpha\beta}$. The combination $\epsilon_{\alpha\beta} = \delta_{\alpha\beta} + 4\pi\chi_{\alpha\beta}$ is the dielectric tensor of the material. Here $\delta_{\alpha\beta}$ is the Kronecker delta function, equal to unity when the two subscripts are the same, and zero otherwise. We shall assume that the reader is familiar with electromagnetic theory at this level, which is covered thoroughly in numerous excellent texts [1.1, 6].[1] However, we will cover below those aspects of the topic essential for the present discussion.

The calculation of the various nonlinear susceptibility tensor elements requires a proper microscopic theory of the material in question. One can write down formal expressions for these quantities,[2] but the resulting expressions are formidable in appearance. At the time of writing, evaluations of these formulae for anything more than schematic models of materials remain at an early stage. The simple models, however, prove most useful as a means of outlining the basic physical properties which control these parameters. We shall be content simply to regard the various coefficients in (1.7) as phenomenological parameters.

We conclude with one final remark. Throughout the present discussion, the electric field $\boldsymbol{E}(\boldsymbol{r}, t)$ is the macroscopic electric field, defined as in Jackson's text [1.1]. Quite often in the literature, one encounters descriptions of the linear or nonlinear response of a material based on a picture which models the atomic or molecular structure explicitly. One then relates the dipole moment per unit volume, our $P_\alpha(\boldsymbol{r}, t)$, to that p_α of an atomic or molecular constituent; p_α may be written as a series similar to (1.6), where now in place of $\chi^{(1)}_{\alpha\beta}$, $\chi^{(2)}_{\alpha\beta\gamma}$, etc., one has the various linear and nonlinear polarizabilities of the individual constituents.

In this case, the electric field which enters the expansion is not the macroscopic electric field here, but the local field which acts on the individual entity; this is the external field, supplemented by that produced by the induced dipole moments which surround the entity in question. Clearly, the local field may differ substantially in value from the macroscopic field. However, it is the case that the local field is always proportional to the macroscopic field; a consequence is that the dipole moment per unit volume may always be expressed as an expansion in powers of the macroscopic field, as we do here. Particular models such as those just described can be very useful, but they can be applied only to materials in which the various atomic and molecular constituents are well separated and well defined (gases, some liquids, rare gas crystals, molecular crystals, alkali halide crystals, . . .). In dense materials, with extended chemical bonds such as those found in semiconductors, it is not clear how to isolate a basic entity, and how to calculate the local field within a simple model that is also meaningful. We thus prefer to phrase our discussion

---

[1] An excellent account of the electrodynamics of crystals, within which the tensor character of the dielectric response is accounted for, is found in [1.6].

[2] A rather general microscopic description of the nonlinear optical response of materials has been given by *Armstrong* et al. [1.7].

entirely within the phenomenological framework where all nonlinear polarizabilities are related to the macroscopic field.

We now turn to a more detailed discussion of the linear response characteristics of materials, and their nonlinear response, before we enter descriptions of nonlinear optical processes.

Problems

1.1 A sphere of radius R is fabricated from ferroelectric material, and has a spatially uniform polarization $\boldsymbol{P} = \hat{\boldsymbol{z}}P_0$ parallel to the z direction everywhere.

(a) Find $\boldsymbol{D}$ and $\boldsymbol{E}$ inside and outside the material.

(b) A positive ion is attracted to, and sticks to one pole of the sphere. Calculate the work required to remove the ion and carry it off to infinity, if it is singly charged and $P_0 = 1.5 \times 10^4$ cgs units.

1.2 A point charge Q is placed at the origin of an isotropic nonlinear dielectric. Thus, both $\boldsymbol{D}$ and $\boldsymbol{E}$ are in the radial direction, by symmetry. One has, with D_r and E_r the radial fields, $D_r = E_r + 4\pi\chi^{(1)}E_r + 4\pi\chi^{(3)}E_r^3$. Assume $\chi^{(3)} > 0$, and discuss the behavior of E_r and D_r, with attention to the limits $r \to \infty$, $r \to 0$.

1.3 In an anisotropic dielectric exposed to static fields, $\chi_{\alpha\beta}^{(1)}$ is symmetric and thus we can always find a set of principal axes $x_0y_0z_0$ within which this tensor is diagonal. Such a material is placed between two parallel metal plates of infinite extent, separated by the distance D. The space between the plates is filled with the dielectric. The z axis is normal to the plates, which are parallel to the xy plane. The z_0 axis makes an angle θ with respect to the z axis, while y and y_0 coincide. Find $\boldsymbol{D}$ and $\boldsymbol{E}$ everywhere, if V is the voltage difference between the metal plates.

2. Linear Dielectric Response of Matter

In this chapter, we discuss aspects of the linear dielectric response of matter; our attention is thus directed toward the contribution to the dipole moment per unit volume displayed in (1.7 b). For much of what follows here, the tensor nature of the dielectric response is of little explicit interest, so we assume for the moment the dielectric susceptibility tensor $\chi^{(1)}_{\alpha\beta}$ introduced in Chap. 1 is diagonal, $\chi^{(1)}_{\alpha\beta} = \delta_{\alpha\beta}\chi$. Then the discussion in Chap. 1 provides the following simple relationship between $\boldsymbol{P}$ and $\boldsymbol{E}$:

$$P_\alpha(\boldsymbol{r}, t) = \chi E_\alpha(\boldsymbol{r}, t) . \tag{2.1}$$

This relationship, which provides the basis for elementary dielectric theory, is clearly unphysical, if taken literally. The macroscopic field $\boldsymbol{E}(\boldsymbol{r}, t)$ may be viewed as a "driving field" to which the electrons and nuclei in the material respond, and rearrange themselves. The result is the induced dipole moment $\boldsymbol{P}(\boldsymbol{r}, t)$. The relationship in (2.1) assumes that the system responds instantaneously to the applied field. In any physical system, a finite time is required for the system to respond to an external driving force. The dipole moment we measure at time t is the consequence of the response of the system to the electric field over some characteristic time interval τ in the recent past. The statement in (2.1) must thus be generalized to incorporate the time lag in the response of the system.

This is accomplished by replacing (2.1) by

$$P_\alpha(\boldsymbol{r}, t) = \int_{-\infty}^{+\infty} dt' \chi(t - t') E_\alpha(\boldsymbol{r}, t') , \tag{2.2}$$

where $\chi(t - t')$ is a function that is nonzero for values of $t - t'$ the order of the characteristic response time τ. The polarizaton we measure at time t clearly is a consequence of the presence of the electric field at past times; the system, of course, does not respond to the future behavior of the electric field. Thus, for any physical system, one has

$$\chi(t - t') \equiv 0 \quad \text{for } t' > t , \tag{2.3}$$

a very simple and obvious statement that has remarkable implications, as we will see below.

The physical significance of $\chi(t - t')$ can be appreciated by supposing the system has been subjected to an impulsive electric field applied at the time t_0:

$$E_\alpha(\boldsymbol{r}, t') = E_\alpha^{(0)}\delta(t - t_0) . \tag{2.4}$$

Then for $t > t_0$,

$$P_\alpha(\boldsymbol{r}, t) = \chi(t - t_0)E_\alpha^{(0)} . \tag{2.5}$$

The function $\chi(t - t_0)$ thus describes the time variation of the dipole moment, after the system has been subjected to a sharp, impulsive "blow." After such a blow, the polarization will decay to zero, possibly ringing or oscillating in the process. For any physical system, a microscopic theory is required to provide us with a description of $\chi(t - t_0)$, though we shall see that, in some circumstances, we can learn a great deal from general principles.

Upon noting (2.3), we can rewrite (2.2) in the form

$$P_\alpha(\boldsymbol{r}, t) = \int_0^\infty dt''\chi(t'')E_\alpha(\boldsymbol{r}, t - t'') , \tag{2.6}$$

where, as remarked earlier, $\chi_\alpha(t'')$ falls to zero when $t'' \gg \tau$.

Suppose $E_\alpha(\boldsymbol{r}, t - t'')$ varies slowly with t'', on the time scale set by τ. We may then replace $E_\alpha(\boldsymbol{r}, t - t'')$ simply with $E_\alpha(\boldsymbol{r}, t)$ to excellent approximation, and (2.6) reduces to (2.1), with

$$\chi = \int_0^\infty dt''\, \chi(t'') . \tag{2.7}$$

Thus, we recover the basic relationship of elementary dielectric theory if the applied electric field varies sufficiently slowly in time. We may refer to such fields as quasi-static in nature.

Our discussion of the inadequacy of (2.1) is not yet complete. In any real material, the dipole moment $P_\alpha(\boldsymbol{r}, t)$ at point $\boldsymbol{r}$ depends not only on the electric field at precisely the point $\boldsymbol{r}$, but on the electric field at other points in space near the point $\boldsymbol{r}$. That this is so may be appreciated from the schematic illustration of the microscopic picture of dense matter given in Fig. 2.1. The figure shows a drawing of a crystal lattice, formed from molecules placed in a regular array. The molecules are linked by chemical bonds, with their origin in the

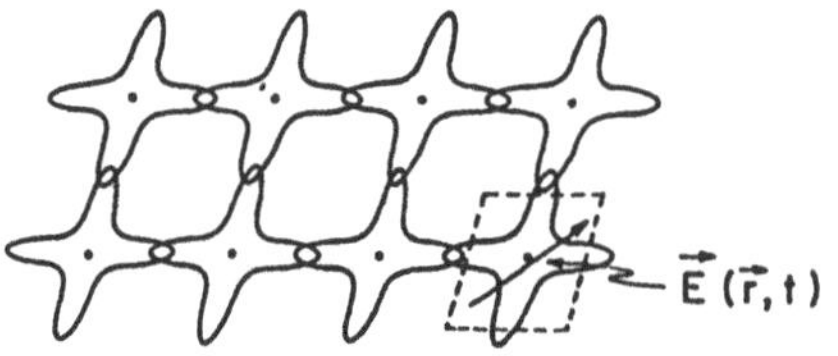

Fig. 2.1. A 2D sketch of molecules in a crystal lattice. The discussion in the text supposes that an electric field is applied to one single molecule of the crystal; the electric field is thus supposed nonzero only within the dotted region

overlap of the electronic wave functions associated with neighboring molecules.

Now imagine an electric field that is well localized in space is applied to the system. We suppose the electric field is nonzero only within the dotted box, within which one of the molecules is located. The electrons within this molecule will be redistributed, and the position of the (positively charged) nuclei will shift. A consequence is that the molecule acquires an electric dipole moment. But since this molecule is bonded to its neighbors, the electronic rearrangement just described will induce modifications in the structure and shifts in nuclear positions in the neighboring molecules. The neighbors thus acquire an electric dipole moment also. An electric field that is highly localized in space not only induces dipole moments within the region where it is nonzero, but polarizes the near vicinity as well.

It follows that the relationship between the electric field and the dipole moment per unit volume must be nonlocal in space, i.e., the dipole moment $\boldsymbol{P}(\boldsymbol{r}, t)$ at point $\boldsymbol{r}$ depends not only on the behavior of the electric field at point $\boldsymbol{r}$, but also on the nature of the electric field in its near vicinity. We thus make one further extension of (2.2) by writing, returning to acknowledge the tensor character of the dielectric response,

$$P_\alpha(\boldsymbol{r}, t) = \sum_\beta \int dt'\, d^3r'\, \chi_{\alpha\beta}(\boldsymbol{r} - \boldsymbol{r}', t - t') E_\beta(\boldsymbol{r}', t') \,. \tag{2.8}$$

If the medium is homogeneous in nature, then $\chi_{\alpha\beta}$ will depend only on the difference $\boldsymbol{r} - \boldsymbol{r}'$, and not on $\boldsymbol{r}$ and $\boldsymbol{r}'$ separately. If there are, for example, variations in density or composition, then a more complex description must be employed.

If the electric field exhibits a slow variation in space, on the length scale that characterizes the range of $\chi_{\alpha\beta}$, and also slow variation in time, we then may replace $E_\beta(\boldsymbol{r}', t')$ in (2.8) by $E_\beta(\boldsymbol{r}, t)$, and we then recover (1.7 b), where now

$$\begin{aligned} \chi_{\alpha\beta} &= \int dt'\, d^3r'\, \chi_{\alpha\beta}(\boldsymbol{r} - \boldsymbol{r}', t - t') \\ &= \int dt''\, d^3r''\, \chi_{\alpha\beta}(\boldsymbol{r}'', t'') \,. \end{aligned} \tag{2.9}$$

We shall frequently employ a Fourier decomposition of the electric fields:

$$E_\beta(\boldsymbol{r}, t) = \int \frac{d^3k\, d\omega}{(2\pi)^4} E_\beta(\boldsymbol{k}, \omega) \mathrm{e}^{i\boldsymbol{k}\cdot\boldsymbol{r}} \mathrm{e}^{-i\omega t} \,. \tag{2.10}$$

It follows from (2.8) that one may write

$$P_\alpha(\boldsymbol{r}, t) = \int \frac{d^3k\, d\omega}{(2\pi)^4} P_\alpha(\boldsymbol{k}, \omega) \mathrm{e}^{i\boldsymbol{k}\cdot\boldsymbol{r}} \mathrm{e}^{-i\omega t} \,, \tag{2.11}$$

where

$$P_\alpha(\boldsymbol{k}, \omega) = \sum_\beta \chi_{\alpha\beta}(\boldsymbol{k}, \omega) E_\beta(\boldsymbol{k}, \omega) \tag{2.12}$$

and

$$\chi_{\alpha\beta}(\boldsymbol{k}, \omega) = \int d^3r\, dt\, \chi_{\alpha\beta}(\boldsymbol{r}, t) \mathrm{e}^{-i\boldsymbol{k}\cdot\boldsymbol{r}} \mathrm{e}^{+i\omega t} \,. \tag{2.13}$$

It is more usual, in linear dielectric theory, to consider the relationship between the displacement field $\boldsymbol{D}$ and the electric field $\boldsymbol{E}$, since it is the former which enters Maxwell's equations directly. From (1.2 a) one has

$$D_\alpha(\boldsymbol{k}, \omega) = \sum_\beta \varepsilon_{\alpha\beta}(\boldsymbol{k}, \omega) E_\beta(\boldsymbol{k}, \omega) \,, \tag{2.14}$$

where

$$\varepsilon_{\alpha\beta}(\boldsymbol{k}, \omega) = \delta_{\alpha\beta} + 4\pi\chi_{\alpha\beta}(\boldsymbol{k}, \omega) \tag{2.15}$$

is the dielectric tensor of the medium.

The expression in (2.15) shows that if we consider the propagation of a plane wave of frequency ω and wave vector $\boldsymbol{k}$ in the material, the dielectric tensor depends separately on the frequency and wave vector of the disturbance. The frequency variation of the dielectric tensor can be overlooked only in the quasi-static limit discussed above. Similarly one sees from (2.13) that the wave vector dependence is negligible in circumstances where the wavelength of the wave is very long compared to the spatial range of $\chi_{\alpha\beta}(\boldsymbol{r}, t)$.

The frequency variation of the dielectric response of materials is a well-known and commonly discussed property, though in some discussions it is not related to the real time response characteristics of the medium as directly as we have done. The wave vector dependence is discussed less commonly because, in many (but by no means in all) circumstances it plays a minor role. A medium for which account must be taken of the wave vector dependence of the dielectric tensor is said to exhibit spatial dispersion. We refer the reader to the text by *Agranovich* and *Ginzburg* [2.1] for a complete discussion of the role of spatial dispersion in the optical response of materials.

The next issue is the nature of the frequency and wave vector dependence of the dielectric tensor.

Any physical system contains characteristic frequencies and time scales. For an atom, we have the frequencies $\omega_{mn} = (E_m - E_n)/\hbar$ that are associated with transitions between quantum states with energy E_m and E_n. These generally lie in the visible or ultra-violet region of the spectrum. In a molecule, we have in addition to these electronic transitions the vibrational normal modes, which lie in the infrared range. In condensed media, such as solids and liquids, the collection of vibrational normal modes and electronic transitions form continuous bands that lie in roughly the same spectral region as those associated with their microscopic constituents. The point of these remarks is that, when we turn to our discusson of nonlinear optical phenomena in materials, the frequencies of the waves of interest are comparable to those characteristic of the internal degrees of freedom of the medium in which the waves propagate. Under these circumstances, we must take due account of the presence of the frequency dependence of the dielectric tensor. We shall see that, in the end, this enters our considerations in a crucial and central manner.

From the discussion which centered around Fig. 2.1, one can appreciate that the spatial range of $\chi_{\alpha\beta}(\boldsymbol{r}, t)$ is microscopic. Application of an electric field within the dotted box of Fig. 2.1 will lead to a disturbance in the crystal that extends well beyond the box, but which nonetheless is confined to a few molecular diameters from the box. Thus, the wavelength of the optical waves of interest to us will be very long compared to the range of $\chi_{\alpha\beta}(\boldsymbol{r}, t)$. The wavelength of visible light, 5000 Å, is a few thousand times longer than intermolecular bond lengths or separations (2–3 Å), and in the infrared wavelengths are many microns. Under these circumstances, we may frequently overlook the influence of the wave vector dependence of the dieletric tensor, and replace $\varepsilon_{\alpha\beta}(\boldsymbol{k}, \omega)$ by $\varepsilon_{\alpha\beta}(0, \omega)$, which for simplicity we shall write as $\varepsilon_{\alpha\beta}(\omega)$.

There are numerous instances, however, where the neglect of spatial dispersion effects is inappropriate, as we discuss later in the present chapter. The role of spatial dispersion in both the linear and nonlinear optical response of materials is a rich and fascinating topic that is being pursued actively at the time of this writing, though we touch on it only lightly here.

2.1 Frequency Dependence of the Dielectric Tensor

In this section, we consider in more detail the general features expected in the frequency dependence of the dielectric tensor of materials. This is a complex topic, so our remarks will be confined to an overview. For simplicity, we also continue to focus our attention on an isotropic material characterized by a simple scalar dielectric function $\varepsilon(\omega)$.

We have, for this case, the relation between $\boldsymbol{D}(\boldsymbol{r}, t)$ and $\boldsymbol{E}(\boldsymbol{r}, t)$ in the form

$$\boldsymbol{D}(\boldsymbol{r},t) = \int_{-\infty}^{+\infty} dt'\,\varepsilon(t - t')\boldsymbol{E}(\boldsymbol{r}, t') , \tag{2.16}$$

where

$$\varepsilon(\omega) = \int_{-\infty}^{+\infty} d\tau\, e^{+i\omega\tau}\varepsilon(\tau) , \tag{2.17}$$

and as discussed above, $\varepsilon(\tau) = 0$ for $\tau < 0$, since the medium responds only to fields at past times.

Since $\boldsymbol{D}$ and $\boldsymbol{E}$ are real, it follows that $\varepsilon(\tau)$ is purely real. However, $\varepsilon(\omega)$ is in general complex. One writes customarily

$$\varepsilon(\omega) = \varepsilon_1(\omega) + \mathrm{i}\varepsilon_2(\omega) , \tag{2.18}$$

where ε_1 and ε_2 are real.

The imaginary part, $\varepsilon_2(\omega)$, has a clear physical interpretation. Suppose the system is subjected to a purely harmonic field with frequency ω:

$$\boldsymbol{E}(\boldsymbol{r}, t) = \boldsymbol{E}_\omega(\boldsymbol{r})\mathrm{e}^{-\mathrm{i}\omega t} + \boldsymbol{E}_\omega(\boldsymbol{r})^*\mathrm{e}^{+\mathrm{i}\omega t} . \tag{2.19}$$

Then

$$\boldsymbol{D}(\boldsymbol{r}, t) = \varepsilon(\omega)\boldsymbol{E}_\omega(\boldsymbol{r})\mathrm{e}^{-\mathrm{i}\omega t} + \varepsilon^*(\omega)\boldsymbol{E}_\omega(\boldsymbol{r})^*\mathrm{e}^{+\mathrm{i}\omega t} , \tag{2.20}$$

where it follows from (2.17) that $\varepsilon(-\omega) = \varepsilon^*(\omega)$.

If U_E is the energy per unit volume stored in the electric field and the polarization it induces in the medium, then the time rate of change of U_E is given by [2.2]

$$\frac{\partial U_E}{\partial t} = \frac{1}{4\pi}\frac{\partial \boldsymbol{D}}{\partial t}\cdot \boldsymbol{E} . \tag{2.21}$$

A short calculation gives the following expression for the time average of $(\partial U_E/\partial t)$, which is the rate at which energy is dissipated in the presence of the electric field:

$$\left\langle \frac{\partial U_E}{\partial t} \right\rangle = -\frac{\mathrm{i}\omega}{4\pi}[\varepsilon(\omega) - \varepsilon^*(\omega)]|\boldsymbol{E}_\omega(\boldsymbol{r})|^2 , \tag{2.22}$$

or

$$\left\langle \frac{\partial U_E}{\partial t} \right\rangle = \frac{\omega}{2\pi}\varepsilon_2(\omega)|\boldsymbol{E}_\omega(\boldsymbol{r})|^2 . \tag{2.23}$$

The presence of the imaginary part of the dielectric constant thus has the consequence that energy is absorbed by the medium, when a time-dependent electric field is present. Clearly, the right-hand side of (2.23) must be positive definite. Thus, if $\omega > 0$, we must also have $\varepsilon_2(\omega) > 0$. From the definition of $\varepsilon(\omega)$ in (2.17), one sees directly that $\varepsilon_2(\omega)$ is an odd function of frequency, while $\varepsilon_1(\omega)$ is even. [Recall $\varepsilon(\tau)$ is real.]

As we have seen, causality requires $\varepsilon(\tau)$ to vanish identically for $\tau < 0$, for any simple physical system. From this fact, and (2.17), it follows that $\varepsilon(\omega)$, considered now a function of complex frequency, is an analytic function of ω in the upper half of the complex ω plane. This simple property leads to a set of remarkable relationships between $\varepsilon_1(\omega)$ and $\varepsilon_2(\omega)$, called the Kramers-Kronig relations.[1] We consider next implications of these statements, which read as follows:

$$\varepsilon_1(\omega) = 1 + \frac{2}{\pi}\mathcal{P}\int_0^\infty \frac{\Omega\varepsilon_2(\Omega)}{(\Omega^2 - \omega^2)}\,d\Omega \tag{2.24 a}$$

and

$$\varepsilon_2(\omega) = \frac{2\omega}{\pi}\mathcal{P}\int_0^\infty \frac{[\varepsilon_1(\Omega) - 1]}{(\omega^2 - \Omega^2)}d\Omega . \tag{2.24 b}$$

In these expressions, the integrals are Cauchy principal values.

These results are quoted in a form appropriate to an insulating material, in which the imaginary part of the dielectric constant vanishes as $\omega \to 0$. If a medium has finite conductivity in the limit $\omega \to 0$, $\varepsilon_2(\omega)$ diverges as $1/\omega$ in this limit [note the relation between the dielectric tensor and the conductivity given in Appendix A, in (A.5)]. It is possible to modify the Kramers-Kronig relations in a manner that applies to conducting media [2.3].

Thus, the real and imaginary parts of the complex dielectric constant are not independent but linked by the relations in (2.24). It follows that any physical picture of matter consistent with causality must produce forms that satisfy the Kramers-Kronig relations. With rather little effort, one may draw conclusions from these expressions of considerable generality.

For example, except at possibly isolated frequencies, the integral on the right-hand side of (2.24 b) will in general be nonzero. Thus, save for the limit $\omega \to 0$, $\varepsilon_2(\omega)$ is in general nonzero. Thus, any physical medium can never be perfectly transparent, since the energy stored in the electric field associated with any electromagnetic wave will be dissipated as the wave propagates through the material, in the manner described by (2.23). We shall use the term nominally transparent to describe materials for which $\varepsilon_2(\omega)$ is very small in the frequency range of interest, so the propagation length of the wave is very long.

The absorption spectrum of a medium provides us with information about the frequency variation of $\varepsilon_2(\omega)$, as we see from (2.23). Once this is known, the frequency variation of $\varepsilon_1(\omega)$ is determined by (2.24 a).

Consider the following example, which is encountered in a variety of different physical systems, as we shall appreciate from the discussion to follow. Suppose the frequency ω lies in the near vicinity of a sharp absorption line, centered at frequency ω_0. We may imagine this line is well separated from other absorption lines or bands of the material, as illustrated in Fig. 2.2. The material may have low-frequency absorption features which lie below ω_m, and

[1] For a derivation of the Kramers-Kronig relation, in a form appropriate to the frequency dependent dielectric constant, see [2.3].

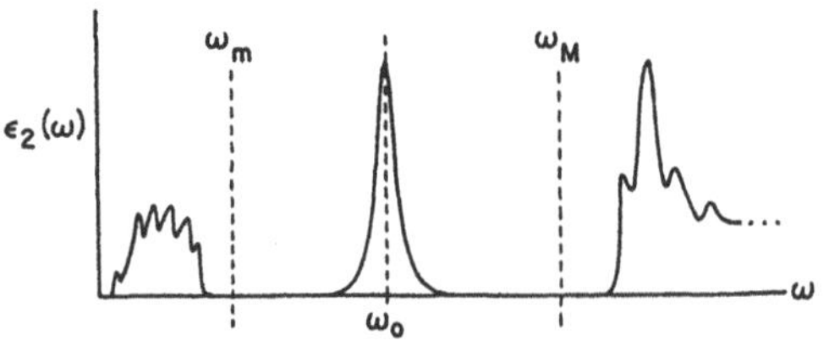

Fig. 2.2. Absorption spectrum of material with a sharp absorption line at frequency ω_0, well separated from low frequency absorptions which lie below ω_m in frequency, and also well separated from high frequency absorption bands which lie above ω_M

high-frequency structures which lie above ω_M. If we consider Na vapor as an example, the D lines which lie in the yellow portion of the visible spectrum lie well below absorption lines that describe transitions to states higher in energy than the $3p$ state that serves as the final state for the transitions associated with the D lines. There are no low-frequency absorption bands in this case. For a vapor formed from molecules with a strong optical transition in the visible, the low-frequency features below ω_m may be the vibrational-rotation bands, which lie in the infrared.

Provided ω lies outside the absorption line centered at ω_0 then to very good approximation, $\varepsilon_1(\omega)$ is given by

$$\varepsilon_1(\omega) = 1 + \frac{2}{\pi}\int_{\omega_M}^{\infty} \frac{\Omega\varepsilon_2(\Omega)}{\Omega^2 - \omega^2}\,d\Omega + \frac{1}{\omega_0^2 - \omega^2}\left[\frac{2}{\pi}\int_{\omega_0} \Omega\varepsilon_2(\Omega)d\Omega\right]$$
$$+ \frac{2}{\pi}\int_0^{\omega_m} \frac{\Omega\varepsilon_2(\Omega)}{\Omega^2 - \omega^2}\,d\Omega\,. \tag{2.25}$$

The integral in square brackets in (2.25) is assumed to range over only the absorption line centered at ω_0.

If ω lies near ω_0, and $\omega_0 \ll \omega_M$, then ω may be set to zero in the first term of (2.25). Furthermore, if $\omega_m \ll \omega_0$, in general the last term will be very small and can be neglected. We then define

$$\varepsilon_\infty = 1 + \frac{2}{\pi}\int_{\omega_M}^{\infty} \frac{\varepsilon_2(\Omega)}{\Omega}\,d\Omega \tag{2.26a}$$

and

$$\Omega_p^2 = \frac{2}{\pi}\int_{\omega_0} \Omega\varepsilon_2(\Omega)d\Omega\,, \tag{2.26b}$$

to find

$$\varepsilon_1(\omega) = \varepsilon_\infty + \frac{\Omega_p^2}{\omega_0^2 - \omega^2}\,. \tag{2.27}$$

The form in (2.27) describes the frequency variation of $\varepsilon_1(\omega)$ for ω close to an isolated narrow absorption line, no matter what the physical origin of the feature. If ω is close to ω_0, but still well outside of the absorption line profile, $\varepsilon_2(\omega)$ will be very small, so $\varepsilon(\omega) \cong \varepsilon_1(\omega)$.

The presence of an absorption line in any physical medium thus introduces dramatic structure in the frequency variation of the dielectric constant. The high frequency transitions, well above ω_0, give a frequency independent "background dielectric" constant ε_∞, and the presence of the line itself leads to the resonant term.

In a variety of texts, one sees expressions such as that in (2.27) derived from very special, often very schematic or naive physical pictures of the microscopic constituents of matter. For example, we may imagine an atom to be composed of an infinitely massive and thus immobile nucleus, to which a negatively charged electron is bound by a harmonic spring. If $\boldsymbol{u}$ is the displacement of the electron produced by an applied electric field $\boldsymbol{E}$, and ω_0 is the resonance frequency associated with the spring, then

$$\boldsymbol{u}'' + \omega_0^2 \boldsymbol{u} = \frac{e}{m} \boldsymbol{E} , \tag{2.28}$$

with e the electron charge and m its mass. Then if the electric field has frequency ω, one has

$$\boldsymbol{u} = \frac{e}{m} \frac{1}{\omega_0^2 - \omega^2} \boldsymbol{E} \tag{2.29}$$

with the dipole moment per unit volume given by

$$\boldsymbol{P} = ne\boldsymbol{u} = \frac{ne^2}{m} \frac{1}{\omega_0^2 - \omega^2} \boldsymbol{E} . \tag{2.30}$$

with n the number of atoms per unit volume. Upon observing that $\boldsymbol{D} = \boldsymbol{E} + 4\pi\boldsymbol{P}$, one has from this picture a frequency dependent dielectric constant from the model whose form is identical to that in (2.27), which we have derived without resort to a particular microscopic model.

Evidently, in the nineteenth century, mechanical models of the atom rather similar to that just sketched were developed, and taken seriously. The motivation was the experimental observation that the real part of the dielectric constant, considered a function of frequency, often exhibited resonant behavior similar to that displayed in (2.27). Such models have been discussed recently in Stone's text on optics [2.4].

The expression in (2.27) applies for frequencies ω which lie close to a sharp absorption line, but still outside the absorption profile of the line. To find the behavior of $\varepsilon_1(\omega)$ as one tunes the frequency ω through the line, we need the explicit form of $\varepsilon_2(\omega)$. Many theories produce Lorentzian profiles for the line shape. With the assumption the line profile is Lorentzian, the Kramers-Kronig integral may be evaluated in closed form, and a complete (but now model dependent) expression for $\varepsilon_1(\omega)$ follows: For example, suppose in the vicinity of the absorption line centered at ω_0, $\varepsilon_2(\omega)$ is given by

$$\varepsilon_2(\omega) = \frac{\Omega_p^2}{2\omega_0}\left[\frac{\gamma/2}{(\omega-\omega_0)^2+(\gamma/2)^2} - \frac{\gamma/2}{(\omega+\omega_0)^2+(\gamma/2)^2}\right], \tag{2.31}$$

where γ is the full width of the absorption line at half maximum, and Ω_p^2 serves as a measure of its integrated strength. We recall, in writing down (2.31), that $\varepsilon_2(\omega)$ is an odd function of frequency. If we absorb the influence of high frequency transitions in a "background" dielectric constant ε_∞, then the combination of (2.31) with (2.24 a) gives

$$\varepsilon_1(\omega) = \varepsilon_\infty + \Omega_p^2\left[\frac{\omega_0^2-\omega^2+(\gamma/2)^2}{[(\omega_0-\omega)^2+(\gamma/2)^2][(\omega_0+\omega)^2+(\gamma/2)^2]}\right], \tag{2.32}$$

which agrees in form with (2.27), if $|\omega_0 - \omega|$ is large compared to the linewidth γ. If the line is very narrow, then for frequencies near ω_0, (2.32) is well approximated by

$$\varepsilon_1(\omega) \cong \varepsilon_\infty + \frac{\Omega_p^2}{2\omega_0}\frac{\omega_0-\omega}{(\omega_0-\omega)^2+(\gamma/2)^2}. \tag{2.33}$$

In Fig. 2.3, we give a sketch of the behavior of $\varepsilon_1(\omega)$ and $\varepsilon_2(\omega)$, near the resonance at ω_0. Notice that above ω_0, $\varepsilon_1(\omega)$ can become negative, if Ω_p^2 is sufficiently large, or the linewidth γ sufficiently small.

In Appendix A, we present a microscopic derivation of the dielectric constant, through use of quantum mechanics. In the limit that the wave vector $\boldsymbol{k} \to 0$, for a material whose dielectric constant is isotropic in this limit, the general expression in (A.40) becomes

$$\varepsilon(\omega) = 1 + \frac{8\pi}{\hbar V}\sum_n \frac{\omega_{n0}\left|\left\langle \psi_n\left|\sum_j e_j r_\alpha^j\right|\psi_0\right\rangle\right|^2}{[\omega_{n0}^2-(\omega+i\eta)^2]}, \tag{2.34}$$

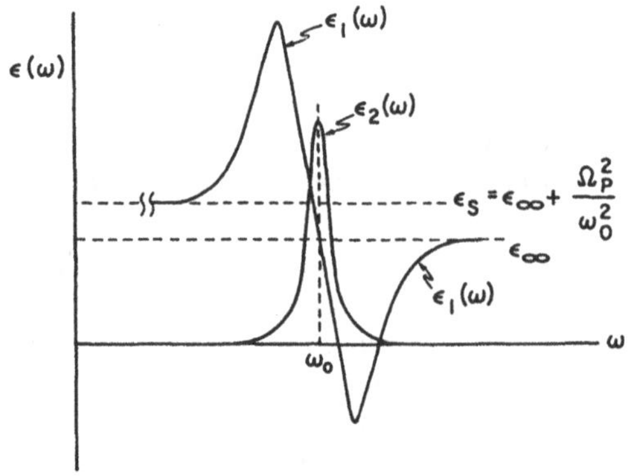

Fig. 2.3. The behavior of the real part, $\varepsilon_1(\omega)$, and imaginary part, $\varepsilon_2(\omega)$, of the complex dielectric constant, in the near vicinity of a sharp absorption line centered at the frequency ω_0

where the sum over n ranges over all the excited states of the medium. Upon recalling the identity, for η infinitesimal [2.5]

$$\frac{1}{x \mp i\eta} = P\frac{1}{x} \pm i\pi\delta(x) ,$$

and writing

$$\frac{\omega_{n0}}{\omega_{n0}^2 - (\omega + i\eta)^2} = \frac{1}{2}\left[\frac{1}{\omega_{n0} - \omega - i\eta} + \frac{1}{\omega_{n0} + \omega + i\eta}\right] .$$

Equation (2.34) gives

$$\varepsilon_1(\omega) = 1 + \frac{8\pi}{\hbar V}\sum_n \frac{\omega_{n0}\left|\left\langle \psi_n \left| \sum_j e_j r_\alpha^j \right| \psi_0 \right\rangle\right|^2}{\omega_{n0}^2 - \omega^2} \tag{2.35 a}$$

and

$$\varepsilon_2(\omega) = \frac{4\pi^2}{\hbar V}\sum_n \left|\left\langle \psi_n \left| \sum_j e_j r_\alpha^j \right| \psi_0 \right\rangle\right|^2 [\delta(\omega - \omega_{n0}) - \delta(\omega + \omega_{n0})] .$$

These expressions (of course!) are compatible with the Kramers-Kronig relations.

We see from (2.35 a) that an excited state that $|\psi_n\rangle$ whose energy is removed from that of other excited states gives a resonant contribution to $\varepsilon_1(\omega)$, with form identical to that displayed in (2.27). In dense matter, as discussed below, the energy spectrum of the excited states in general forms a continuum, so we have absorption bands that extend over various frequency ranges characteristic of the material.

We pause to comment on the behavior of electromagnetic waves which propagate in a medium described by the dielectric constant in (2.27).

The electromagnetic waves that propagate in an isotropic dielectric are purely transverse in nature, as they are in free space. Later on in this chapter, we shall see that in annisotropic dielectrics, where the tensor nature of the dielectric response asserts itself, the character of the waves is more complex.

We may seek solutions of Maxwell's equations for which all field components exhibit the space and time variation $\exp[i(\boldsymbol{k}\cdot\boldsymbol{r} - \omega t)]$. For such waves, once again transverse in character, the wave equation provides a connection between the wave vector $\boldsymbol{k}$ and the frequency. It is a short exercise to see that

$$\frac{c^2k^2}{\omega^2} = \varepsilon(\omega) . \tag{2.36}$$

Suppose for the moment that the dielectric constant ε is independent of frequency. Then (2.36) describes a plane wave whose phase fronts advance at the speed $c/\sqrt{\varepsilon}$; the quantity $\sqrt{\varepsilon}$ is thus the index of refraction of the medium, in this example. The ratio ω/k, known as the phase velocity v_p, is independent of frequency.

We may construct a localized packet of electromagnetic energy, a pulse, by superimposing various plane waves, in a manner familiar from discussions of wave packet propagation in quantum mechanics [2.6]:

$$E_\alpha(\boldsymbol{r}, t) = \int \frac{d^3k}{(2\pi)^3} E_\alpha(\boldsymbol{k}) e^{i(\boldsymbol{k}\cdot\boldsymbol{r} - ckt/\sqrt{\varepsilon})} . \tag{2.37}$$

Since the phase fronts all advance at precisely the same speed (in the special case that ε is frequency independent), the wave packet propagates at the speed $c/\sqrt{\varepsilon}$, undistorted in shape.

When the dielectric constant is frequency dependent (observe that the Kramers-Kronig relations never allow ε to be strictly independent of frequency over a finite domain), the above picture changes. For instance, when (2.27) is a suitable representation of $\varepsilon(\omega)$, we have

$$\frac{c^2k^2}{\omega^2} = \varepsilon_\infty + \frac{\Omega_p^2}{\omega_0^2 - \omega^2} . \tag{2.38}$$

The phase velocity, $v_p(\omega) = \omega/k(\omega)$ now depends on frequency. Thus, the wave packet will spread as it propagates through the material. Later in this volume in Chap. 7, we shall see that the "center of mass" of the wave packet propagates at the group velocity $v_g(\omega) = \partial\omega/\partial k$, which as we shall see momentarily, can differ greatly from the phase velocity.

From (2.38), it is straightforward to find an explicit expression for ω as a function of $\boldsymbol{k}$; this functional relationshp is called often the dispersion relation of the wave in the medium. For each choice of wave vector, there are two frequencies allowed. There is thus an upper branch, and a lower branch to the dispersion relation:

$$\omega_\pm^2(k) = \frac{1}{2}\left(\frac{c^2k^2}{\varepsilon_\infty} + \omega_0^2 + \frac{\Omega_p^2}{\varepsilon_\infty}\right) \pm \frac{1}{2}\left[\left(\frac{c^2k^2}{\varepsilon_\infty} + \omega_0^2 + \frac{\Omega_p^2}{\varepsilon_\infty}\right)^2 - 4\omega_0^2\frac{c^2k^2}{\varepsilon_\infty}\right]^{1/2} . \tag{2.39}$$

The dispersion relations are plotted in Fig. 2.4; here $\varepsilon_s = \varepsilon_\infty + \Omega_p^2/\omega_0^2$ is the static dielectric constant of the model. In the frequency band between ω_0 and $\omega_0(\varepsilon_s/\varepsilon_\infty)^{1/2}$, $\varepsilon(\omega)$ is negative, and electromagnetic waves cannot propagate in the medium. The wave vector is purely imaginary.

On the lower branch, described by $\omega_-(k)$, the phase velocity is always less than $c/\sqrt{\varepsilon_s}$, while on the upper branch described by $\omega_+(k)$, it is always greater than $c/\sqrt{\varepsilon_\infty}$. Most particularly, as $k \to 0$, the phase velocity becomes infinite; there is a finite region of wave vector near $k = 0$ where $v_p(\omega)$ is greater than the vacuum velocity of light! The group velocity, $v_g(\omega)$, is always less than the velocity of light. Thus, an electromagnetic pulse, which transports energy

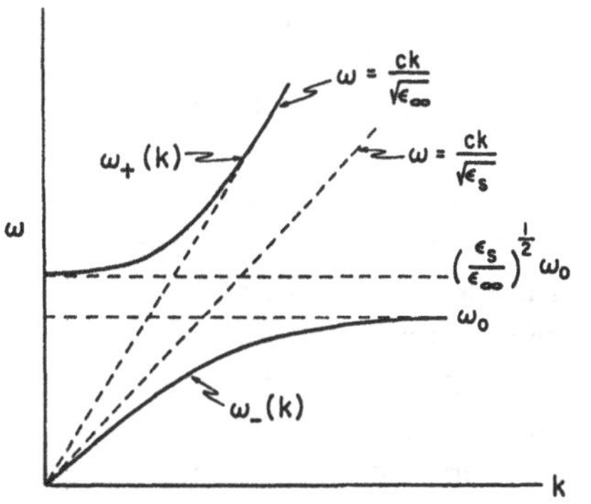

Fig. 2.4. The dispersion relation for electromagnetic waves which propagate in a medium with a frequency dependent dielectric constant such as that displayed in (2.27). The quantity ε_s is the static dielectric constant, $\varepsilon_s = \varepsilon_\infty + \Omega_p^2/\omega_0^2$

from one point to another, can never propagate faster than the velocity of light, since its speed is controlled by the group velocity. This statement leaves one unsettled, however, because the notion that the group velocity controls the propagation of a wave packet rests on a set of approximations as we shall see in Chap. 7; one does not explore in detail the behavior of the leading edge of the pulse. One must inquire further to insure that no information of any sort is propagated through the medium faster than the vacuum velocity of light. This question was addressed and resolved several decades ago by Sommerfeld, in a most elegant analysis.[2] No piece of the pulse, even the leading tip, is transported from point to point faster than the velocity of light in vacuum, despite the behavior of the phase velocity.

The above description of electromagnetic wave propagation in a medium where $\varepsilon(\omega)$ has a resonance such as that displayed in (2.27) is encountered often. For frequencies near ω_0, one often employs a picture which treats in an explicit manner the excitation of the medium by the electric field in the wave. This begins by considering the electromagnetic waves which propagate in a "sterile" dielectric with dielectric constant ε_∞. One then introduces an explicit description of the excitation responsible for the resonant term in $\varepsilon(\omega)$ (vibrational normal mode, electronic excited state), and the coupling of the electric field to this excitation. The resulting entity, fully equivalent to the electromagnetic waves discussed here, is described by the dispersion relation in (2.38) in the end. One then has a picture which views the electromagnetic mode in the medium as a "bare" electromagnetic wave in the background dielectric ε_∞, "dressed" or "clothed" by the excitation in the medium, dragged along with it in the process. Such a decomposition of the electromagnetic wave is useful to introduce, when one desires a more microscopic picture of the entity that propagates in the material. In the literature on condensed matter physics, when such a picture of electromagnetic waves is employed, the modes are referred to as polaritons, a term that is often confusing, and in fact one that crept into

[2] This issue is discussed in depth in the monograph by *Brillouin* [2.7].

present usage by means of a misunderstanding [2.8]. In this text, we shall not use the polariton picture, though it can prove exceedingly useful. We refer the reader to a review article which describes this approach, and summarizes a number of applications of it [2.9].

We conclude this subsection with comments on the nature of the frequency dependencies found in $\varepsilon(\omega)$, in various forms of matter. There are certain general properties that will be useful to keep in mind as we proceed.

The form for $\varepsilon(\omega)$ displayed in (2.27) applies to any substance within which one encounters a sharp absorption line, well separated in frequency from nearby features in $\varepsilon_2(\omega)$. This picture can apply to the gas phase, where the atoms are well separated, so the excitation energies of any one atom are uninfluenced by the presence of others. In this case, an explicit expression for $\varepsilon(\omega)$ follows by letting $\boldsymbol{k} \to 0$ in (A.41).

In dense matter, such as a solid or liquid, the wave functions of neighboring atoms overlap, and the spectrum of eigenvalues of the electronic degrees of freedom is more complex than in the gas phase, where in the notation of Appendix A a given excitation frequency $\omega_{n0} = (E_n - E_0)/\hbar$ has its origin in an excited state E_n that is N fold degenerate, where N is the number of atoms in the sample. One excited level is associated with each atom in the limit of high dilution. When the wave functions overlap, the N fold degenerate level evolves into a band of N states, with width and detailed nature controlled by the interatomic distances and geometry of the material, as illustrated in Fig. 2.5 a.

There are two broad classes of materials, insulators which fail to carry electrical current in response to an applied static electric field, and metals within which one realizes current flow in this circumstance. The properties of any material are controlled by the manner in which the energy levels in Fig. 2.5 a

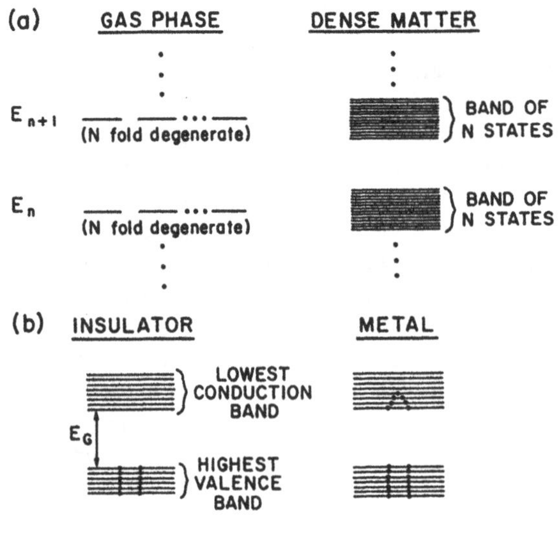

Fig. 2.5. **a** The spectrum of energy levels is schematically illustrated for a collection of N atoms in the gas phase, and for such a collection in a dense phase such as a solid or liquid. **b** An illustration of the occupancy of the energy levels in an insulator and a metal

are occupied by electrons. The Pauli principle allows us to place two electrons in each state, one with spin up and one with spin down. If each atom contains an even number of electrons, then simple counting shows that we will precisely fill up the first $(n/2)$ energy bands, with n the number of electrons per atom. The situation is illustrated in the left panel of Fig. 2.5 b. The first excited state then lies a distance E_g above the highest energy occupied band, called the valence band. One refers to E_g as the energy gap. To promote an electron from the valence band to the lowest state in the first unoccupied band (the conduction band) requires absorption of energy from a field whose frequency is larger than $E_g/\hbar$; for frequencies with frequency lower than this value, the electrons remain "frozen" in their original state. The term conduction band has its origin in the fact that electrons which reside in it are readily accelerated by even a static electric field, since the energy levels form a continuum.

If there is an odd number of electrons per atom (or per formula unit, in more complex solids), then the highest occupied band is only partially filled, as illustrated in the right panel of Fig. 2.5 b. There are states arbitrarily close in energy to the highest filled state. A static field applied to such a system induces electron motion, and a conduction current, very much as in a classical electron fluid. This is a metal.

The rule that materials with an even number of electrons per formula unit form an insulator, but those with an odd number are metals is obeyed widely. But there are important exceptions. For example, energy bands associated with atomic energy levels nearby in energy may overlap, rather than be separate and distinct as illustrated in Fig. 2.5. It is then possible to have an even number of electrons per atom, while at the same time the highest occupied state remains within a continuum in the energy spectrum of the material. We then have a metal.

Consider next the behavior of $\varepsilon_2(\omega)$ from electronic transitions, for an insulating material. For $\omega < E_g/\hbar$, no transitions can be induced, so in this simple picture, $\varepsilon_2(\omega) = 0$ for such frequencies. The material is then transparent; the wave vector of an electromagnetic wave is thus perfectly real, and it propagates infinitely far without attenuation. For $\omega > E_g/\hbar$, we have allowed transitions, $\varepsilon_2(\omega) \neq 0$, and an electromagnetic wave is absorbed, usually quite strongly within the first few electron volts of the onset of absorption. The frequency $E_g/\hbar$ is referred to as the absorption edge.

In a metal, we have similar contributions to $\varepsilon_2(\omega)$ from transitions from the highest (partially) occupied band, also a conduction band, and higher energy unfilled bands. In a similar manner, we have transitions into empty states in the conduction band, from filled bands lower in energy. These contributions, referred to as interband transitions, are quite similar in nature to those found in an insulator. In addition, we must consider the transitions between states in the highest partially occupied conduction band, the intra band transitions. We can treat these in a purely quantum mechanical fashion by evaluating their contribution to the dielectric tensor through use of the formal results derived in Appendix A. It is much easier, however, to realize that these low energy transitions within the conduction band are, in fact, the origin of the conduction current which is described quite easily by treating the conducting electrons in a classical manner. We proceed with a brief sketch of how this is done.

Consider the i-th electron, exposed to a spatially uniform electric field of frequency ω. We shall suppose, as is the case in many of the highly conducting simple metals, that the mobile electron has a mass m close to the free electron mass. The equation of motion can be written

$$\frac{d\boldsymbol{v}_i}{dt} = \frac{e}{m}\boldsymbol{E}\,\mathrm{e}^{-\mathrm{i}\omega t} + \left(\frac{d\boldsymbol{v}_i}{dt}\right)_{\mathrm{coll}}, \tag{2.40}$$

where the second term on the right-hand side has its origin in collisions experienced by the electron. The average velocity $\langle\boldsymbol{v}\rangle = \left(\sum_{i=1}^{N}\boldsymbol{v}_i\right)/N$ then obeys

$$\frac{d\langle\boldsymbol{v}\rangle}{dt} = \frac{e}{m}\boldsymbol{E}\,\mathrm{e}^{-\mathrm{i}\omega t} + \left(\frac{d\langle\boldsymbol{v}\rangle}{dt}\right)_{\mathrm{coll}}, \tag{2.41}$$

where for the influence of the collisions, one adopts a relaxation time approximation,

$$\left(\frac{d\langle\boldsymbol{v}\rangle}{dt}\right)_{\mathrm{coll}} = -\frac{1}{\tau}\langle\boldsymbol{v}\rangle\,. \tag{2.42}$$

The current density is $\boldsymbol{J} = ne\langle\boldsymbol{v}\rangle$, with n the number of electrons per unit volume. One sees easily that

$$\boldsymbol{J} = \sigma(\omega)\boldsymbol{E}\,, \tag{2.43}$$

where the frequency dependent conductivity is given by

$$\sigma(\omega) = \frac{ne^2\tau}{m}\frac{1}{1-\mathrm{i}\omega\tau} = \frac{\sigma_0}{1-\mathrm{i}\omega\tau} \tag{2.44}$$

with $\sigma_0 = ne^2\tau/m$ being the dc conductivity, appropriate to the limit $\omega\tau \ll 1$. For simple metals at room temperature (Cu, Ag, Al), one has[3] $\tau \cong 10^{-14}$ s, so the limit $\omega\tau < 1$ applies to infrared frequencies. In the visible, $\omega\tau \gg 1$.

The above picture is referred to often as the Drude model; it works well for a wide range of conducting materials, not only the simple metals, though the effective mass m of the electrons can differ substantially from the free electron mass.

In the presence of the conduction current, the "$\nabla \times \boldsymbol{H}$" Maxwell equation now reads, for fields of frequency ω,

$$\nabla \times \boldsymbol{H} = -\mathrm{i}\frac{\omega}{c}\boldsymbol{D} + \frac{4\pi}{c}\boldsymbol{J} = -\mathrm{i}\frac{\omega}{c}\,\varepsilon_{\mathrm{inter}}(\omega)\boldsymbol{E} + \frac{4\pi}{c}\sigma(\omega)\boldsymbol{E}\,, \tag{2.45 a}$$

[3] Estimates of the value of τ for simple metals are given in [2.10].

where $\varepsilon_{\text{inter}}(\omega)$ is the contribution to the dielectric constant from interband transitions. We may rewrite (2.45) as

$$\nabla \times \boldsymbol{H} = -\mathrm{i}\frac{\omega}{c}\,\varepsilon(\omega)\boldsymbol{E}\,, \tag{2.45 b}$$

where

$$\varepsilon(\omega) = \varepsilon_{\text{inter}}(\omega) + \frac{4\pi\mathrm{i}\sigma(\omega)}{\omega} \tag{2.46 a}$$

or, using (2.44),

$$\varepsilon(\omega) = \varepsilon_{\text{inter}}(\omega) + \frac{4\pi\mathrm{i}\sigma_0}{\omega(1 - \mathrm{i}\omega\tau)} \tag{2.46 b}$$

which is the same as

$$\varepsilon(\omega) = \varepsilon_{\text{inter}}(\omega) - \frac{\omega_{\mathrm{p}}^2}{\omega^2(1 + \mathrm{i}/\omega\tau)}\,, \tag{2.46 c}$$

where $\omega_{\mathrm{p}}^2 = 4\pi ne^2/m$ is called the plasma frequency.

Thus, the presence of the free electrons may be accounted for by simply adding the appropriate term to our complex, frequency dependent dielectric constant. In essence, there is no physical difference between a time dependent dipole moment density $\boldsymbol{P}$, and the presence of a current density $\boldsymbol{J}$, provided one regards $-\mathrm{i}\omega\boldsymbol{P}$ as equivalent to $\boldsymbol{J}$. For the purposes of our discussion, the only difference between a metal and an insulator is the form of $\varepsilon(\omega)$. Note that in the presence of conduction current, $\varepsilon_2(\omega)$ diverges as $\omega \to 0$ inversely with ω; the presence of such a divergence requires the Kramers-Kronig relations to be modified, as mentioned earlier, and discussed explicitly by *Landau* and *Lifshitz* [2.3].

Anytime there is a time dependent dipole moment per unit volume $\boldsymbol{P}(\boldsymbol{r}, t)$, charges in the system are in motion, and there is an internal current density $\boldsymbol{J}(\boldsymbol{r}, t)$. We may then write $\varepsilon_{\text{inter}}(\omega) = 1 + 4\pi\mathrm{i}\sigma_{\text{inter}}(\omega)/\omega$, and write $\varepsilon(\omega) = 1 + 4\pi\mathrm{i}\sigma_{\mathrm{T}}(\omega)/\omega$, where $\sigma_{\mathrm{T}}(\omega)$ is the total conductivity, which is the sum of that from the mobile conducting electrons (if present), and from the currents with origin in the electron motions associated with the interband processes. In our microscopic derivation of the dielectric tensor presented in Appendix A, we calculate the total current density from all electron motions, then form the wave vector and frequency dependent dielectric tensor from the appropriate generalization of the statement $\varepsilon(\omega) = 1 + 4\pi\mathrm{i}\sigma_{\mathrm{T}}(\omega)/\omega$.

While we have discussed the nature of the dielectric response of insulators and metals, there is a third and very important class of materials, the semiconductors.

A perfectly pure, stoichiometric semiconductor is in fact an insulator at the absolute zero of temperature, within the above classification scheme. The band gap E_g is sufficiently small, however (1–2 eV, or possibly considerably less) that at room temperature, a non-negligible fraction of the electrons in the highest valence band are thermally excited into the conduction band. The material is thus a conductor at room temperature, because of these thermally excited electrons. This is an example of an intrinsic semiconductor. Such materials may have their conductivity altered by the addition of impurities. For example, P has one more electron than Si. When P is added to Si as an impurity, four of its outer electrons are locked into tetrahedrally arranged chemical bonds with the four silicon neighbors, and the fifth electron is released into the conduction band, where it moves about freely, to contribute to the conduction current. A semiconductor whose electrical conductivity is controlled predominantly by electrons introduced by impurities is referred to as an extrinsic semiconductor. The dielectric response of semiconducting materials is described through use of the concepts introduced above.

We have one final aspect of the electronic contributions to the dielectric constant of insulators that is of great importance in the literature on nonlinear optics. Consider an insulator, and suppose an electron is excited across the energy gap into the conduction band. The electron will wander away from the site where the excitation process occurs; this site is positively charged, since an electron is missing from it. One refers to the positively charged region as a hole in the valence band. The electron and hole do not move independently, but are attracted by means of their mutual Coulomb attraction. They will in fact form bound states, of hydrogen like character, in which the electron orbits around the hole. These bound states are referred to as excitons. The excitons lie in the band gap between the highest occupied valence band, and the lowest unoccupied conduction band: one excites an electron from the top of the valence band to the bottom of the conduction band, at the energy cost E_g. The two then form a bound state, and their energy is lowered by Δ_n, the binding energy of the n-th exciton level. The energy of the exciton is thus

$$E_n^{(ex)} = E_g - \Delta_n \,. \tag{2.47 a}$$

Excitons can be excited by electromagnetic radiation with frequency $E_n^{(ex)}/\hbar < E_g/\hbar$. The excitons are discrete levels, and one thus finds a sequence of sharp absorption lines associated with exciton formation, below the absorption edge, above which one has a continuous absorption band, produced by the interband transitions. For frequencies very near those of the exciton levels, their discrete nature and associated sharp absorption lines allows (2.27) to describe the frequency variation of the dielectric constant.

Electrons and holes move in solids with effective masses that differ from free electron masses. Also, the Coulomb attraction between them is reduced by the static dieletric constant ε_s of the material. If we treat the exciton as a hydrogen-like object, the exciton series has Rydberg character, with binding energies given by

$$\Delta_n = \frac{1}{\varepsilon_s^2}\left(\frac{\mu^*}{m}\right)\frac{R_H}{n^2}, \tag{2.47b}$$

where μ^* is the reduced mass of the electron-hole combination, m the free electron mass, and $R_H = 13.6$ eV is the Rydberg. For semiconductors, $\varepsilon_s \cong 10$, $\mu^* \cong 0.1$ m, so the exciton binding energies are in the range of 100 meV. The effective Bohr radius is $a_0^* = \varepsilon_s(m/\mu^*)a_0$, with a_0 the atomic Bohr radius (0.528 Å), so the separation between electron and hole is in the range of 50 Å, justifying the simple model just employed. Since the exciton radius is large compared to the lattice constant, the medium can indeed be treated as a continuum. In many materials where the band gap is larger, ε_s is smaller, μ^* is larger, so Δ_n may be in the electron-volt range. Then a_0^* is small enough that a proper microscopic theory of the exciton is required.

This completes our survey of the contributions to $\varepsilon_2(\omega)$ from the electronic excitations of materials. We have assumed all atoms are fixed rigidly in various positions, while in fact they engage in vibrational motions about their equilibrium positions. Electromagnetic radiation may be absorbed by exciting vibrational normal modes of the system, called phonons in solids. A vibrational normal mode of frequency ω_v has its energy quantized in units of $\hbar\omega_v$, and if the nature of the atomic motions are appropriate can lead to a sharp absorption line at the frequency ω_v.

While typical electronic absorptions give rise to features in $\varepsilon_2(\omega)$ in the energy range of 1–10 eV in typical materials (excluding metals, where the absorption from conducting electrons gives absorption from frequencies the order of $1/\tau$ down to zero frequency), the vibrational frequencies lie in the infrared frequency range, from a few tens of cm^{-1} to perhaps 1000 cm^{-1} (from a few meV to perhaps 200 meV) depending on the nature of the material. Thus, the vibrational normal modes lead to absorption lines and bands from the infrared into the far infrared.

Sharp vibrational absorption lines occur in ionic materials. The chemical unit cell must be electrically neutral, and thus contains both positive and negative ions, whose relative motion is excited by the presence of an electric field of frequency ω. In such a material, there can be one or perhaps several sharp lines, one associated with each normal mode of the basic unit cell whose symmetry is such that the mode is excited by the field. The number of these modes, each of which must have an oscillatory electric dipole moment when excited, is controlled by the number of atoms in the unit cell, and their geometrical arrangement [2.11]. Once again, in the near vicinity of such a sharp absorption line, (2.27) (possibly generalized to tensor form in an anisotropic crystal) serves to describe the frequency variation of $\varepsilon(\omega)$. Note also that the discussion in Appendix A is very general, and the final formula may be used to analyze the contributions to the dielectric tensor from vibrationally excited states, as well as electronic excited states.

We thus have sharp absorption lines in the far infrared, in ionic materials. As the frequency ω increases, to the point where it exceeds the maximum vibration frequency ω_M of the crystal, we enter a regime where energy conservation considerations require 2, 3 or more vibrational quanta to interact with the incoming photon or electromagnetic wave, since energy cannot be conserved in a process that involves only a single vibrational excitation. In this frequency range, referred to as the multiphonon absorption regime, $\varepsilon_2(\omega)$ decreases dramatically with increasing frequency, exhibiting an exponential fall-off [2.12].

At finite temperatures, the existence of thermally excited lattice motions affects the electronic absorption, in a variety of ways. For example, a photon whose energy lies below the absorption edge ($E_g/\hbar$) can be absorbed by lifting an electron from the highest occupied valence band into the conduction band, provided a sufficient number of vibrational quanta are absorbed simultaneously, to provide the energy necessary to promote the electron into the conduction band. The onset of optical absorption is thus not sharp, but one has a tail which extends below the absorption edge. This temperature dependent feature, which falls off exponentially in the variable (E_g-$\hbar\omega$), is referred to as the Urbach tail.

A schematic illustration of the various contributions to $\varepsilon_2(\omega)$ is given in Fig. 2.6. We plot here the absorption constant $\alpha(\omega) = (\omega/c)\varepsilon_2(\omega)/\sqrt{\varepsilon_1}$. The

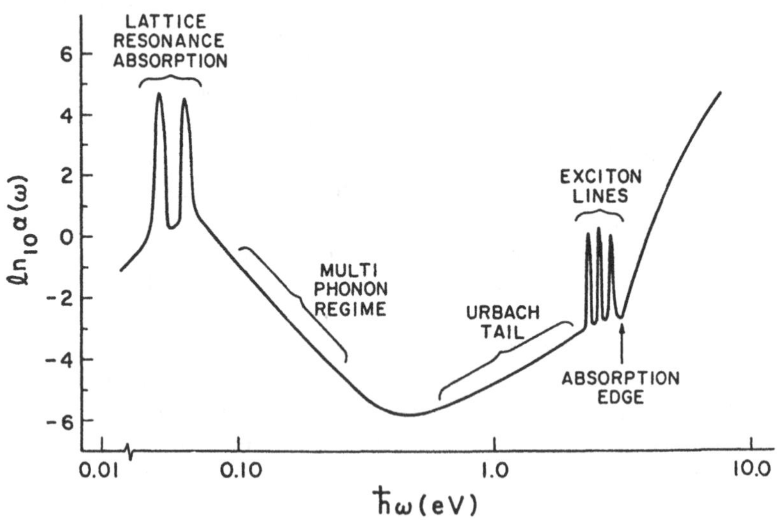

Fig. 2.6. The absorption constant $\alpha(\omega) = (\omega/c)\varepsilon_2(\omega)/\sqrt{\varepsilon_1(\omega)}$ is schematically shown for a typical insulating material, illustrating the various features discussed in the text. It is assumed the units of $\alpha(\omega)$ are cm^{-1}; the inverse of this quantity is the attenuation length of an electromagnetic wave of frequency ω in the medium

inverse of $\alpha(\omega)$ is the attenuation length of an electromagnetic wave of frequency ω. As we proceed with our discussion, we will want to keep this picture in mind, along with the influence of the various structures on the frequency variation of $\varepsilon_1(\omega)$. We see, as required by the expression in (2.24 b), that the material is never perfectly transparent. There is always some process that leads to absorption.

There is a minimum in the absorption constant $\alpha(\omega)$, as one passes from the multiphonon regime into the Urbach tail. In glasses from which optical fibers are constructed, this minimum lies in the near vicinity of 1 eV, or the 1 μm wavelength regime. Optical fibers typically operate in this frequency range.

2.2 Wave Vector Dependence of the Dielectric Tensor

We turn next to a brief discussion of the wave vector dependence of the dielectric tensor, which as we have seen enters as a consequence of the nonlocal response of the system of interest, in coordinate space.

In Appendix A, we have presented a microscopic theory of the wave vector and frequency dependent dielectric tensor of materials. The wave vector $\boldsymbol{k}$ enters the final expression in (A.27) in two distinctly different ways:

(i) In the energy denominator, the excitation energy ω_{n0} is in fact a function of wave vector, $\omega_{n0}(\boldsymbol{k})$. This is because, if the ground state wave function $|\psi_0(0)\rangle$ is an eigenstate total momentum with eigenvalue zero, the eigenstate $|\psi_n(\boldsymbol{k})\rangle$ which enters as the intermediate state has the total momentum $\hbar\boldsymbol{k}$. One says the system has temporarily absorbed a photon with momentum $\hbar\boldsymbol{k}$, and acquired its momentum in the process in this virtual transition. The energy denominator is thus $\omega_{n0}(\boldsymbol{k}) = [E_n(\boldsymbol{k}) - E_0(0)]/\hbar$, and the resonance is shifted away from its zero wave vector value $[E_n(0) - E_0(0)]/\hbar$ by the recoil energy associated with the center of mass motion of the system.

(ii) In the numerator, of course $J^p_\alpha(\boldsymbol{k})$ depends on the wave vector $\boldsymbol{k}$. If one lets $\boldsymbol{k} \rightarrow 0$ in the matrix elements which enter the numerator of (A.27), as we see from the discussion which leads to (A.36), these matrix elements become those associated with the electric dipole allowed transitions of the quantum theory of radiative transitions [2.13]. There are selection rules that govern which of these matrix elements are nonzero. For instance, if the Hamiltonian is invariant under the parity operation, all eigenstates have even or odd parity. The ground state has even parity, so when the matrix elements in (A.27) are replaced by their limit as $\boldsymbol{k} \rightarrow 0$, as we have done to obtain (A.36), only odd parity states enter the sum on n. The even parity states are thus "silent," in this approximation; note that the wave vector dependence of the denominator remains, and can become important, as we shall see.

If the wave vector $\boldsymbol{k}$ is retained as nonzero in the matrix elements which remain in the numerator, the electric dipole selection rule breaks down, and it is possible for even parity intermediate states to contribute to the dielectric tensor, connected to the ground state wave function by the electric quadrupole operator, magnetic dipole operator, or a structure higher order in the multipole expansion of radiation theory [2.13]. If the wavelength of the radiation is large compared to the size of the fundamental entity (atom, exciton, . . .) in the system of interest, these matrix elements are small compared to their electric dipole counterparts. But if the frequency lies very close to $\omega_{n0}(\boldsymbol{k})$, a possibility that can be achieved readily in a variety of systems with tunable dye lasers, then such dipole forbidden transitions can produce large contributions to $\varepsilon_{\alpha\beta}(\boldsymbol{k}, \omega)$. Spatial dispersion effects can thus be substantial, for frequencies very near resonance.

We may appreciate this point further by considering the behavior of $\varepsilon_{\alpha\beta}(\boldsymbol{k}, \omega)$ near a dipole allowed transition, where the wave vector dependence of the denominator in (A.27) enters importantly. As a simple example, consider (A.41), which describes the contribution from electric dipole allowed transitions to $\varepsilon_{\alpha\alpha}(\boldsymbol{k}, \omega)$ from an array of noninteracting atoms. If we define $d_\alpha(n0) = \langle\phi_n|\Sigma_{j=1}^{n} e_j r^j_\alpha|\phi_0\rangle$, the function $\chi_{\alpha\alpha}(\boldsymbol{k}, \omega)$ introduced in (2.13) has the form, with $n = N/V$,

$$\chi_{\alpha\alpha}(\boldsymbol{k}, \omega) = \frac{2n}{\hbar} \sum_n \frac{\omega_{n0}(0)|d_\alpha(n0)|^2}{[\omega_{n0}^2(\boldsymbol{k}) - (\omega + \mathrm{i}\eta)^2]} . \tag{2.48 a}$$

Suppose the frequency ω lies very close to one particular transition, so a single term on the right-hand side of (2.48 a) provides the dominant contribution to $\chi_{\alpha\alpha}(\boldsymbol{k}, \omega)$. To excellent approximation

$$\chi_{\alpha\alpha}(\boldsymbol{k}, \omega) = \frac{n}{\hbar} \frac{|d_\alpha(n0)|^2}{\omega_{n0}(k) - \omega - \mathrm{i}\eta} . \tag{2.48 b}$$

Since the wave vector dependence of the excitation energy $\omega_{n0}(\boldsymbol{k})$ arises from the recoil effect discussed above, we have

$$\omega_{n0}(\boldsymbol{k}) = \omega_{n0}(0) + \frac{\hbar k^2}{2M} , \tag{2.49}$$

with M the mass of the recoiling object. Then (2.48 b) may be rewritten to read

$$\chi_{\alpha\alpha}(\boldsymbol{k}, \omega) = \frac{2nM|d_\alpha(n0)|^2}{\hbar^2} \frac{1}{k^2 + \Gamma^2(\omega)} , \tag{2.50}$$

where $\Gamma(\omega) = (2M/\hbar)^{1/2}[\omega_{n0}(0) - \omega - \mathrm{i}\eta]^{1/2}$. We shall always choose $\Gamma(\omega)$ so that $\mathrm{Re}\{\Gamma(\omega)\} > 0$ in the limit $\eta \to 0$.

From (2.10–13), one sees that the function

$$\chi_{\alpha\alpha}(\boldsymbol{r} - \boldsymbol{r}', \omega) = \int \frac{d^3k}{(2\pi)^3} \chi_{\alpha\alpha}(\boldsymbol{k}, \omega) \mathrm{e}^{\mathrm{i}\boldsymbol{k}\cdot(\boldsymbol{r}-\boldsymbol{r}')} \tag{2.51}$$

describes the spatial range of the response of the medium to a monochromatic electric field of frequency ω. That is, if an electric field whose variation in space and time is given by

$$E_\alpha(\boldsymbol{r}, t) = E_\alpha \delta(\boldsymbol{r} - \boldsymbol{r}_0)\mathrm{e}^{-\mathrm{i}\omega t}, \tag{2.52}$$

which is localized at the point $\boldsymbol{r}_0$, then the α-th Cartesian of the dipole moment per unit volume induced by the field is

$$P_\alpha(\boldsymbol{r}, t) = \chi_{\alpha\alpha}(\boldsymbol{r} - \boldsymbol{r}_0, \omega)E_\alpha \mathrm{e}^{-\mathrm{i}\omega t}. \tag{2.53}$$

With use of (2.50), a simple integration gives

$$\chi_{\alpha\alpha}(\boldsymbol{r} - \boldsymbol{r}'; \omega) = \frac{nM|d_\alpha(n0)|^2}{2\pi\hbar^2} \frac{\mathrm{e}^{-\Gamma(\omega)|\boldsymbol{r}-\boldsymbol{r}'|}}{|\boldsymbol{r} - \boldsymbol{r}'|}. \tag{2.54}$$

First consider frequencies ω which lie below $\omega_{n0}(0)$. We have

$$\lim_{\eta\to 0} \Gamma(\omega) = \left(\frac{2M}{\hbar}\right)^{1/2} [\omega_{n0}(0) - \omega]^{1/2}, \quad \omega < \omega_{n0}(0). \tag{2.55}$$

Thus, $\chi_{\alpha\alpha}(\boldsymbol{r} - \boldsymbol{r}'; \omega)$ has the Yukawa form, and falls off exponentially with increasing $|\boldsymbol{r} - \boldsymbol{r}'|$. Notice, however, that as $\omega \to \omega_{n0}(0)$ from below, $\Gamma(\omega) \to 0$. Thus, for frequencies very close to resonance, $\chi_{\alpha\alpha}(\boldsymbol{r} - \boldsymbol{r}'; \omega)$ becomes very long ranged. It follows that close to resonance, one always has to be concerned about spatial dispersion effects, in that the spatial range of the response function becomes very long, and local dielectric theory breaks down.

Above the resonance, $\omega > \omega_{n0}(0)$, $\Gamma(\omega)$ becomes pure imaginary,

$$\Gamma(\omega) = \mathrm{i}\kappa(\omega), \quad \omega > \omega_{n0}(0) \tag{2.56a}$$

where

$$\kappa(\omega) = \left(\frac{2M}{\hbar}\right)^{1/2} [\omega - \omega_{n0}(0)]^{1/2}. \tag{2.56b}$$

Thus,

$$\chi_{\alpha\alpha}(\boldsymbol{r} - \boldsymbol{r}', \omega) = \frac{nM|d_\alpha(n0)|^2}{2\pi\hbar^3} \frac{\mathrm{e}^{-\mathrm{i}\kappa(\omega)|\boldsymbol{r}-\boldsymbol{r}'|}}{|\boldsymbol{r} - \boldsymbol{r}'|}. \tag{2.57}$$

Just a bit above resonance, $\kappa(\omega)$ is very small, and the wave length of the spatial oscillations in $\chi_{\alpha\alpha}(\boldsymbol{r} - \boldsymbol{r}', \omega)$ is very long. Spatial dispersion again asserts itself importantly. As ω increases, $\kappa(\omega)$ becomes large, the spatial oscillations in $\chi_{\alpha\alpha}(\boldsymbol{r} - \boldsymbol{r}', \omega)$ are very rapid, and detailed analysis shows that once again only the near vicinity of the point $\boldsymbol{r}'$ matters significantly. One recovers the local theory once again, though how this happens is not clear from what can be presented here.

The above discussion, based on the near resonant response of a noninteracting array of atoms or molecules is rather academic. The mass M is so very large that one has to be very close to resonance indeed for the spatial range of

$\chi_{\alpha\alpha}(\boldsymbol{r} - \boldsymbol{r}', \omega)$ to be appreciable; on physical grounds one expects spatial dispersion effects to become important when the spatial range of $\chi_{\alpha\alpha}(\boldsymbol{r} - \boldsymbol{r}', \omega)$ becomes a non-negligible fraction of the wavelength of light in the medium.

However, discussions of the nature of the optical response of solids [2.1] shows that for frequencies very close to the exciton absorption lines, a form identical to (2.48 b) and (2.50) describes the wave vector and frequency variation of the dielectric susceptibility tensor. The mass M which enters (2.49) is now the total mass of the bound electron-hole pair, which is the order of the free electron mass, rather than that of an atom or molecule. The latter is larger by a factor in the range of 10^4–10^5. If the frequency ω lies within a few meV of the exciton resonance, $\Gamma(\omega)$ lies in the range of 10^6 cm^{-1}. The index of refraction of a typical semiconductor lies in the range of 3.5 or 4, so the wavelength of visible radiation in the medium is in the range of 10^{-5} cm. Thus $\Gamma(\omega)$ is an appreciable fraction of the wavelength. Theoretical and experimental studies show that the optical response of such materials is influenced in a rich and dramatic manner, for frequencies in the near vicinity of the exciton resonances [2.1].

While we shall not explore spatial dispersion effects further in the present text, the above discussion serves to acquaint us with the limitations of local dielectric theory, in which it is presumed the dipole moment per unit volume $\boldsymbol{P}(\boldsymbol{r}, t)$ is proportional to the electric field $\boldsymbol{E}(\boldsymbol{r}, t)$ at the same spatial point.[4]

2.3 Electromagnetic Waves in Anisotropic Dielectrics

While we have placed emphasis on the tensor character of the dielectric response of crystalline materials, we have yet to explore the consequences of this feature.

If the material is isotropic in nature, such as a liquid or a gas, then the dielectric tensor reduces to a simple scalar quantity, in the limit of zero wave vector, $\boldsymbol{k} \to 0$. We may write $\varepsilon_{\alpha\beta} = \varepsilon\delta_{\alpha\beta}$, where for the moment, we make no explicit reference to the frequency dependence of the dielectric tensor in this section. One may establish easily that in a material whose macroscopic symmetry is cubic, the dielectric tensor again reduces to a scalar. In such a medium, where $\boldsymbol{D} = \varepsilon\boldsymbol{E}$, the electromagnetic waves are purely transverse as they are in free space; they propagate with phase velocity $c/\sqrt{\varepsilon}$, as we have discussed above.

[4] It should be remarked that in metals of high purity, and at low temperatures, the electron mean free path is long compared to the skin depth. Under these conditions, the simple picture in which the current density $\mathbf{J}(\boldsymbol{r}, t)$ is proportional to the electric field $\boldsymbol{E}$ at the same point breaks down in a particularly dramatic manner. One refers to this as the anomalous skin effect regime. The electrodynamics of metals can be influenced strongly by spatial dispersion effects, even though the frequency is not close to any resonance in the material [2.14].

In an anisotropic medium, the polarization and propagation characteristics are more complex in nature. We may illustrate with a simple example. Many crystals have a unique direction or growth axis called often the optic axis. One sees this by examining single crystals of a material such as quartz. The crystals are (if growth conditions are stable) in the form of elongated cylinders with hexagonal cross section. The optic axis is the direction parallel to the long axis of the material. Such a material can be described by dielectric tensor of the form, if the symmetry in the basal plane is sufficiently high (the existence of three-, four- or six-fold symmetry suffices)

$$\varepsilon = \begin{pmatrix} \varepsilon_\perp & 0 & 0 \\ 0 & \varepsilon_\perp & 0 \\ 0 & 0 & \varepsilon_\parallel \end{pmatrix}, \tag{2.58}$$

where $\varepsilon_\parallel$ and $\varepsilon_\perp$ describe the dielectric response to an electric field applied parallel to and perpendicular to the optic axis, respectively.

We have, as assumed throughout this text, $\boldsymbol{B} = \boldsymbol{H}$. If we consider a plane wave in which all field components exhibit space and time dependence proportional to $\exp[\mathrm{i}(\boldsymbol{k}\cdot\boldsymbol{r} - \omega t)]$, then Maxwell's equations (1.1) assume the form

$$\boldsymbol{k}\cdot\boldsymbol{D} = 0, \quad \boldsymbol{k}\cdot\boldsymbol{B} = 0\,, \tag{2.59 a}$$

$$\boldsymbol{k}\times\boldsymbol{E} = +\frac{\omega}{c}\boldsymbol{B}, \quad \boldsymbol{k}\times\boldsymbol{B} = -\frac{\omega}{c}\boldsymbol{D}\,. \tag{2.59 b}$$

The statements in (2.59 a) require both $\boldsymbol{D}$ and $\boldsymbol{B}$ to lie in the plane perpendicular to the wave vector $\boldsymbol{k}$. This is true not only for a material described by the dielectric tensor displayed in (2.58), but for any arbitrary dielectric. Furthermore, the second equation in (2.59 b) requires $\boldsymbol{D}$ and $\boldsymbol{B}$ to be perpendicular to each other, within the plane perpendicular to $\boldsymbol{k}$. Thus, if we consider the three vectors $\boldsymbol{k}$, $\boldsymbol{D}$ and $\boldsymbol{B}$, their relationship is exactly the same as the vectors $\boldsymbol{k}$, $\boldsymbol{E}$ and $\boldsymbol{B}$ associated with an electromagnetic wave in free space.

In the anisotropic dielectric, however, the electric field $\boldsymbol{E}$ is not, in general, perpendicular to the wave vector, but may contain a component parallel to $\boldsymbol{k}$ (a longitudinal component) as well as one transverse to $\boldsymbol{k}$. To obtain a description of these waves, one requires an explicit form for the dielectric tensor. We thus turn to a material described by the dielectric tensor in (2.58).

If we align the z axis of a coordinate system along the optic axis, then the dielectric tensor in (2.58) is invariant in form under any arbitrary rotation about the z axis. Thus, there is no loss of generality if we choose $\boldsymbol{k}$ to lie in the xz plane, as illustrated in Fig. 2.7. We can decompose $\boldsymbol{E}$ and $\boldsymbol{B}$ into components within and perpendicular to the xz plane. We write

$$\boldsymbol{E} = \boldsymbol{E}^{(\mathrm{e})} + \hat{\boldsymbol{y}}E^{(\mathrm{o})} \tag{2.60 a}$$

and

$$\boldsymbol{B} = \boldsymbol{B}^{(\mathrm{o})} + \hat{\boldsymbol{y}}B^{(\mathrm{e})}\,, \tag{2.60 b}$$

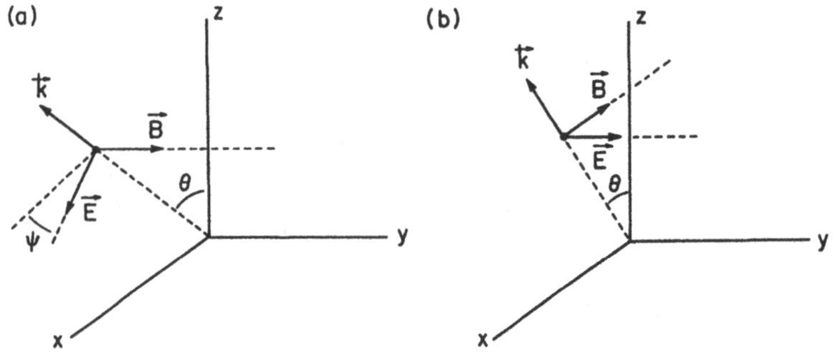

Fig. 2.7. The electric and magnetic fields associated with **a** the extraordinary wave, and **b** the ordinary wave in a uniaxial dielectric described by the dielectric tensor in (2.58). In **a**, the $\boldsymbol{B}$ field is parallel to the xy plane, and the electric field lies in the xz plane, which also contains $\boldsymbol{k}$. For a general value of θ, $\boldsymbol{E}$ is not perpendicular to $\boldsymbol{k}$, and $\psi \neq 0$. In **b**, $\boldsymbol{B}$ and $\boldsymbol{E}$ are both perpendicular to $\boldsymbol{k}$, with $\boldsymbol{B}$ in the xz plane and $\boldsymbol{E}$ perpendicular to it

where $\boldsymbol{E}^{(e)}$ and $\boldsymbol{B}^{(o)}$ each lie in the xz plane. The motivation for the notation will be clear shortly. We have also, with θ the angle between $\boldsymbol{k}$ and the z axis, $k_x = k \sin\theta$ and $k_z = k\cos\theta$. The Maxwell equations then break down into the following, noting that $\boldsymbol{B}^{(o)}$ must necessarily be perpendicular to $\boldsymbol{k}$:

$$k|\boldsymbol{B}^{(o)}| = \frac{\omega}{c}\varepsilon_\perp E^{(o)}, \tag{2.61 a}$$

$$kE^{(o)} = \frac{\omega}{c}|\boldsymbol{B}^{(o)}|; \tag{2.61 b}$$

and

$$\varepsilon_\perp E_x^{(e)} \sin\theta + \varepsilon_\parallel E_z^{(e)} \cos\theta = 0, \tag{2.62 a}$$

$$k[E_z^{(e)} \sin\theta - E_x^{(e)} \cos\theta] = -\frac{\omega}{c}B^{(e)}, \tag{2.62 b}$$

$$k[\hat{\mathbf{x}} \cos\theta - \hat{\mathbf{z}} \sin\theta]B^{(e)} = \frac{\omega}{c}[\hat{\mathbf{x}}\varepsilon_\perp E_x^{(e)} + \hat{\mathbf{z}}\varepsilon_\parallel E_z^{(e)}]. \tag{2.62 c}$$

The equations thus decouple into two sets, one linking the two fields $(\boldsymbol{B}^{(o)}, \hat{\mathbf{y}}E^{(o)})$, and one linking the pair $(\boldsymbol{E}^{(e)}, \hat{\mathbf{y}}B^{(e)})$.

The first wave, involving the pair $(\boldsymbol{B}^{(o)}, \hat{\mathbf{y}}E^{(o)})$, is called the ordinary wave. By construction, $\hat{\mathbf{y}}E^{(o)}$ is perpendicular to $\boldsymbol{k}$, and we have established that $\boldsymbol{B}^{(o)}$ is also. This is thus a transverse wave, very similar in nature to the electromagnetic wave in the isotropic dielectric. Equation (2.61) easily provides the dispersion relation for the wave:

$$\frac{c^2k^2}{\omega^2} = \varepsilon_\perp \equiv n_0^2, \tag{2.63}$$

which is identical in form to (2.36). The quantity $n_o = \sqrt{\varepsilon_\perp}$ is called the index of refraction of the ordinary wave.

The second wave, linking the pair $(\boldsymbol{E}^{(e)}, \hat{\boldsymbol{y}}B^{(e)})$ is called the extraordinary wave. While $\hat{\boldsymbol{y}}B^{(e)}$ is perpendicular to the wave, from (2.62 a) it follows that $\boldsymbol{E}^{(e)}$ is not. Upon eliminating $B^{(e)}$ from (2.62 c) using (2.62 b), one finds

$$\left[\cos^2\theta - \frac{\omega^2\varepsilon_\perp}{c^2k^2}\right]E_x^{(e)} - \sin\theta\cos\theta\, E_z^{(e)} = 0\ , \tag{2.64 a}$$

$$-\sin\theta\cos\theta\, E_x^{(e)} + \left[\sin^2\theta - \frac{\omega^2\varepsilon_\parallel}{c^2k^2}\right]E_z^{(e)} = 0\ , \tag{2.64 b}$$

which admit a nontrivial solution when

$$\frac{c^2k^2}{\omega^2} = \left[\frac{\sin^2\theta}{\varepsilon_\parallel} + \frac{\cos^2\theta}{\varepsilon_\perp}\right]^{-1} \equiv n_e^2(\theta)\ , \tag{2.65}$$

where $n_e(\theta)$ is the index of refraction of the extraordinary wave.

The index of refraction of the ordinary wave is, in this example, independent of the propagation direction of the wave, while the extraordinary wave has an index of refraction dependent on the direction of propagation. Later in this volume, we shall see how these two features may be exploited to advantage when we appreciate aspects of the nonlinear mixing of electromagnetic waves.

The angle ψ in Fig. 2.7 a is a measure of the degree to which the electric field in the extraordinary wave is tilted out of the plane perpendicular to $\boldsymbol{k}$. One may show that

$$\tan(\psi) = (\varepsilon_\parallel - \varepsilon_\perp)\left[\frac{\sin\theta\cos\theta}{\varepsilon_\parallel\cos^2\theta + \varepsilon_\perp\sin^2\theta}\right]\ . \tag{2.66}$$

Once again, the electric and magnetic fields in the two waves are illustrated in Fig. 2.7.

Problems

2.1 We have seen that free electrons are characterized by a frequency dependent conductivity

$$\sigma(\omega) = \frac{ne^2}{m}\frac{1}{1/\tau - \mathrm{i}\omega}$$

and they contribute a term to the frequency dependent susceptibility of the form

$$\chi(\omega) = -\frac{ne^2}{m}\frac{1}{\omega(\omega + \mathrm{i}/\tau)}\ .$$

(a) Find expressions for $\sigma(t - t')$, the "real time conductivity" which relates the current density $J(t)$ to the past behavior of the electric field $E(t)$. Also, find $\chi(t - t')$ that relates the polarization $P(t)$ to $E(t')$. (To handle the pole at $\omega = 0$, replace ω by $\omega + i\eta$ and let $\eta \to 0$. In essence, one must always examine the response of the system to a field switched on adiabatically from time $t = -\infty$.)

Suppose the electrons are subjected to a pulsed electric field, which vanishes as the time $t \to +\infty$. Notice that $J(t)$ vanishes as $t \to +\infty$, but $P(t)$ does not.

(b) The electrons in this problem have an average velocity v described by

$$\frac{dv}{dt} = \frac{e}{m}E(t) - \frac{v}{\tau} .$$

By directly integrating this equation, for general $E(t)$, show that when the electrons respond to a field pulse,

$$\lim_{t\to\infty} P(t) = ne\Delta x ,$$

where Δx is the displacement of the electron gas in response to the pulse.

2.2 The electrons discussed in problem 2.1 are placed in a static magnetic field of strength B_0, parallel to the z direction.

(a) Find the form of the frequency dependent polarizability tensor $\chi_{\alpha\beta}(\omega)$ for the electrons, in the limit the relaxation time $\tau \to \infty$.

(b) Use this tensor to discuss the dispersion relation and polarization of electromagnetic waves which propagate parallel to the z direction. Show that as the frequency $\omega \to 0$, one mode has frequency proportional to the square of the wave vector. In solid state physics, such waves are called helicons, and in magnetospheric physics they are referred to as whistlers.

2.3 In texts on electromagnetic theory, one encounters the notion that energy is stored in the electromagnetic field. It is demonstrated, that for nonmagentic media where $\mathbf{B} = \mathbf{H}$, the energy density U has the time derivative

$$\frac{dU}{dt} = \frac{1}{4\pi}\left[\frac{\partial \mathbf{D}}{\partial t}\cdot \mathbf{E} + \frac{\partial \mathbf{B}}{\partial t}\cdot \mathbf{B}\right] .$$

(a) Consider an isotropic dielectric with frequency dependent dielectric constant $\varepsilon(\omega)$, assumed here to be real and positive. An electromagnetic wave propagates through the medium. Its electric field, at a selected point in space, has the form

$$\mathbf{E}(t) = \hat{\mathbf{e}}[E_\omega(t)e^{-i\omega t} + E_\omega^*(t)e^{+i\omega t}] ,$$

where $E_\omega(t)$ is an envelope function that varies slowly in time, changing very little over one period $2\pi/\omega$ of the wave, and also changing little over the characteristic response times of the material.

Show that to good approximation one has

$$D(t) = \hat{e}\left[e^{-i\omega t}\varepsilon(\omega)E_\omega(t) + ie^{-i\omega t}\left(\frac{\partial \varepsilon}{\partial \omega}\right)\left(\frac{\partial E_\omega}{\partial t}\right) + \text{c.c.}\right].$$

(b) Let $\langle dU/dt\rangle$ be the time average of the quantity dU/dt, averaged over a period $2\pi/\omega$ of the "carrier," during which $E_\omega(t)$ changes little. Show that

$$\left\langle \frac{dU}{dt}\right\rangle = \frac{1}{4\pi}\left[2\varepsilon(\omega) + \omega\frac{\partial \varepsilon}{\partial \omega}\right]\frac{\partial}{\partial t}|E_\omega(t)|^2 .$$

Hence, on the time average, we may say the wave contains the energy density

$$\langle U\rangle = \frac{1}{4\pi}\left[2\varepsilon(\omega) + \omega\frac{\partial \varepsilon}{\partial \omega}\right]|E_\omega(t)|^2 .$$

(c) The Poynting vector $S = cE \times B/4\pi$ provides the energy/unit area-unit time transported by the wave. Express the magnitude $\langle S\rangle$ of the time averaged Poynting vector in terms of $\varepsilon(\omega)$, and $E_\omega(t)$.

The energy transport velocity $v_E(\omega)$ is defined by the relation

$$\langle S\rangle = v_E(\omega)\langle U\rangle .$$

Find an expression for $v_E(\omega)$ and show it to be identical to the group velocity calculated from the dispersion relation in (2.36).

When the imaginary part of $\varepsilon(\omega)$ is nonzero, the energy transport and group velocities may differ substantially. See [2.7] of the present chapter for a discussion.

2.4 A particular crystal has a preferred axis, parallel to the z direction. Furthermore the crystal is left unchanged by 120° rotations about the z direction (one says the z axis is a three-fold axis). Show that the dielectric tensor has the form given in (2.58), i.e., the dielectric tensor is invariant in form under arbitrary rotations around the z axis.

2.5 A point charge is placed at the origin of the coordinate system, in an anisotropic dielectric described by the dielectric tensor in (2.58). Find the electrostatic potential everywhere, and sketch the electric field lines for the two cases $\varepsilon_\perp > \varepsilon_\parallel$, and $\varepsilon_\perp < \varepsilon_\parallel$.

3. Nonlinear Dielectric Response of Matter

In this chapter, we explore general aspects of the nonlinear dielectric response of materials. Many of the principles and notions follow along lines very similar to those discussed in Chap. 2. As a consequence, we can be quite a bit briefer here.

Our interest is in the structure of the nonlinear contribution to the dipole moment per unit volume displayed in (1.7 c) This expression is subject to the same criticisms that we leveled against (1.7 b), and also (2.1). In any real medium, the dipole moment per unit volume $P_\alpha(\boldsymbol{r}, t)$ is not simply controlled by the instantaneous value of the electric field at the point $\boldsymbol{r}$, but in fact we must realize that there is a time lag in the response of the medium. Also, the response is nonlocal in space, for the same reasons discussed in Chap. 2. If we consider the extension of (1.7 c) that incorporates these features, then for a homogeneous medium, we may write

$$\begin{aligned} P_\alpha^{(\mathrm{NL})}(\boldsymbol{r}, t) = &\sum_{\beta\gamma} \int d^3r_1\, dt_1\, d^3r_2\, dt_2 \chi^{(2)}_{\alpha\beta\gamma}(\boldsymbol{r} - \boldsymbol{r}_1, t - t_1; \boldsymbol{r} - \boldsymbol{r}_2, t - t_2) \\ &\times E_\beta(\boldsymbol{r}_1, t_1) E_\gamma(\boldsymbol{r}_2, t_2) + \sum_{\beta\gamma\delta} \int d^3r_1\, dt_1\, d^3r_2\, dt_2\, d^3r_3\, dt_3 \\ &\times \chi^{(3)}_{\alpha\beta\gamma\delta}(\boldsymbol{r} - \boldsymbol{r}_1, t - t_1; \boldsymbol{r} - \boldsymbol{r}_2, t - t_2; \boldsymbol{r} - \boldsymbol{r}_3, t - t_3) \\ &\times E_\beta(\boldsymbol{r}_1, t_1) E_\gamma(\boldsymbol{r}_2, t_2) E_\delta(\boldsymbol{r}_3, t_3) + \cdots . \end{aligned} \tag{3.1}$$

We can proceed by introducing a Fourier decomposition of the electric field, as in (2.10). Consider, for simplicity, just the second order contribution the nonlinear electric dipole moment proportional to $\chi^{(2)}_{\alpha\beta\gamma}$. The generalization to the case of the higher order terms will be clear after we examine the structure of this term. We introduce the wave vector and frequency dependent second order susceptibility $\chi^{(2)}_{\alpha\beta\gamma}(\boldsymbol{k}_1\omega_1; \boldsymbol{k}_2\omega_2)$, defined by

$$\begin{aligned} \chi^{(2)}_{\alpha\beta\gamma}(\boldsymbol{k}_1\omega_1; \boldsymbol{k}_2\omega_2) = &\int \frac{d^3k_1 d\omega_1 d^3k_2 d\omega_2}{(2\pi)^8} \mathrm{e}^{-i\boldsymbol{k}_1\cdot(\boldsymbol{r}-\boldsymbol{r}_1)} \mathrm{e}^{+i\omega_1(t-t_1)} \mathrm{e}^{-i\boldsymbol{k}_2\cdot(\boldsymbol{r}-\boldsymbol{r}_2)} \mathrm{e}^{+i\omega_2(t-t_2)} \\ &\times \chi^{(2)}_{\alpha\beta\gamma}(\boldsymbol{r} - \boldsymbol{r}_1, t - t_1; \boldsymbol{r} - \boldsymbol{r}_2, t - t_2) . \end{aligned} \tag{3.2}$$

One then finds that

$$P_\alpha^{(\mathrm{NL})}(\boldsymbol{r},t) = \int \frac{d^3k_1 d\omega_1 d^3k_2 d\omega_2}{(2\pi)^3} \mathrm{e}^{\mathrm{i}(\boldsymbol{k}_1+\boldsymbol{k}_2)\cdot\boldsymbol{r}} \mathrm{e}^{-\mathrm{i}(\omega_1+\omega_2)t}$$
$$\times \sum_{\beta\gamma} \chi_{\alpha\beta\gamma}^{(2)}(\boldsymbol{k}_1\omega_1;\boldsymbol{k}_2\omega_2) E_\beta(\boldsymbol{k}_1\omega_1) E_\gamma(\boldsymbol{k}_2\omega_2) + \cdots . \qquad (3.3)$$

The physical interpretation of the right-hand side of (3.3) is as follows: If we isolate the contributions from a particular wave vector and frequency $(\boldsymbol{k}_1\omega_1)$, combined with that from $(\boldsymbol{k}_2\omega_2)$, we see that the two plane waves $\exp[\mathrm{i}(\boldsymbol{k}_1\cdot\boldsymbol{r}_1 - \omega_1 t_1)]$ and $\exp[\mathrm{i}(\boldsymbol{k}_2\cdot\boldsymbol{r}_2 - \omega_2 t_2)]$ "mix" together in the nonlinear material, to produce an output wave at the frequency $\omega_1 + \omega_2$, and whose wave vector $\boldsymbol{k}$ is the sum $\boldsymbol{k}_1 + \boldsymbol{k}_2$ of the wave vector of the two input waves. If we were to consider the third order terms generated by $\chi_{\alpha\beta\gamma\delta}^{(3)}$, then three waves with wave vector and frequency combinations $(\boldsymbol{k}_1\omega_1)$, $(\boldsymbol{k}_2\omega_2)$ and $(\boldsymbol{k}_3\omega_3)$ will generate output waves with wave vectors $\boldsymbol{k}_1 + \boldsymbol{k}_2 + \boldsymbol{k}_3$, at frequency $\omega_1 + \omega_2 + \omega_3$.

It is evident from construction that $\chi_{\alpha\beta\gamma}^{(2)}(\boldsymbol{k}_1\omega_1;\boldsymbol{k}_2\omega_2)$ must be left invariant under simultaneous interchange of $(\beta\boldsymbol{k}_1\omega_1)$ and $(\gamma\boldsymbol{k}_2\omega_2)$. A similar remark applies to $\chi_{\alpha\beta\gamma\delta}^{(3)}(\boldsymbol{k}_1\omega_1;\boldsymbol{k}_2\omega_2;\boldsymbol{k}_3\omega_3)$, not displayed here in the interest of brevity. We shall invoke this property in some of our later discussions.

If we consider just simple plane waves which propagate in a medium, then we must look a bit more closely to see the nature of the output. The point is that a real plane wave is not described by a single complex exponential $\exp[\mathrm{i}(\boldsymbol{k}\cdot\boldsymbol{r} - \omega t)]$, but by a superposition of this form with $\exp[-\mathrm{i}(\boldsymbol{k}\cdot\boldsymbol{r} - \omega t)]$. For example, if we have the plane wave whose electric field has the form

$$E_\beta^{(a)}(\boldsymbol{r},t) = E_\beta^{(a)} \cos(\boldsymbol{k}_a\cdot\boldsymbol{r} - \omega_a t + \phi_a) , \qquad (3.4)$$

this can be written

$$E_\beta^{(a)}(\boldsymbol{r},t) = \frac{1}{2} E_\beta^{(a)} \mathrm{e}^{\mathrm{i}\phi_a} \mathrm{e}^{\mathrm{i}(\boldsymbol{k}_a\cdot\boldsymbol{r}-\omega_a t)} + \frac{1}{2} E_\beta^{(a)} \mathrm{e}^{-\mathrm{i}\phi_a} \mathrm{e}^{-\mathrm{i}(\boldsymbol{k}_a\cdot\boldsymbol{r}-\omega_a t)} . \qquad (3.5)$$

If we represent this wave by (2.10), we must choose

$$E_\beta(\boldsymbol{k},\omega) = \frac{(2\pi)^4}{2} E_\beta^{(a)} \mathrm{e}^{\mathrm{i}\phi_a} \delta(\boldsymbol{k} - \boldsymbol{k}_a)\delta(\omega - \omega_a)$$
$$+ \frac{(2\pi)^4}{2} E_\beta^{(a)} \mathrm{e}^{-\mathrm{i}\phi_a} \delta(\boldsymbol{k} + \boldsymbol{k}_a)\delta(\omega + \omega_a) . \qquad (3.6)$$

Suppose we have two plane waves which propagate in the medium, so the total "input" electric field is given by

$$E_\beta(\boldsymbol{r},t) = E_\beta^{(a)} \cos(\boldsymbol{k}_a\cdot\boldsymbol{r} - \omega_a t + \phi_a) + E_\beta^{(b)} \cos(\boldsymbol{k}_b\cdot\boldsymbol{r} - \omega_b t + \phi_b) . \qquad (3.7)$$

Then if we consider the contributions to the dipole moment per unit volume from the second order susceptibility $\chi_{\alpha\beta\gamma}^{(2)}$, there will be three classes of terms that we may describe as follows:

(i) Self interaction terms associated with wave (a):
These will be proportional to $E_\beta^{(a)}E_\gamma^{(a)}$ and will have Fourier components proportional to $\boldsymbol{k}_a + \boldsymbol{k}_a \equiv 2\boldsymbol{k}_a$, and $\omega_a + \omega_a = 2\omega_a$; this wave is called the second harmonic. Also, we have electric fields with Fourier components $\boldsymbol{k}_a + (-\boldsymbol{k}_a) = 0$, and $\omega_a + (-\omega_a) = 0$. Thus, the self interactions within the wave also produce spatially uniform, dc electric fields. One refers to this as optical rectification.

(ii) Self interaction terms associated with wave (*b*):
The nature of these terms is identical to those associated with wave (*a*). We thus get the second harmonic of wave (*b*), along with optical retification from this wave.

(iii) Terms which describe the interaction between wave (*a*) and wave (*b*):
These terms are proportional to $E_\beta^{(a)}E_\gamma^{(b)}$, and $E_\beta^{(b)}E_\gamma^{(a)}$. We have contributions to the nonlinear dipole moment at the sum and differences $\pm(\boldsymbol{k}_a \pm \boldsymbol{k}_b)$, and $\pm(\omega_a \pm \omega_b)$. We thus find sum frequency generation and difference frequency generation, through the second order terms, in the presence of two input waves.

When we examine the contribution from the third order term proportional to $\chi^{(3)}_{\alpha\beta\gamma\delta}$, when one or more waves is present, we find a spectrum of nonlinear waves that can be described in terms of self interactions of a wave with itself, and nonlinear interactions between the various waves. We shall encounter specific examples later in the text.

3.1 Frequency Variation of the Nonlinear Susceptibilities

We see from (3.2) that $\chi^{(2)}_{\alpha\beta\gamma}$ depends on the frequencies ω_1 and ω_2 of the two interacting waves. Consider, for example, just the variation with ω_1, while the second frequency ω_2 is held fixed. For fixed ω_2, $\chi^{(2)}_{\alpha\beta\gamma}(\boldsymbol{k}_1\omega_1;\boldsymbol{k}_2\omega_2)$ considered a function of ω_1 obeys a Kramers-Kronig relation similar to that followed by the frequency dependent dielectric constant, as discussed in the previous chapter. That this is so follows simply from causality, which requires the real time representation of $\chi^{(2)}_{\alpha\beta\gamma}$, the functions $\chi^{(2)}_{\alpha\beta\gamma}(\boldsymbol{r} - \boldsymbol{r}_1, t - t_1; \boldsymbol{r} - \boldsymbol{r}_2, t - t_2)$ to vanish when $t_1 > t$. Then, from the definitions of $\chi^{(2)}_{\alpha\beta\gamma}(\boldsymbol{k}_1\omega_1; \boldsymbol{k}_2\omega_2)$ in (3.2), one sees this function is an analytic function of ω_1, in the upper half ω_1 plane. On physical grounds, this object must vanish as $\omega_1 \to \infty$. One may then write down the relevant Kramers-Kronig relation at once, following the discussion given by *Landau* and *Lifshitz* [3.1].

In the case of the dielectric constant, we saw that the Kramers-Kronig relation proved most valuable indeed, in our effort to understand general aspects of the frequency variation of the complex dielectric constant $\varepsilon(\omega)$. We know that $\varepsilon_2(\omega)$ is directly related to the frequency dependent absorption constant of a material; from our understanding of the frequency dependence of the absorption constant, we can thus draw inferences about the behavior of $\varepsilon_1(\omega)$. To the experimentalist, the Kramers-Kronig relation for $\varepsilon(\omega)$ is particularly valuable. If the absorption rate, from which $\varepsilon_2(\omega)$ can be inferred, is measured over a wide spectral range, then $\varepsilon_1(\omega)$ can be calculated directly. It is possible also to measure both $\varepsilon_1(\omega)$ and $\varepsilon_2(\omega)$ separately, in certain kinds of reflectivity studies. The Kramers-Kronig relations then serve as a consistency check on the results, if data is available over a sufficiently wide spectral range.

The Kramers-Kronig relations for the nonlinear susceptibilities are far less useful, and are encountered infrequently. The reason is that even at one pair of frequencies ω_1 and ω_2, the accurate measurement of a nonlinear susceptibility element like $\chi^{(2)}_{\alpha\beta\gamma}(\boldsymbol{k}_1\omega_1; \boldsymbol{k}_2\omega_2)$ proves rather difficult. If and when this can be done, say for fixed ω_2, the measurement will cover a limited range of ω_1. The Kramers-Kronig integrals are then of little practical assistance in generating useful information. As we shall appreciate shortly, the dependence of the various nonlinear susceptibility elements on ω_1 and ω_2 is rather more complex than realized in the linear dielectric response, where only a single frequency is involved, and one encounters singularities at the various excitation frequencies of the material (see Appendix A).

A simple model used frequently in the literature serves as a guide to the frequency variation of the nonlinear susceptibilities of a material [3.2].

Consider first a simple charged particle, with charge e, bound by a perfectly harmonic potential to an infinitely massive center. It is driven by an electric field $E(t)$; we consider one dimensional motion only, and the electric field is applied parallel to the direction of the motion. We write the potential energy $V_2(x)$ in the form

$$V_2(x) = \frac{1}{2} M\omega_0^2 x^2 , \tag{3.8}$$

where M is the mass of the particle, ω_0 its resonance frequency, and x the displacement from equilibrium. The equation of motion is of course

$$\ddot{x} + \Gamma\dot{x} + \omega_0^2 x = \frac{e}{M} E(t) \tag{3.9}$$

with Γ a damping rate. For a field of frequency ω, $E(t) = E(\omega)\exp(-\mathrm{i}\omega t)$, then $x(t) = x(\omega)\exp(-\mathrm{i}\omega t)$, where

$$x(\omega) = \frac{eE(\omega)}{M} \frac{1}{\omega_0^2 - \mathrm{i}\omega\Gamma - \omega^2} . \tag{3.10}$$

The electric dipole moment is given by $p(\omega) = ex(\omega)$. We write $p(\omega) = \alpha(\omega)E(\omega)$, where $\alpha(\omega)$ is the polarizability, given by

$$\alpha(\omega) = \frac{e^2}{M} \frac{1}{\omega_0^2 - \mathrm{i}\omega\Gamma - \omega^2} . \tag{3.11}$$

If there are n such charges per unit volume, and we can ignore interactions between them, then the frequency dependent dielectric constant is $\varepsilon(\omega) = 1 + 4\pi n\alpha(\omega)$. We arrive once again at a familiar frequency dependence.

The above simple model can be applied quite directly to a dilute array of linear diatomic molecules. Then M is the reduced mass of each molecule, ω_0 its vibrational frequency, and e the effective charge that describes its coupling to the external electric field. (Symmetry considerations require e to vanish for a diatomic molecule with inversion symmetry, such as O_2 or N_2).

If we now drive the above molecule with two superimposed electric fields, one with frequency ω_1 and one with frequency ω_2, we have for the total field felt by the molecule

$$E(t) = E(\omega_1)e^{-i\omega_1 t} + E(\omega_2)e^{-i\omega_2 t} + \text{complex conjugate} . \tag{3.12}$$

The fact that (3.9) is a linear differential equation means that the resulting displacement is just a simple superposition of that at frequency ω_1 and at frequency ω_2:

$$x(t) = x(\omega_1)e^{-i\omega_1 t} + x(\omega_2)e^{-i\omega_2 t} + \text{complex conjugate} . \tag{3.13}$$

We have no second harmonics, or interactions between the two input waves, for this very simple case.

For any actual physical system such as a linear diatomic molecule, the motion is not harmonic, but there are anharmonic terms in the total potential energy which supplement the principal terms $V_2(x) = M\omega_0^2 x^2/2$. For small amplitude displacements, we may expand the full potential $V(x)$ in a power series, to obtain in addition to $V_2(x)$, the anharmonic terms that will be written

$$V_3(x) + V_4(x) + \cdots = \frac{1}{3}Max^3 + \frac{1}{4}Mbx^4 + \cdots . \tag{3.14}$$

Retaining the first two terms gives

$$\ddot{x} + \Gamma\dot{x} + \omega_0^2 x + ax^2 + bx^3 = \frac{e}{M}E(t) . \tag{3.15}$$

We now obtain a model description of the nonlinear interactions discussed earlier from a general perspective. Suppose we now drive the system with an electric field (necessarily real) that is a superposition of frequency ω_1 and ω_2:

$$E(t) = \frac{1}{2}(E_1 e^{-i\omega_1 t} + E_1^* e^{+i\omega_1 t}) + \frac{1}{2}(E_2 e^{-i\omega_2 t} + E_2^* e^{+i\omega_2 t}) . \tag{3.16}$$

We can analyze the response of the oscillator by expanding the displacement in powers of the electric field $E(t)$. Thus, we shall write

$$x(t) = x^{(1)}(t) + x^{(2)}(t) + x^{(3)}(t) + \cdots , \tag{3.17}$$

where $x^{(i)}(t)$ is proportional to the i-th power of the field $E(t)$. If one inserts (3.17) into (3.15), and equates like power of $E(t)$, one finds the hierarchy

$$\ddot{x}^{(1)} + \Gamma\dot{x}^{(1)} + \omega_0^2 x^{(1)} = \frac{e}{M}E(t)\,, \tag{3.18 a}$$

$$\ddot{x}^{(2)} + \Gamma\dot{x}^{(2)} + \omega_0^2 x^{(2)} + a(x^{(1)})^2 = 0\,, \tag{3.18 b}$$

$$\ddot{x}^{(3)} + \Gamma\dot{x}^{(3)} + \omega_0^2 x^{(3)} + 2ax^{(1)}x^{(2)} + b(x^{(1)})^3 = 0\,, \tag{3.18 c}$$

where we ignore higher order terms.

The solution of (3.18 a) can be written at once. We define

$$d(\omega) = \frac{1}{\omega_0^2 - \mathrm{i}\omega\Gamma - \omega^2} \tag{3.19}$$

and very much as above we have

$$x^{(1)}(t) = \frac{e}{2M}[E_1 d(\omega_1)\mathrm{e}^{-\mathrm{i}\omega_1 t} + E_1^* d^*(\omega_1)\mathrm{e}^{+\mathrm{i}\omega_1 t} + E_2 d(\omega_2)\mathrm{e}^{-\mathrm{i}\omega_2 t}$$
$$+ E_2^* d^*(\omega_2)\mathrm{e}^{\mathrm{i}\omega_2 t}]\,. \tag{3.20}$$

One proceeds by moving the term $a(x^{(1)})^2$ to the right-hand side of (3.18 b), then treating it as an inhomogeneous driving term. The contributions to $x^{(2)}(t)$ are then the following:

(i) The second harmonic of the wave at frequency ω_1:

$$x^{(2)}(t) = -\frac{ae^2}{4M^2}E_1^2 d(2\omega_1)d^2(\omega_1)\mathrm{e}^{-\mathrm{i}2\omega_1 t} + \text{complex conjugate}\,; \tag{3.21 a}$$

(ii) The second harmonic of the wave at frequency ω_2:

$$x^{(2)}(t) = -\frac{ae^2E_2^2}{4M^2}d(2\omega_2)d^2(\omega_2)\mathrm{e}^{-\mathrm{i}2\omega_2 t} + \text{complex conjugate}\,; \tag{3.21 b}$$

(iii) The time independent (dc) component of the displacement:

$$x^{(2)}(t) = -\frac{ae^2}{2M^2}(|d(\omega_1)|^2|E_1|^2 + |d(\omega_2)|^2|E_2|^2)\,; \tag{3.21 c}$$

(iv) Generation of the sum frequency $\omega_1 + \omega_2$:

$$x^{(2)}(t) = -\frac{ae^2}{2M^2}d(\omega_1 + \omega_2)d(\omega_1)d(\omega_2)E_1E_2\mathrm{e}^{-\mathrm{i}(\omega_1+\omega_2)t}$$
$$+ \text{complex conjugate}\,; \tag{3.21 d}$$

(v) Generation of the difference frequency $\omega_1 - \omega_2$:

$$x^{(2)}(t) = -\frac{ae^2}{2M^2} d(\omega_1 - \omega_2) d(\omega_1) d^*(\omega_2) E_2 E_2^* e^{-i(\omega_1 - \omega_2)t}$$

$$+ \text{complex conjugate} . \tag{3.21 e}$$

The dipole moment $p(t)$ of our model oscillator is just $ex(t)$. Hence the various contributions to the second order nonlinear polarizability are obtained by multiplying the equations of (3.21) by e.

The third order terms are treated in a manner very similar to the second order terms. In (3.18 c), one moves the terms $b(x^{(1)})^3 + 2ax^{(1)}x^{(2)}$ to the right-hand side, and treats them as inhomogeneous driving terms, utilizing (3.20) and (3.21). The output $x^{(3)}(t)$ contains many terms, as one can appreciate from the expressions above. We shall quote only one contribution to the output that will prove useful later on, that at frequency $2\omega_1 - \omega_2$. At this frequency, we have

$$x^{(3)}(t) = -\frac{e^2}{4M^3} d(2\omega_1 - \omega_2) d^2(\omega_1) d^*(\omega_2)$$

$$\times \left[\frac{3}{2} b - 2a^2 d(\omega_1 - \omega_2) - a^2 d(2\omega_1)\right] E_1^2 E_2^* e^{-i(2\omega_1 - \omega_2)t}$$

$$+ \text{complex conjugate} . \tag{3.22}$$

The various expressions given above have a common feature, which can be illustrated by examining (3.21 d). We have the interaction between two input waves, one with frequency ω_1, and one with frequency ω_2. Note that there is a factor $d(\omega)$, associated with each input wave which exhibits a resonance when the frequency ω is "tuned" to the frequency ω_0 of the oscillator. Thus, we obtain resonant enhancement of the output if either input wave lies close in frequency to ω_0, not surprisingly.

The output is at the frequency $(\omega_1 - \omega_2)$, for the term displayed in (3.21 e). An array of oscillators, each with a dipole moment with time dependence $\exp[-i(\omega_1 - \omega_2)t]$, will generate an electromagnetic wave at frequency $(\omega_1 - \omega_2)$, as one may appreciate by inserting the nonlinear dipole moment density into the Maxwell equations; we shall see how this occurs in Chap. 4, for an important example. Notice that there appears a factor $d(\omega_1 - \omega_2)$ in (3.21 e). Thus, we also get resonant enhancement if the frequency $(\omega_1 - \omega_2)$ lies close to the resonance frequency of the oscillator. Similarly, for the case of sum frequency generation, (3.21 d) shows a resonance when $(\omega_1 + \omega_2)$ lies near ω_0.

We have here an example of a three wave interaction: the wave of frequency ω_1 "mixes" with that of frequency ω_2 to, among other things, produce a third output wave at $(\omega_1 - \omega_2)$. If any one of the waves, including the output wave, is in resonance with the physical system (the oscillator in this case) that provides the nonlinearity responsible for their interaction, we obtain resonance enhancement. This is a general principle of nonlinear wave mixing phenomena.

For the nonlinear susceptibilities we encounter in our study of nonlinear optics, one may appreciate this by examining the general structure of the quantum mechanical formulae for the nonlinear susceptibilities [3.3].

If we turn to (3.22), we have a description of an interaction between four waves. There are two interactions between waves of frequency ω_1, a second with the wave at ω_2, and there is then an output wave at frequency $2\omega_1 - \omega_2$. First suppose the cubic term ax^3 vanished, while the quartic term bx^4 remains nonzero. We then have the direct interaction between four waves, as we see from the $b(x^{(1)})^3$ term in (3.18 c). Upon noting $x^{(1)}$ is directly proportional to the total electric field $E(t)$ in (3.16), we see the response of the oscillator has its origin in a driving term proportional to $(E_1 e^{-i\omega_1 t})^2(E_2^* e^{i\omega_2 t})$. We thus have two enhancement factors associated with the frequency ω_1, $d(\omega_1)^2$, and a single resonant enhancement factor associated with ω_2. There is finally an enhancement factor associated with the output wave at $2\omega_1 - \omega_2$. The term proportional to b describes a direct four wave interaction, as illustrated diagramatically in Fig. 3.1 a.

When $a \neq 0$, so the cubic terms enter, we obtain additional contributions to the amplitude of the output at $(2\omega_1 - \omega_2)$, as we see in (3.22). These have the same overall enhancement factors as the direct four wave mixing promoted by the bx^4 term. However, there are additional resonant enhancement factors when $a \neq 0$, inside the square bracket in (3.22).

We have seen that a, acting by itself, promotes the three wave interactions responsible for the "outputs" displayed in (3.21). An effective four wave interaction results from a succession of two three wave interactions, as illustrated

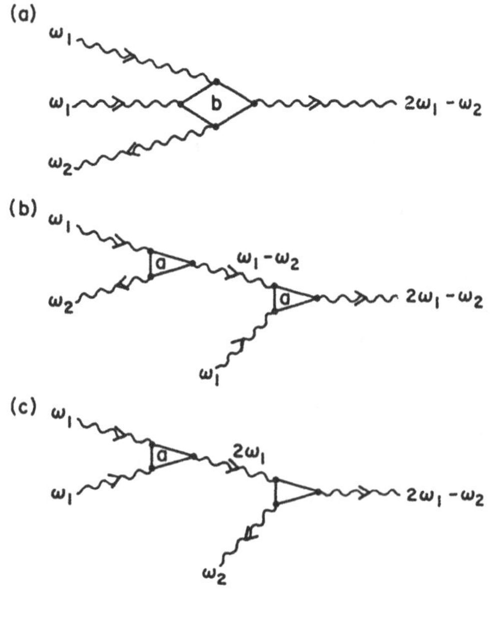

Fig. 3.1. An illustration of the processes which contribute to the four wave mixing which leads to the output at frequency $2\omega_1 - \omega_2$ displayed in (3.22). The Fourier component of a wave with frequency dependence $\exp(-i\omega t)$ is indicated by a left to right directed arrow, while the Fourier component of this wave which exhibits the frequency dependence $\exp(+i\omega t)$ is illustrated by a right to left directed arrow. We illustrate the interactions responsible for **a** the first, **b** the second, and **c** the third term of (3.22)

in Figs. 3.1 b and 3.1 c. The second term in (3.22) is the consequence of the interaction displayed in Fig. 3.1 b. We have an intermediate wave of frequency $(\omega_1 - \omega_2)$ generated by mixing of the input waves at ω_1 and ω_2, which then immediately mixes with the ω_1 wave, to produce output at $(2\omega_1 - \omega_2)$. The presence of the intermediate wave, not part of the output, is responsible for the additional enhancement factor in (3.22). The third term comes from the process illustrated in Fig. 3.1 c; the enhancement factor $d(2\omega_1)$ comes from the intermediate wave with frequency $2\omega_1$.

The anharmonic oscillator model discussed above can be applied in a literal manner to a simple molecule, where coupling to a single vibrational normal mode dominates the response of the above systems. Thus, the model applies in the infrared region of the spectrum. In fact, discrete, isolated electronic levels provide contributions to the nonlinear susceptibilities very similar in overall structure to those obtained for the anharmonic oscillator model. One may apply formulae for the nonlinear contributions to the dipole moment quite similar to those just derived to electronic excitations in gaseous media (the levels are quite sharp for such systems), and also to the exciton levels of semiconductors. Indeed, dye lasers can be tuned very close to exciton transitions in semiconducting crystals and films, with the consequence that the intensity of a desired nonlinear output can be increased by several orders of magnitude over those appropriate to off-resonance radiation. We have also seen that solids and other forms of dense matter contain absorption bands which extend over a broad frequency range. The nonlinear susceptibilities are enhanced substantially also for frequencies which lie close to the edges of such bands.

We can now appreciate that the frequency variations of the nonlinear susceptibilities are more complex than that of the linear dielectric constant. The higher order susceptibilities describe nonlinear interactions between several waves, and we can have resonant enhancement of the output if any of the participating waves has frequency close to an internal excitation energy of the system which promotes the mixing.

3.2 Wave Vector Dependence of the Nonlinear Susceptibilities

We shall offer only brief remarks on this topic. When the perturbation theoretic expressions for the nonlinear susceptibilities are examined [3.3], one sees wave vector dependencies that enter the formulae in a manner similar to (A.32). The resonant denominators acquire excitation energies which depend on one or more of the wave vectors of the interacting waves. If one is operating at a frequency very close to that of an isolated resonance, this wave vector dependence can shift its apparent frequency. Also, the presence of the nonzero wave vector within the matrix elements in the numerator can "activate" particular resonances, whose excitation matrix element vanishes in the long wavelength limit.

Finite wave vector effects can also modify selection rules deduced in the limit of very long wavelengths. Consider, for example, the second order electric susceptibility tensor $\chi^{(2)}_{\alpha\beta\gamma}(\boldsymbol{k}_1\omega_1; \boldsymbol{k}_2\omega_2)$ defined in (3.2). We have seen that, in the limit $\boldsymbol{k}_1, \boldsymbol{k}_2 \to 0$ symmetry requires this tensor to vanish identically, for any material which possesses an inversion center.[1] When $\chi^{(2)}_{\alpha\beta\gamma}$ vanishes, phenomena such as second harmonic generation, and the generation of the sum and difference frequencies by virtue of the interaction of a wave with frequency ω_1 with a wave of frequency ω_2, will be absent.

However, as we know $\chi^{(2)}_{\alpha\beta\gamma}$ depends on the wave vectors $\boldsymbol{k}_1$ and $\boldsymbol{k}_2$. A Taylor series expansion then yields, for a material with an inversion center,

$$\chi^{(2)}_{\alpha\beta\gamma}(\boldsymbol{k}_1\omega_1; \boldsymbol{k}_2\omega_2) = \sum_\delta \left(\frac{\partial\chi^{(2)}_{\alpha\beta\gamma}}{\partial k_1^\delta}\right)_0 k_1^\delta + \sum_\delta \left(\frac{\partial\chi^{(2)}_{\alpha\beta\gamma}}{\partial k_2^\delta}\right)_0 k_2^\delta + \cdots . \tag{3.23}$$

Under rotations and inversions of the coordinate system, derivatives such as $(\partial\chi^{(2)}_{\alpha\beta\gamma}/\partial k_1^\delta)_0$ are fourth rank tensors. There are nonzero elements of fourth rank tensors in any material, even if it is completely isotropic.

Thus, by exploiting the wave vector dependence of $\chi^{(2)}_{\alpha\beta\gamma}$, one may "activate" nonlinear processes forbidden by symmetry in the limit of long wavelengths. In the usual circumstance, where the wavelength of the radiation is long compared to the underlying microscopic lengths in the system, the terms proportional to the wave vector will be quite small in magnitude. However, as discussed in the previous section, their magnitude can be enhanced substantially by operating near resonance.

Suppose we consider the second order term in (3.1), and endow $\chi^{(2)}_{\alpha\beta\gamma}$ with the wave vector dependence displayed in (3.23). We may define

$$E_\alpha(\boldsymbol{r}_1, \omega_1) = \int \frac{d^3k_1}{(2\pi)^3} E_\alpha(\boldsymbol{k}_1, \omega_1) e^{i\boldsymbol{k}_1\cdot\boldsymbol{r}_1} , \tag{3.24}$$

which describes the spatial variation of the component of the electric field with frequency ω_1. The first term in (3.1) can then be cast into the form

$$P_\alpha^{(\mathrm{NL})}(\boldsymbol{r}, t) = \frac{1}{i} \sum_{\beta\gamma\delta} \int \frac{d\omega_1 d\omega_2}{(2\pi)^2} e^{-i\omega_1 t} e^{-i\omega_2 t} \left\{ \left(\frac{\partial\chi^{(2)}_{\alpha\beta\gamma}}{\partial k_1^\delta}\right)_0 \left[\frac{\partial E_\beta(\boldsymbol{r}, \omega_1)}{\partial r^\delta}\right] E_\gamma(\boldsymbol{r}, \omega_2) \right.$$
$$\left. + \left(\frac{\partial\chi^{(2)}_{\alpha\beta\gamma}}{\delta k_2^\delta}\right)_0 \left[\frac{\partial E_\gamma}{\partial r^\delta}(\boldsymbol{r}, \omega_2)\right] E_\beta(\boldsymbol{r}, \omega_1) \right\} . \tag{3.25}$$

In (3.25), the nonlinear electric dipole moment per unit volume, $\boldsymbol{P}_\alpha^{(\mathrm{NL})}(\boldsymbol{r}, t)$, is related to quantities evaluated at the same point in space. We see, however, that it is the spatial gradients of the various components of the electric field that enter, in addition to the electric field itself.

[1] A compilation of the magnitude of nonzero elements of $\chi^{(2)}_{\alpha\beta\gamma}$ measured for various insulating materials can be found in [Ref. 3.4, Table 7.1]. For more details see [3.5].

If we return to (1.6), before we encountered the notion that there is time lag in the response of the system in the time domain, and also the response is nonlocal in space, we see that $\chi^{(2)}_{\alpha\beta\gamma}$ may be viewed as endowing the electric susceptibility of the medium with a dependence on electric field; in general, the dielectric susceptibility tensor is defined by writing $\chi_{\alpha\beta} = \partial P_\alpha/\partial E_\beta$, with the derivative evaluated at finite, nonzero applied field. When the nonlocal response of the medium is evaluated to leading order in the wave-vector dependent corrections to a quantity such as $\chi^{(2)}_{\alpha\beta\gamma}(\boldsymbol{k}_1\omega_1; \boldsymbol{k}_2\omega_2)$, as in (3.23), the dipole moment per unit volume may be regarded to be a function of the gradient tensor $\partial E_\beta/\partial r_\gamma$, in addition to the components of $\boldsymbol{E}$ itself. Then $\partial P_\alpha/\partial E_\beta$ evaluated at finite fields contains terms proportional to $\partial E_\beta/\partial r_\gamma$. Electric field gradients in general interact with the quadrupolar components of the charge distribution of a system. One can refer to such contributions to the nonlinear dielectric response as quadrupole contributions to the nonlinear susceptibility.

3.3 Remarks on the Order of Magnitude of the Nonlinear Susceptibilities

In Sect. 3.1 of the present chapter, we explored the frequency variation of the nonlinear dipole moment for an anharmonic oscillator driven by a superposition of electric fields, one with frequency ω_1 and one with frequency ω_2. We could continue on to use this example to estimate the size of the various contributions to the nonlinear susceptibility. If, however, we take this model literally, we must recognize that in practice, nearly all experiments in nonlinear optics are carried out in or near the visible range of frequencies. Such frequencies lie very far above those characteristic of the internal vibrations of even very small, tightly bound molecules. A consequence is that it is the electronic degrees of freedom, and not the vibrational degrees of freedom that are the source of the nonlinear response characteristic of most materials. Thus, while our anharmonic oscillator model proved to be a most useful means of illustrating the frequency dependencies of the various contributions to the nonlinear susceptibilities, we have to proceed differently to estimate their order of magnitude in nearly all actual experiments. The reader will note that microscopic theories of the electronic contributions provide rather complex expressions for $\chi^{(2)}_{\alpha\beta\gamma}$, and $\chi^{(2)}_{\alpha\beta\gamma\delta}$, but the energy denominators one encounters are similar in structure to those in the anharmonic oscillator example.

However, a simple argument serves to set the order of magnitude of $\chi^{(2)}_{\alpha\beta\gamma}$ and $\chi^{(3)}_{\alpha\beta\gamma\delta}$. We see that to calculate these tensors, we need to expand the electric dipole moment per unit volume in powers of an externally applied electric field $\boldsymbol{E}$. It is the response of the most weakly bound, outermost electrons in the atoms, molecule, or formula unit from which the material of interest is fabricated that provides the dominant contribution to the nonlinear susceptibilities.

Let E_a be the strength of the average electric field from the nucleus and nearby electrons experienced by such an outer electron, and E the strength of the external electric field, assumed for the moment to be either static, or with frequency spectrum far from resonance. Our basic expansion, (1.6), may then be viewed as an expansion in powers of the dimensionless parameter E/E_a. Thus, it follows that the typical magnitude of elements of the second order susceptibility $\chi^{(2)}_{\alpha\beta\gamma}$, denoted for the moment by $\langle\chi^{(2)}\rangle$, are equal to $\langle\chi^{(1)}\rangle/E_a$, where $\langle\chi^{(1)}\rangle$ is an average of the elements $\chi^{(1)}_{\alpha\beta}$. Similarly, in an obvious notation, $\langle\chi^{(3)}\rangle \approx \langle\chi^{(1)}\rangle/E_a^2$.

For most insulating solid materials, values of the dielectric constant lie in the range of from 2 to 10; the former value is appropriate for wide gap transparent insulators, and the latter for semiconductors. Thus, one has $\langle\chi^{(1)}\rangle \sim 0.1$ to 1.0. We may estimate E_a as follows: Suppose the binding energies of the outer electrons lie in the range of ten electron volts, and they reside on the average $1\,\text{Å} = 10^{-8}$ cm from the nucleus. We may then estimate $E_\alpha \cong 10^9$ V/cm, or in esu we have $E_a \cong 3 \times 10^6$ statvolts/cm. Hence, we estimate

$$\langle\chi^{(2)}\rangle \sim 3 \times 10^{-8} \text{ to } 3 \times 10^{-7} \text{ esu}$$

and

$$\langle\chi^{(3)}\rangle \sim 10^{-14} \text{ to } 10^{-13} \text{ esu}.$$

These estimates are indeed in the range of experimentally measured nonlinear susceptibilities of solid materials, at frequencies away from electronic resonances [3.4], though one surely finds examples of materials that have susceptibilities that lie outside these ranges.

Most dense insulating materials suffer dielectric breakdown for external fields not very far in excess of 10^6 V/cm as we remarked in Chap. 1. Thus, in practice, the largest values of E/E_a we may expect to encounter lie in the range of 10^{-3}. Under these conditions, the basic notion that underlies our discussion so far, that we may begin by expanding the electric dipole moment $\boldsymbol{P}$ in powers of the external electric field $\mathbf{E}$ as in (1.6) and in its generalizations such as (3.1), is very well justified indeed.

However, there are cases where the nonlinear response coefficients can substantially exceed the above estimate. This can be the case where a medium is excited at frequencies very close to resonance. For example, alkali vapors have electronic excited states in the visible, as one knows from the Na D lines, that provide very strong absorption within the yellow portion of the visible spectrum. In the vapor phase, these lines can be very narrow, and with tunable dye lasers one can operate very close to resonance. Even though the atomic density is much lower than in solids, very large values of $\langle\chi^{(3)}\rangle$ can be realized. The perturbation expansion in powers of $\boldsymbol{E}$ may then break down.

We shall use the following terminology in what follows: The regimes in which the expansion of the dipole moment per unit volume in powers of the external field $\boldsymbol{E}$ suffices will be referred to as the regime of weak nonlinearities; under such circumstances, the material's electronic structure is perturbed only very modestly by the applied field. The regime of strong nonlinearity, encountered only for frequencies very close to a narrow resonance, is where the perturbation approach cannot be used as a framework for the analysis. In this circumstance, the electronic structure of the system is perturbed strongly by the applied field.

We now turn to the descripton of selected but important examples of nonlinear optical phenomena.

Problems

3.1 For the anharmonically bound charge explored after (3.15), find an expression for the nonlinear electric dipole moment at the frequency $3\omega_1$. Discuss the interpretation of the result in terms of diagrams such as those in Fig. 3.1.

3.2 We have a linear triatomic molecule of the type AB_2. Each B atom has the charge Q, and for the purposes of this problem the mass of atom A is infinite. The atoms are coupled by anharmonic springs described by the potential function found by combining (3.8) with (3.14), where x is the difference between the instantaneous and the equilibrium bond lengths. The system is thus described by a model Hamiltonian

$$H = \frac{1}{2m_B}(P_1^2 + P_2^2) + V(x_1) + V(-x_2) ,$$

with x_1 and x_2 the displacement from equilibrium of the two B atoms. One is located at $+x_0$, and the other at $-x_0$, and motions are confined to the x axis.

The molecule is exposed to an electric field $E_0 \cos(kx - \omega t)$. Find the component of electric dipole moment at the frequency 2ω, and show it vanishes as $k \to 0$.

3.3 In an isotropic medium, we have seen that $\chi^{(2)}_{\alpha\beta\gamma}(\boldsymbol{k}_1\omega_1; \boldsymbol{k}_2\omega_2)$ vanishes, in the limit that the wave vectors $\boldsymbol{k}_1$ and $\boldsymbol{k}_2$ both vanish.

Consider, in such an isotropic medium, the influence of the terms linear in wave vector, as described in (3.23). We have a wave with frequency ω_1, wave vector $\boldsymbol{k}_1$ parallel to z, and electric field $\boldsymbol{E}_1$ is parallel to x. We wish to inquire if through these terms, the wave may "mix" with a transverse wave of frequency ω_2, and wave vector $\boldsymbol{k}_2$, to produce output at $\omega_1 \pm \omega_2$. Explore this possibility (a) for the case where $\boldsymbol{k}_2$ is parallel to z also, and (b) $\boldsymbol{k}_2$ is directed along y.

3.4 A quantum mechanical system is described by the Hamiltonian H_0, and has the set of eigenfunctions $\{|\psi_n\rangle\}$. It is subjected to a perturbation of the form

$$V(t) = v_1 e^{-i(\omega_1 + i\eta)t} + v_1^{+} e^{+i(\omega_1 - i\eta)t} + v_2 e^{-i(\omega_2 + i\eta)t} + v_2^{+} e^{+i(\omega_2 - i\eta)t} ,$$

which we may write in the form

$$V(t) = \sum_{\sigma=+,-} \sum_{j=1}^{2} v_j(\sigma) e^{-i\sigma\omega_j t} e^{\eta t} ,$$

where $v_1(+) = v_1$, $v_1(-) = v_1^{+}$, etc.

Calculate the expectation value $\langle\psi|O|\psi\rangle$ correct through terms second order in $V(t)$, i.e., extend (A.23) in Appendix A to the second order of perturbation theory.

This result can be used as the basis for microscopic calculation of $\chi^{(2)}_{\alpha\beta\gamma}(\boldsymbol{k}_1\omega_1; \boldsymbol{k}_2\omega_2)$ after the appropriate choices are made for the operators O, v_1, v_2^{+}, etc. Notice your expression contains resonant denominators that are singular when $\omega_1 = \omega_{n0}$, $\omega_2 = \omega_{n0}$, and the output frequency $\sigma_i\omega_i + \sigma_j\omega_j = \omega_{n0}$, where $\hbar\omega_{n0}$ is an excitation energy of the system.

4. Basic Principles of Nonlinear Wave Interactions: Second Harmonic Generation and Four Wave Mixing

We now turn from our discussion of the nature of a material's linear and nonlinear response to an external electric field, to the consequences of the latter. We have seen that we can decompose the electric dipole moment per unit volume into a linear portion, and a nonlinear portion, as in (1.7 a). When we Fourier transform all quantities with respect to time by writing

$$E_\alpha(\boldsymbol{r}, t) = \int_{-\infty}^{+\infty} \frac{d\omega}{2\pi} E_\alpha(\boldsymbol{r}, \omega) \mathrm{e}^{-\mathrm{i}\omega t} , \tag{4.1}$$

then introduce the frequency dependent dielectric tensor, (1.3 a,b) become

$$\nabla \cdot [\boldsymbol{\varepsilon}(\omega) \cdot \boldsymbol{E}(\boldsymbol{r}, \omega)] = -4\pi \nabla \cdot \boldsymbol{P}^{(\mathrm{NL})}(\boldsymbol{r}, \omega) \tag{4.2a}$$

and

$$\nabla^2 \boldsymbol{E}(\boldsymbol{r}, \omega) - \nabla(\nabla \cdot \boldsymbol{E}(\boldsymbol{r}, \omega)) + \frac{\omega^2}{c^2} \boldsymbol{\varepsilon}(\omega) \cdot \boldsymbol{E}(\boldsymbol{r}, \omega) = -\frac{4\pi\omega^2}{c^2} \boldsymbol{P}^{(\mathrm{NL})}(\boldsymbol{r}, \omega) . \tag{4.2b}$$

We shall ignore the influence of the wave vector dependence of the dielectric tensor in what follows.

Quite clearly, the nonlinear polarization present at the frequency ω can be viewed as a source of electromagnetic radiation at that frequency. We shall focus our attention on the theory of second harmonic generation, treated first within a perturbation theoretic framework. We shall then turn to a nonperturbative analysis of the process which, as we shall come to appreciate, is required in certain circumstances.

4.1 Perturbation Theoretic Analysis of Second-Harmonic Generation

Suppose the material is illuminated with a laser beam, here taken to be a simple plane wave, of frequency ω_1 and wave vector $\boldsymbol{k}_1$. If we assume the intensity of the second harmonic radiation is very weak, then we can ignore the depletion of the primary wave due to a conversion of a portion of its energy into second harmonic. We may then calculate $\boldsymbol{P}^{(\mathrm{NL})}(\boldsymbol{r}, \omega)$ by simply inserting the expression for the amplitude of the initial wave into the appropriate terms in the power series expansion of the dipole moment per unit volume in powers of electric field. Let the incident field be given by

$$\boldsymbol{E}(z,t) = \hat{\boldsymbol{e}}E(\omega_1)\mathrm{e}^{\mathrm{i}k_1 z}\mathrm{e}^{-\mathrm{i}\omega_1 t} + \hat{\boldsymbol{e}}E^*(\omega_1)\mathrm{e}^{-\mathrm{i}k_1 z}\mathrm{e}^{+\mathrm{i}\omega_1 t}\,, \tag{4.3}$$

where we orient the z axis along the propagation direction of the beam. The components of the dipole moment with the frequency $2\omega_1$ are

$$P_\alpha^{(\mathrm{NL})} = E^2(\omega_1)\sum_{\beta\gamma}\chi^{(2)}_{\alpha\beta\gamma}\hat{e}_\beta\hat{e}_\gamma \mathrm{e}^{\mathrm{i}2k_1 z}\mathrm{e}^{-\mathrm{i}2\omega_1 t} + \mathrm{c.c.}\,. \tag{4.4}$$

If we suppose the dielectric is a simple isotropic dielectric, or if we suppose the propagation direction of the incident wave is aligned with a principal axis of the dielectric tensor in the more general case, the electric field of the incident wave will lie in the xy plane. The incident wave is then transverse. However, in general, even in such a simple case the nonlinear polarization in (4.4) may have a component parallel to the z axis, as well as in the xy plane, because of the tensor character of $\chi^{(2)}_{\alpha\beta\gamma}$. It is convenient to break $\boldsymbol{P}^{(\mathrm{NL})}$ into two pieces, one parallel to $\hat{\boldsymbol{z}}$ and one in the xy plane:

$$\boldsymbol{P}^{(\mathrm{NL})} = \hat{\boldsymbol{z}}P_\parallel^{(\mathrm{NL})} + \boldsymbol{P}_\perp^{(\mathrm{NL})}\,. \tag{4.5}$$

The amplitude of the second harmonic field will have amplitude dependent on only z, for an incident wave of plane wave character. We write, for the second harmonic field at the frequency $\omega_2 = 2\omega_1$

$$\boldsymbol{E}(z,\omega_2) = \hat{\boldsymbol{z}}E_\parallel(z,\omega_2) + \boldsymbol{E}_\perp(z,\omega_2)\,. \tag{4.6}$$

One may show rather easily that $\nabla^2\boldsymbol{E} - \nabla(\nabla\cdot\boldsymbol{E}) = \partial^2\boldsymbol{E}_\perp/\partial z^2$. Furthermore, it will simplify our discussion to assume the z direction is a principal axis, and in fact that the dielectric material is isotropic in its response to electric fields in the xy plane. The discussion would then be applicable to second harmonic generation in a crystal such as quartz, for the case where the incident beam propagates parallel to the optic axis.

We combine the decompositions described above with the model, (4.2 a) and (4.2 b) become

$$\frac{\partial^2}{\partial z^2}\boldsymbol{E}_\perp + \omega_2^2\frac{\varepsilon_\perp(\omega_2)}{c^2}\boldsymbol{E}_\perp + \frac{4\pi\omega_2^2}{c^2}\boldsymbol{P}_\perp^{(\mathrm{NL})} + \left[\frac{\omega_2^2\varepsilon_\parallel(\omega_2)}{c^2}E_\parallel + \frac{4\pi\omega_2^2}{c^2}P_\parallel^{(\mathrm{NL})}\right]\hat{\boldsymbol{z}} = 0 \tag{4.6 a}$$

and

$$\frac{\partial}{\partial z}\left[\varepsilon_{\|}(\omega_2)E_{\|} + 4\pi P_{\|}^{(\mathrm{NL})}\right] = 0 , \tag{4.6b}$$

where $\varepsilon_{\perp}$ and $\varepsilon_{\|}$ describe the dielectric response perpendicular and parallel to the optic axis, respectively.

Satisfaction of (4.6 a) requires the two conditions:

$$E_{\|}(\omega_2, z) = -\frac{4\pi}{\varepsilon_{\|}(\omega_2)} P_{\|}^{(\mathrm{NL})}(\omega_2, z) \tag{4.7a}$$

and

$$\left(\frac{\partial}{\partial z^2} + k_2^2\right)\boldsymbol{E}_{\perp}(\omega_2, z) = -\frac{4\pi\omega_2^2}{c^2}\boldsymbol{P}_{\perp}^{(\mathrm{NL})}(\omega_2, z) . \tag{4.7b}$$

Note that satisfaction of (4.7 a) insures that (4.6 b) is obeyed. We have introduced

$$k_2 = \frac{\omega_2}{c}\sqrt{\varepsilon_{\perp}(\omega_2)} \equiv \frac{2\omega_1}{c}\sqrt{\varepsilon_{\perp}(2\omega_1)} , \tag{4.8}$$

which is the wave vector of a wave of frequency $\omega_2 = 2\omega_1$, as it propagates freely in the medium.

The component of the second harmonic wave parallel to the optic axis, and to the direction of propagation of the second harmonic wave, is given by the simple expression in (4.7 a).

The analysis of $\boldsymbol{E}_{\perp}$ will prove of more interest. While it is not a difficult matter to solve (4.7 b) as it stands, in fact we can turn to an approximate treatment based on a scheme used commonly in situations such as the present. As one progresses along the propagation direction, the amplitude of the second harmonic builds up very slowly, as a consequence of the smallness of $\chi^{(2)}_{\alpha\beta\gamma}$. The amplitude changes very little, if we move just one wavelength. Thus, for the amplitude of the second harmonic wave, we write

$$\boldsymbol{E}_{\perp}(\omega_2, z) = \boldsymbol{E}(\omega_2, z)\mathrm{e}^{\mathrm{i}k_2 z} , \tag{4.9}$$

where the spatial variation of $\exp(\mathrm{i}k_2 z)$ is assumed rapid compared to that of $\boldsymbol{E}(\omega_2, z)$. Then we have

$$\begin{aligned}\frac{\partial^2 \boldsymbol{E}_{\perp}}{\partial z^2} &= -\left(k_2^2\boldsymbol{E} - 2\mathrm{i}k_2\frac{\partial \boldsymbol{E}}{\partial z} - \frac{\partial^2 \boldsymbol{E}}{\partial z^2}\right)\mathrm{e}^{\mathrm{i}k_2 z} \\ &\cong -\left(k_2^2\boldsymbol{E} - 2\mathrm{i}k_2\frac{\partial \boldsymbol{E}}{\partial z}\right)\mathrm{e}^{\mathrm{i}k_2 z} .\end{aligned} \tag{4.10}$$

When this is inserted into (4.7 b), and we write $\boldsymbol{E} = \hat{\mathbf{e}}^{(2)}E(\omega_2, z)$ where $\hat{\mathbf{e}}^{(2)}$ is a unit vector parallel to $\boldsymbol{P}_{\perp}^{(\mathrm{NL})}$, we may write (4.7 b) in the form

$$\frac{\partial E(\omega_2, z)}{\partial z} = \frac{2\pi \mathrm{i}\omega_2^2}{c^2 k_2}\bar{\chi}^{(2)}E^2(\omega_1)\mathrm{e}^{\mathrm{i}(2k_1-k_2)z}\,, \tag{4.11}$$

where

$$\bar{\chi}^{(2)} = \sum_{\alpha\beta\gamma} \chi^{(2)}_{\alpha\beta\gamma}\hat{e}^{(2)}_{\alpha}\hat{e}_{\beta}\hat{e}_{\gamma}\,. \tag{4.12}$$

The scheme used to obtain (4.11) is called the slowly varying envelope approximation.

It is an elementary matter to integrate (4.11). We shall assume the surface of the material is at $z = 0$. The second harmonic field vanishes there, and builds in intensity as one moves into the material. We thus integrate (4.11) subject to the boundary condition $E(\omega_2, 0) = 0$. The result may be arranged to read, with $\Delta k = 2k_1 - k_2$,

$$E(\omega_2, z) = \frac{4\pi \mathrm{i}\omega_2^2\bar{\chi}^{(2)}E^2(\omega_1)}{c^2 k_2}\mathrm{e}^{\mathrm{i}\Delta kz/2}\frac{\sin(\Delta kz/2)}{\Delta k}\,. \tag{4.13}$$

The energy per unit area per unit time carried by the second harmonic is found by evaluating the Poynting vector. In the slowly varying envelope approximation, we have for the magnitude S of the Poynting vector $S = c^2k_2|E(\omega_2, z)|^2/2\pi\omega_2$, or

$$S = \frac{8\pi\omega_2^3|\bar{\chi}^{(2)}|^2}{c^3k_2}|E(\omega_1)|^4\frac{\sin^2(\Delta kz/2)}{(\Delta k)^2}\,. \tag{4.14}$$

The crucial parameter which controls the intensity of the second harmonic output is Δk. Recall that $k_1 = \omega_1\sqrt{\varepsilon(\omega_1)}/c$ is the wave vector of the primary wave of frequency ω_1, while $k_2 = 2\omega_1\sqrt{\varepsilon(2\omega_2)}/c$ is that of a freely propagating wave with frequency $2\omega_1$. If the dielectric constant were to be independent of frequency, then $\varepsilon(2\omega_1) = \varepsilon(\omega_1)$, and we have $\Delta k = 0$. Upon noting that

$$\lim_{\Delta k\to 0}\frac{\sin(\Delta kz/2)}{\Delta k} = \frac{z}{2}\,, \tag{4.15}$$

in this limit the field envelope $E_{\perp}(\omega_2, z)$ grows linearly with z, and the power flow in the second harmonic increases as z^2. Clearly, at large values of z our perturbation theory which ignores depletion of the first harmonic, breaks down, though it is quite clear that we wish to achieve the condition $\Delta k = 0$ to generate an intense second harmonic wave. As we have seen earlier, the dielectric constant of any medium depends on frequency, unfortunately. Thus, in general, the condition $\Delta k = 0$ is not realized. We shall discuss shortly how one may arrange to satisfy this condition.

A direct experimental test of the predictions of (4.14) can be found in the work of *Maker* et al. [4.1]. For example, as the path length z is varied, one expects oscillations with distance of travel in the second harmonic output, for $\Delta k \neq 0$. In [4.1], the path length is varied by using a thin film to generate the second harmonic, and rotating the film. The path length in the medium is then $d/\cos\theta$, where d is the film thickness and θ the angle between the pump beam (in the crystal) and the normal to the film. In Fig. 4.1, we show the comparison between theory (solid line) and experiment (dots), taken from [4.1].

When the condition $\Delta k = 0$ is obeyed, the interaction which leads to second harmonic generation is said to be phase matched. Physically, the reason why a phase matched interaction leads to intense output is the following: The nonlinear dipole moment exhibits the spatial variation $\exp(i2k_1 z)$, as we have seen. If we consider two small regions of space separated by the distance d, the phase difference between the dipoles in each responsible for generating the second harmonic radiation is thus $2k_1 d$. The radiation emitted by each set of dipoles has the frequency $\omega_2 = 2\omega_1$, and will propagate through the medium with the wave vector k_2. If $k_2 = 2k_1$, the radiation field emitted from the set of dipoles in the first volume is exactly in phase with that emitted by the dipoles in the second. The fields reinforce coherently. If we add up the fields radiated by all the dipoles in the strip which lies between 0 and z, since all the dipole fields reinforce coherently, the amplitude of the second harmonic is linear with z, and its intensity varies as z^2.

If $\Delta k \neq 0$, the length $\ell_c = 1/|\Delta k|$ has the following interpretation: A strip of "second harmonic" dipoles with width ℓ_c radiate second harmonic fields which reinforce constructively. Thus, for $z \ll \ell_c$, the second harmonic fields grow linearly with z. As z increases beyond ℓ_c, we encounter destructive interference, the field no longer grows monotonically, and in fact the combined effects of constructive and destructive interference lead to the oscillatory behavior for the field envelope displayed in (4.13). The length ℓ_c is called the coherence length of the nonlinear interaction.

Some comments follow from the above remarks. First notice that if the process of second harmonic generation is phase matched, then no matter how small $\bar{\chi}^{(2)}$ is, our perturbation theoretic treatment of the phenomenon breaks

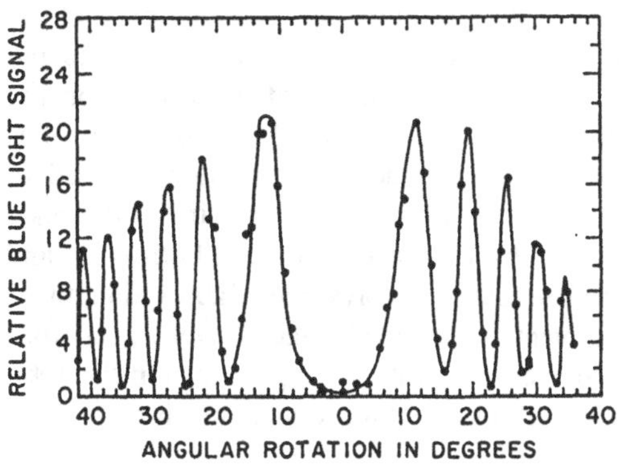

Fig. 4.1. The intensity of the second harmonic, as a function of rotation angle for a laser beam incident on a film of KDP. The experimental data (dots) is compared with theoretical prediction of (4.14) (solid line). This tests the dependence of the intensity of the second harmonic signal with path length in the medium; in (4.41), $z = d/\cos\theta$, with θ the angle between the pump beam inside the crystal, and the normal to the film. The figure has been reproduced from [4.1]

down if the optical path length in the medium is sufficiently long. Validity of the perturbation treatment requires the path length L be shorter than $L_c = c^2k_2/[2\pi\omega_2^2|\bar{\chi}^{(2)}E(\omega_1)|]$, a criterion which follows upon comparing the intensity of the second harmonic wave with that of the first harmonic, if $L \gtrsim L_c$, the intensity of the second harmonic is comparable to that of the pump beam, even though $\bar{\chi}^{(2)}$ is small, and $P^{(NL)}$ is everywhere small compared to the contribution to the dipole moment per unit volume linear in the external field.

It follows also from the above that if we have a medium whose thickness is small compared to the coherence length ℓ_c, then we need not be concerned about the question of phase matching. The second harmonic intensity is then independent of ℓ_c, and the wavevector mismatch Δk.

There is another means of understanding the phenomenon of phase matching. While our treatment here is based entirely on classical physics, we may also adopt a quantum mechanical view point, which treats the incident beam as a collection of photons, each of energy $\hbar\omega_1$. When $\chi^{(2)}_{\alpha\beta\gamma} \neq 0$, the photons may interact with each other. This may be appreciated by noting that the presence of $\chi^{(2)}_{\alpha\beta\gamma}$ leads to a term in the energy density of the medium proportional to $\Sigma_{\alpha\beta\gamma}\chi^{(2)}_{\alpha\beta\gamma}E_\alpha E_\beta E_\gamma$ (these remarks assume $\chi^{(2)}_{\alpha\beta\gamma}$ is real, and ignore complications which arise from its frequency dependence). When the electric field is expressed in terms of the photon annihilation and creation operators, one sees this term leads to three photon interactions. Two photons may "fuse" to form a third. Thus, two quanta in the pump beam may fuse to form a single quantum of energy $2\hbar\omega_1$. While such an interaction clearly conserves energy, a photon of wave vector $\boldsymbol{k}$ also carries momentum $\hbar\boldsymbol{k}$. Thus, unless $k_2 = 2k_1$, momentum is not conserved in the interaction. The full classical treatment given below of the problem of phase matched second harmonic generation will show that under such conditions, all the pump beam is converted to second harmonic. Thus the photons of frequency ω_1 fuse until the supply is exhausted.

It is quite possible to give a theoretical treatment of second harmonic generation, with use of the photon annihilation and creation operators; of course the final answer must agree with that provided by our classical theory, when the occupation numbers of the states involved are large compared to unity. This is insured by the correspondence principle of quantum theory. Full quantum theoretical treatments are to be found in the literature. In the experience of the present author, the classical approach to such problems is far more flexible and powerful. It is very tricky to incorporate the influence of absorption on the nonlinear interactions, within quantum theory, for example. Most such treatments ignore its role as a consequence, though in fact it is important in practice. The treatment presented here, while very simple, in fact is valid in the presence of absorption as it stands, though in our discussion we regarded both $\varepsilon(\omega_1)$ and $\varepsilon(2\omega_1)$ as real. All we need to do is realize that in the presence of absorption, these are complex. Also, the classical treatment is extended rather easily to incorporated boundary conditions and finite size effects (within the perturbation theoretic framework), while full quantum treatments of such influences are much more cumbersome, in the view of this writer.

While we have encountered the notion of phase matching in the context of second harmonic generation, in fact the concept enters crucially into a diverse array of nonlinear interactions. A key ingredient critical to achieving an intense output is the realization of phase matching in a nonlinear interaction of interest. We next turn to a discussion of the means of achieving phase matching, in the specific case of second harmonic generation.

4.2 Methods of Achieving the Phase Matching Condition

In general, second harmonic generation experiments are carried out in crystalline media since, as we have seen, for $\chi^{(2)}_{\alpha\beta\gamma}$ to be nonzero, we require a material within which inversion symmetry is lacking. Gasses and liquids are isotropic, of course, and this third rank tensor thus vanishes in the limit of zero-wave vector. It should be remarked that $\chi^{(2)}_{\alpha\beta\gamma}$ is nonzero in the near vicinity of the surface of any medium, since atoms in the surface occupy positions that lack inversion symmetry. Since such a small fraction of the total number of atoms reside in the surface, the intensity of second harmonic radiation generated from the surface region is quite small. We explore surface nonlinear effects in Chap. 8.

We need a long optical path length for intense second harmonic signals to be generated; one must thus operate well below the absorption edge of the medium. We have seen in Chap. 2 that $\varepsilon_1(\omega)$ increases with frequency in this spectral range, for dielectric materials, so $\varepsilon_1(2\omega) > \varepsilon_1(\omega)$ and one cannot achieve phase matching.

However, in anisotropic crystals, by means of a trick, the condition can be realized. In Sect. 2.3, we discussed the electromagnetic modes of a uniaxial crystal, with dielectric constant $\varepsilon_{\parallel}(\omega)$ for electric fields applied parallel to the optic axis, and $\varepsilon_{\perp}(\omega)$ applied in the basal plane. There is the ordinary wave, with index of refraction $n_o(\theta,\omega) = \sqrt{\varepsilon_{\perp}(\omega)}$ independent of propagation angle, and the extraordinary wave with index of refraction $n_e(\theta,\omega)$ that does depend on the angle between the wave vector of the wave, and the optic axis. The expression for $n_e(\theta,\omega)$ is given in (2.65).

If the input wave is an ordinary wave, and the output wave is an extraordinary wave, then there is one particular angle θ where precise phase matching may be achieved, provided the inequality $\varepsilon_{\parallel}(2\omega) < \varepsilon_{\perp}(\omega)$ is satisfied. The situation is illustrated in Fig. 4.2. The crystal KDP satisfies the required condition, and *Maker* et al. [4.1] verified the critical role played by the phase matching condition. We reproduce their data in Fig. 4.3; one sees the dramatic variation of the intensity of the second harmonic output (blue light) as the angle θ is swept through the critical angle θ_0. There is a much more modest variation with azimuthal angle ϕ. While the phase matching condition is in fact independent of ϕ in a uniaxial dielectric, factors such as the coefficient of transmission through the film surfaces and $\bar{\chi}^{(2)}$ depend on azimuth. The intensity of the second harmonic exhibits a dependence on ϕ as a consequence.

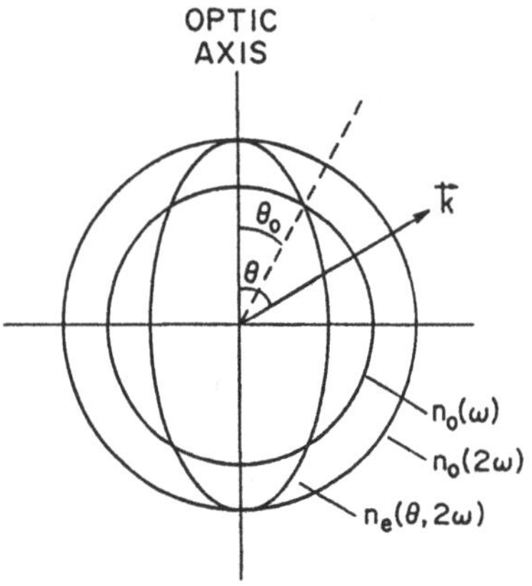

Fig. 4.2. An illustration of how one satisfies the phase matching condition in a uniaxial dielectric. We show the index of refraction $n_o(\omega)$ and $n_o(2\omega)$ for the ordinary ray, and $n_e(\theta,2\omega)$ as a function of propagation angle θ for the case $\varepsilon_{\|}(2\omega) < \varepsilon_{\perp}(\omega)$. One satisfies the phase matching condition at the particular angle θ_0

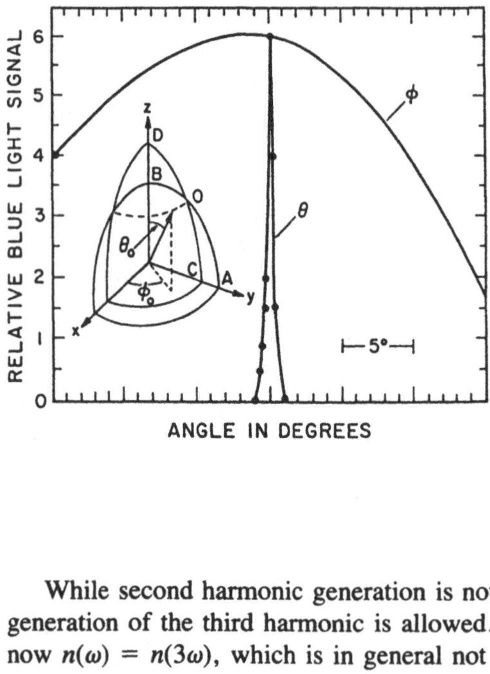

Fig. 4.3. The variation of second harmonic intensity (blue light) with polar angle θ and azimuthal angle ϕ, in KDP. The data is reproduced from [4.1]

While second harmonic generation is not possible in gasses or liquids, the generation of the third harmonic is allowed. The phase matching condition is now $n(\omega) = n(3\omega)$, which is in general not satisfied. It is, however, possible to "tune" the frequency variation of the dielectric constant of some gases, so that phase matching can be achieved at one desired frequency. Consider a gas of atoms, called species 1, where the constituents have an excited state that produces a resonance in the dielectric constant at a frequency ω_0 that lies between ω and 3ω. Then as we see from Fig. 2.3, $n_1(\omega) > n_1(3\omega)$. Now mix in atoms of type 2, whose first resonance lies above 3ω. For a gas of type 2 atoms, $n_2(\omega) < n_2(3\omega)$. For the mixture, the index of refraction is $(1 - f)n_1(\omega) + fn_2(\omega)$, where f is the fraction of type 2 atoms present. One may always choose f so that the phase matching condition is satisfied at selected frequencies. Generation of the third harmonic in the gas phase has been reported by *Bloom* et al. [4.2,3], under conditions where phase matching has been achieved.

4.3 Evolution of the Second-Harmonic Wave under Phase Matched Conditions

We have seen that when the process of second-harmonic generation is perfectly phased matched, the perturbation theory breaks down if the path length is sufficiently long, for arbitrarily small $\bar{\chi}^{(2)}$.

To proceed, we must realize that under these conditions, the pump beam at the frequency ω_1 is depleted, as the second harmonic is generated. The first step is thus to allow the amplitudes $E(\omega_1)$ and $E^*(\omega_1)$ in (4.3) to vary with distance of travel z. We thus begin by replacing $E(\omega_1)$ by $E(\omega_1, z)$ in (4.11), after setting $k_2 = 2k_1$. One then has

$$\frac{\partial E(\omega_2, z)}{\partial z} = \frac{4\pi i \omega_1^2}{c^2 k_1} \bar{\chi}^{(2)} E^2(\omega_1, z) . \tag{4.16}$$

We now require a second statement which describes the evolution with z of $E(\omega_1, z)$. There is, in fact, a contribution to the nonlinear polarization which exhibits the time variation $\exp(-i\omega_1 t)$. This arises, in our general expression (3.3), from the interaction of the Fourier component of the second harmonic with time variation $\exp(-i2\omega_1 t)$ with that of the pump beam with time variation $\exp(+i\omega_1 t)$. The former has amplitude proportional to $E(\omega_2, z)$, and the latter $E^*(\omega_1, z)$ [(4.3)]. We may write this contribution to the magnitude of the nonlinear polarization at frequency ω_1 in the form[1]

$$P^{(NL)}(\omega_1, z) = 2\bar{\chi}^{(2)} E(\omega_2, z) E^*(\omega_1, z) . \tag{4.17}$$

When this expression is inserted into the inhomogeneous wave equation for $\boldsymbol{E}_\perp(\omega_1, z)$, and the slowly envelope approximation is invoked, one finds

$$\frac{\partial E(\omega_1, z)}{\partial z} = \frac{4\pi i \omega_1^2}{c^2 k_1} \bar{\chi}^{(2)} E(2\omega_1, z) E^*(\omega_1, z) . \tag{4.18}$$

These two differential equations are to be solved with the boundary conditions $E(2\omega_1, 0) = 0$, and $E(\omega_1, 0) = E(\omega_1)$, the amplitude of the pump beam at the surface. It is possible to solve this set in closed form. We next proceed with the solution.

---

[1] See remarks on the permutation symmetry of the second order susceptibility given in [4.4]. It is quite obvious from the structure of our (3.3) that $\chi^{(2)}_{\alpha\beta\gamma}(\boldsymbol{k}_1\omega_1, \boldsymbol{k}_2\omega_2)$ is invariant under simultaneous exchange of the combinations $(\alpha \boldsymbol{k}_1\omega_1)$ and $(\beta \boldsymbol{k}_2\omega_2)$ as we have remarked in Chap. 3.

We begin by writing

$$E(\omega_1, z) = E(\omega_1) f_1(z) \tag{4.19a}$$

and

$$E(\omega_2, z) = E(\omega_1) f_2(z), \tag{4.19b}$$

where $f_1(z)$ and $f_2(z)$ obey the boundary conditions

$$f_1(0) = 1 \tag{4.19c}$$

and

$$f_2(0) = 0\,. \tag{4.19d}$$

We also change to a dimensionless measure of length:

$$z = L_c \xi\,,$$

where $L_c = c^2 k_1 / 4\pi\omega_1^2 \bar{\chi}^{(2)} E(\omega_1)$ is the critical length which entered our earlier discussion.

Our equations then become

$$\frac{\partial f_2}{\partial \xi} = \mathrm{i} f_1^2 \tag{4.20a}$$

and

$$\frac{\partial f_1}{\partial \xi} = \mathrm{i} f_2 f_1^*\,. \tag{4.20b}$$

One may proceed by breaking f_1 and f_2 into amplitudes and phases

$$f_{1,2}(\xi) = u_{1,2}(\xi) \mathrm{e}^{\mathrm{i}\phi_{1,2}(\xi)}\,, \tag{4.21}$$

then separating (4.20 a,b) into real and imaginary parts. This leads to a set of four coupled differential equations:

$$\frac{\partial u_2}{\partial \xi} = -u_1^2 \sin(2\phi_1 - \phi_2), \tag{4.22a}$$

$$u_2 \frac{\partial \phi_2}{\partial \xi} = u_1^2 \cos(2\phi_1 - \phi_2), \tag{4.22b}$$

$$\frac{\partial u_1}{\partial \xi} = u_1 u_2 \sin(2\phi_1 - \phi_2)\,, \tag{4.22c}$$

and

$$\frac{\partial \phi_1}{\partial \xi} = u_2 \cos(2\phi_1 - \phi_2)\,. \tag{4.22d}$$

These may be reduced to a set of three equations, by noting the phase angles enter only in the combination $\psi = 2\phi_1 - \phi_2$. Thus,

$$\frac{\partial u_2}{\partial \xi} = -u_1^2 \sin \psi \, , \tag{4.23 a}$$

$$\frac{\partial u_1}{\partial \xi} = u_1 u_2 \sin \psi \, , \tag{4.23 b}$$

and

$$\frac{\partial \psi}{\partial \xi} = \left(2u_2 - \frac{u_1^2}{u_2}\right) \cos \psi \, . \tag{4.23 c}$$

Through use of (4.23 a,b), (4.23 c) may be arranged to read

$$\tan \psi \frac{\partial \psi}{\partial \xi} - \frac{2}{u_1} \frac{\partial u_1}{\partial \xi} - \frac{1}{u_2} \frac{\partial u_2}{\partial \xi} = 0 \tag{4.24}$$

which is in fact a statement of a conservation law:

$$\frac{\partial}{\partial \xi} \ln[u_2 u_1^2 \cos \psi] = 0 \tag{4.25}$$

or

$$u_2(\xi) u_1^2(\xi) \cos[\psi(\xi)] = \text{const} \, . \tag{4.26}$$

Now as $\xi \to 0$, $u_2(\xi) \to 0$. Hence the constant of integration on the right-hand side of (4.26) vanishes and we must have $\psi(\xi) = +\pi/2$ or $-\pi/2$, independent of ξ everywhere. We choose the minus sign so that $\partial u_2/\partial \xi > 0$ in (4.23 a) and the second harmonic grows with increasing ξ.

We are then left with the pair of equations

$$\frac{\partial u_2}{\partial \xi} = +u_1^2 \, , \tag{4.27 a}$$

$$\frac{\partial u_1}{\partial \xi} = -u_1 u_2 \, , \tag{4.27 b}$$

which require

$$u_2 \frac{\partial u_2}{\partial \xi} + u_1 \frac{\partial u_1}{\partial \xi} = 0 \tag{4.28}$$

or

$$u_1^2(\xi) + u_2^2(\xi) = 1 \, . \tag{4.29}$$

The result in (4.29) is a statement of energy conservation. From the time average of the Poynting vector, the energy per unit area per unit time carried by the pump beam is (in our slowly varying envelope approximation)

$$S_1(\xi) = \frac{ck_1}{2\pi\omega_1}|E(\omega_1)|^2 u_1^2(\xi) \tag{4.30 a}$$

while that carried by the second harmonic is

$$S_2(\xi) = \frac{ck_2}{2\pi\omega_2}|E(\omega_1)|^2 u_2^2(\xi)\,. \tag{4.30 b}$$

Under the conditions of perfect matching, $k_1/\omega_1 = k_2/\omega_2$, so (4.29) is equivalent to the statement $S_1(\xi) + S_2(\xi) = S_1(0)$.

It is now a straightforward matter to integrate (4.27 a), eliminating u_1^2 from (4.27 a) by use of (4.29). One finds

$$u_2(\xi) = \tanh \xi \tag{4.31 a}$$

and

$$u_1(\xi) = \frac{1}{\cosh \xi}\,, \tag{4.31 b}$$

or returning to our original variables,

$$E(\omega_2, z) = \mathrm{i}E(\omega_1)\tanh(z/L_c) \tag{4.32 a}$$

and

$$E(\omega_2, z) = \frac{E(\omega_1)}{\cosh(z/L_c)}\,, \tag{4.32 b}$$

where we note that the choice $\phi_1 = 0$ leads to $\phi_2 = \pi/2$.

For small values of z, $z \ll L_c$, the result in (4.32 a) agrees with the perturbation theoretic expression in (4.13).

We see that as z increases, all of the intensity of the pump beam is converted to the second harmonic. We have seen that even for the most intense electric fields to which matter may be exposed without breakdown, the nonlinear polarization $\boldsymbol{P}^{(\mathrm{NL})}$ is in fact very small compared to the linear contribution to the dipole moment per unit volume (we assume one is not operating very close to a resonance in $\chi^{(2)}_{\alpha\beta\gamma}$ with this remark). A weak perturbation acting over a very long distance, or over a very long time may have dramatic effects, under suitable conditions. Although we are operating in the regime of weak nonlinearity, in a sense defined earlier, the end effect of the perturbation is very large.

We saw that second harmonic generation, when phase matched, can be viewed in quantum theory as the consequence of photon-photon interactions, with origin in the nonlinear response of the medium to light. Once again, the photons in the pump beam interact to form the second harmonic until the supply is exhausted.

We direct the reader's attention to a most dramatic photograph, the cover of the July 1963 issue of the jounal *Scientific American*. One sees a transparent KDP crystal, struck from the left by a red laser beam. A blue beam emerges from the right![2] This is a dramatic illustration of the behavior described by the equation of (4.32).

This discussion serves to illustrate a general principle of nonlinear optics, and in fact of nonlinear wave interactions in matter quite generally. The key to generating intense output from any particular nonlinear interaction, possibly controlled by a modest or small nonlinear coupling constant, is to achieve phase matching. This in combination with a sufficiently long path length or interaction volume, will allow an appreciable fraction of the pump beam or beams to be converted to output. In our remarks in Chap. 3, we have seen that the mixing of two laser beams, one of frequency ω_1 and one of frequency ω_2, produces radiation at the frequencies $2\omega_1$, $2\omega_2$, $\omega_1 + \omega_2$, and $\omega_1 - \omega_2$ through action of $\chi^{(2)}_{\alpha\beta\gamma}$. An additional "zoo" of frequencies arises from the presence of the third order susceptibility $\chi^{(3)}_{\alpha\beta\gamma\delta}$. All of these radiations will be very weak, but one or more will become very intense if phase matching is achieved.

4.4 Other Examples of Nonlinear Wave Interactions

4.4.1 Four Wave Mixing Spectroscopy

There are a variety of other nonlinear interactions one may explore, as the remarks of the previous paragraph suggest. In this section, we comment on two that prove particularly interesting, in the view of the present author. Both of these are referred to as four wave interactions, and are mediated by the third order susceptibility $\chi^{(3)}_{\alpha\beta\gamma\delta}$.

The first is the process diagrammed in Fig. 3.1. We have the interaction between two laser beams, one with frequency ω_1 and wave vector $\boldsymbol{k}_1$, and one with frequency ω_2 and wave vector $\boldsymbol{k}_2$. The output is at frequency $2\omega_1 - \omega_2$, and with wave vector $2\boldsymbol{k}_1 - \boldsymbol{k}_2$.

A description of this process must recognize that the output can be generated in various steps, similar in character to the processes diagrammed in Fig. 3.1. Three waves may mix directly through $\chi^{(3)}_{\alpha\beta\gamma\delta}$, to produce a nonlinear dipole moment $P^{(\mathrm{NL})}_{\alpha}$ with the desired character. The input electric field is

$$E_\alpha(\boldsymbol{r}, t) = E_\alpha(\boldsymbol{k}_1\omega_1)\mathrm{e}^{\mathrm{i}(\boldsymbol{k}_1\cdot\boldsymbol{r}-\omega_1 t)} + E_\alpha(\boldsymbol{k}_2\omega_2)\mathrm{e}^{\mathrm{i}(\boldsymbol{k}_2\cdot\boldsymbol{r}-\omega_2 t)} + \mathrm{c.c.}\,, \tag{4.33}$$

[2] The red light, when doubled in frequency, produces radiation which lies in the near ultra violet and which is thus invisible as a consequence. Such radiation, however, leaves a blue image on the film used to take the photograph shown on the cover of the journal Scientific American.

and when due account is taken of the fact that $\chi^{(3)}_{\alpha\beta\gamma\delta}(\boldsymbol{k}_1\omega_1, \boldsymbol{k}_2\omega_2, \boldsymbol{k}_3\omega_3)$ is necessarily invariant under interchange of any pair of the collection $(\boldsymbol{k}_1\omega_1\beta)$, $(\boldsymbol{k}_2\omega_2\gamma)$, $(\boldsymbol{k}_3\omega_3\delta)$, one has the contribution to the nonlinear moment given by

$$P^{(\mathrm{NL})}_\alpha(\boldsymbol{r}t) = 3\sum_{\beta\gamma\delta} \chi^{(3)}_{\alpha\beta\gamma\delta} E_\beta(\boldsymbol{k}_1\omega_1)E_\gamma(\boldsymbol{k}_1\omega_1)E^*_\delta(\boldsymbol{k}_2\omega_2) \times \mathrm{e}^{\mathrm{i}(2\boldsymbol{k}_1-\boldsymbol{k}_2)\cdot\boldsymbol{r}}\mathrm{e}^{-\mathrm{i}(2\omega_1-\omega_2)t}\,. \tag{4.34}$$

This may be inserted into the basic wave equation, to generate an expression for the output electric field. Within the framework of perturbation theory the analysis proceeds along lines very similar to our discussion of second harmonic generation.

However, in a medium which lacks an inversion center, by means of $\chi^{(2)}_{\alpha\beta\gamma}$ a wave of frequency ω_1 and wave vector $\boldsymbol{k}_1$ may mix with a wave of frequency ω_2 and wave vector $\boldsymbol{k}_2$, to produce a dipole moment per unit volume, and hence radiation at frequency $\omega_1 - \omega_2$, and wave vector $\boldsymbol{k}_1 - \boldsymbol{k}_2$. The electric field so generated may then "mix" with the wave of frequency ω_1 and wave vector $\boldsymbol{k}_1$, in the original beam again through action of $\chi^{(2)}_{\alpha\beta\gamma}$, to produce the output of interest. A second two step process involves the generation first of the second harmonic at $2\omega_1$, $2\boldsymbol{k}_1$ of the input wave at frequency ω_1, then a second interaction with the "ω_2" wave to produce the output. We thus have a second contribution to the nonlinear dipole moment per unit volume that may be written as

$$\begin{aligned} P^{(\mathrm{NL})}_\alpha(\boldsymbol{r}, t) = 2\sum_{\beta\gamma} [&\chi^{(2)}_{\alpha\beta\gamma}(\boldsymbol{k}_1 - \boldsymbol{k}_2, \omega_1 - \omega_2; \boldsymbol{k}_1\omega_1) \\ &\times E_\beta(\boldsymbol{k}_1 - \boldsymbol{k}_2, \omega_1 - \omega_2)E_\gamma(\boldsymbol{k}_1, \omega_1) \\ &+ \chi^{(2)}_{\alpha\beta\gamma}(2\boldsymbol{k}_1, 2\omega_1; -\boldsymbol{k}_2 - \omega_2)E_\beta(2\boldsymbol{k}_1, 2\omega_1)E^*_\gamma(\boldsymbol{k}_2, \omega_2)] \\ &\times \mathrm{e}^{\mathrm{i}(2\boldsymbol{k}_1-\boldsymbol{k}_2)\cdot\boldsymbol{r}}\mathrm{e}^{-\mathrm{i}(2\omega_1-\omega_2)t}\,. \end{aligned} \tag{4.35}$$

The total output at the frequency $2\omega_1 - \omega_2$ must be found by adding the dipole moment per unit volume in (4.35) to that in (4.34), then inserting the sum into the wave equation.

We can calculate the intermediate fields $E_\beta(\boldsymbol{k}_1 - \boldsymbol{k}_2, \omega_1 - \omega_2)$ and $E_\beta(2\boldsymbol{k}_1, 2\omega_1)$ through use of (4.2 b) which, for simplicity we apply to an isotropic dielectric. Suppose on the right-hand side of this expression, we let $P^{(\mathrm{NL})}_\alpha(\boldsymbol{r}, \omega)$ have the spatial variation $\exp(\mathrm{i}\boldsymbol{Q}\cdot\boldsymbol{r})$. The solution of the equation will have $E_\alpha(\boldsymbol{r}, \omega) = E_\alpha(\boldsymbol{Q}, \omega)\exp(\mathrm{i}\boldsymbol{Q}\cdot\boldsymbol{r})$, and one finds $E_\alpha(\boldsymbol{Q}, \omega)$ by inverting a 3×3 matrix:

$$\sum_\beta \left\{\delta_{\alpha\beta}\left[Q^2 - \frac{\omega^2}{c^2}\,\varepsilon(\omega)\right] - Q_\alpha Q_\beta\right\}E_\beta(\boldsymbol{Q}, \omega) = \frac{4\pi\omega^2}{c^2}\,P^{(\mathrm{NL})}_\alpha(\boldsymbol{Q}, \omega)\,. \tag{4.36}$$

The result of the matrix inversion may be written, with $\hat{\mathbf{Q}}$ a unit vector in the direction of $\boldsymbol{Q}$,

$$E_\alpha(Q,\omega) = -\frac{4\pi}{\varepsilon(\omega)}\hat{Q}_\alpha[\hat{\mathbf{Q}}\cdot P^{(\mathrm{NL})}(Q,\omega)] + \frac{4\pi\omega^2}{c^2}\frac{\{P_\alpha^{(\mathrm{NL})}(Q,\omega) - Q_\alpha[Q\cdot P^{(\mathrm{NL})}(Q,\omega)]\}}{Q^2 - \frac{\omega^2}{c^2}\varepsilon(\omega)} . \tag{4.37}$$

We may now choose the combination $Q = k_1 - k_2$, $\omega = \omega_1 - \omega_2$, and then the combination $Q = 2k_1$ and $\omega = 2\omega_1$, to generate the Fourier components $E_\beta(k_1 - k_2, \omega_1 - \omega_2)$ and $E_\beta(2k_1, 2\omega_1)$ which appear in (4.35). We are to take for this purpose, respectively,

$$P_\alpha^{(\mathrm{NL})}(k_1 - k_2, \omega_1 - \omega_2) = 2\sum_{\beta\gamma}\chi^{(2)}_{\alpha\beta\gamma}(k_1\omega_1; -k_2\omega_2)E_\beta(k_1\omega_1)E^*_\gamma(k_2\omega_2) \tag{4.38 a}$$

and

$$P_\alpha^{(\mathrm{NL})}(2k_1, 2\omega_1) = 2\sum_{\beta\gamma}\chi^{(2)}_{\alpha\beta\gamma}(k_1\omega_1; k_1\omega_1)E_\beta(k_1\omega_1)E_\gamma(k_1\omega_1) . \tag{4.38 b}$$

When these expressions are combined, in a notation that overlooks the wave vector dependence of the various elements of the nonlinear susceptibility tensor, we have a rather complex but rich expression for the nonlinear moment:

$$\begin{aligned} P_\alpha^{(\mathrm{NL})}(r,t) = 3\mathrm{e}^{\mathrm{i}(2k_1-k_2)\cdot r}\mathrm{e}^{-\mathrm{i}(2\omega_1-\omega_2)t}\sum_{\beta\gamma\delta}\Bigg[&\chi^{(3)}_{\alpha\beta\gamma\delta}(\omega_1;\omega_1;-\omega_2)\\ &+\frac{16\pi}{3}\sum_{\mu\nu}\chi^{(2)}_{\alpha\nu\beta}(\omega_1-\omega_2;\omega_1)\left\{\frac{(\omega_1-\omega_2)^2[\delta_{\mu\nu}-\hat{n}_\nu(12)\hat{n}_\mu(12)]}{c^2|k_1-k_2|^2-(\omega_1-\omega_2)^2\varepsilon(\omega_1-\omega_2)}\right.\\ &\left.-\frac{\hat{n}_\nu(12)\hat{n}_\mu(12)}{\varepsilon(\omega_1-\omega_2)}\right\}\chi^{(2)}_{\mu\gamma\delta}(\omega_1;-\omega_2)\\ &+\frac{16\pi}{3}\sum_{\mu\nu}\chi^{(2)}_{\alpha\nu\delta}(2\omega_1;-\omega_2)\left\{\frac{(2\omega_1)^2(\delta_{\mu\nu}-\hat{k}_{1\nu}\hat{k}_{1\mu})}{c^2(2k_1)^2-(2\omega_1)^2\varepsilon(2\omega_1)}-\frac{\hat{k}_{1\nu}\hat{k}_{1\mu}}{\varepsilon(2\omega_1)}\right\}\\ &\chi^{(2)}_{\mu\beta\gamma}(\omega_1;\omega_1)\Bigg]E_\beta(k_1\omega_1)E_\gamma(k_1\omega_1)E^*_\delta(k_2\omega_2) . \end{aligned} \tag{4.39}$$

In these expressions, $\hat{\mathbf{n}}(12)$ is a unit vector in the direction of $k_1 - k_2$.

The output of the four wave mixing process is proportional to the square of the modulus of the magnitude of $P_\alpha^{(\mathrm{NL})}(r,t)$; this is controlled by the quantity in the large square brackets, which may be viewed as an effective third order susceptibility, in which the direct contribution $\chi^{(3)}_{\alpha\beta\gamma\delta}$ is supplemented by that from $\chi^{(2)}_{\alpha\nu\beta}$ taken to second order in perturbation theory.

Of particular interest are the energy denominators in the second order terms, which lead to resonances in the output of the four wave mixing experiment, under suitable conditions. In a typical experiment, ω_1 and ω_2 lie in the visible range, but the difference $\omega_1 - \omega_2$ may lie in the infrared. If one of the two laser sources is a dye laser, then one of the two frequencies may be "tuned," thus allowing the difference $(\omega_1 - \omega_2)$ to be swept through resonances in the second and third term of (4.39). If ω_1 and ω_2 are fixed, the angle between the two beams may be varied, so one may scan the variable $|\boldsymbol{k}_1 - \boldsymbol{k}_2|$ and tune through resonance by this means.

We have seen in Chap. 2 that the dispersion relation of an electromagnetic wave of frequency Ω and wave vector $\boldsymbol{Q}$ is given in an isotropic dielectric by

$$c^2Q^2 = \Omega^2\varepsilon(\Omega) . \tag{4.40}$$

Thus, the resonance in the second term of (4.39) occurs when $|\boldsymbol{k}_1 - \boldsymbol{k}_1|$ and $(\omega_1 - \omega_2)$ are "tuned" to the wave vector and frequency of an electromagnetic normal mode of the material. One is driving the system at the difference frequency $(\omega_1 - \omega_2)$ and wave vector $(\boldsymbol{k}_1 - \boldsymbol{k}_2)$, through the contribution to the nonlinear dipole moment in (4.38 a). The response, which contains the resonance just described, is monitored in the visible frequency range at the frequency $2\omega_1 - \omega_2$, through the electromagnetic wave generated by the third order dipole moment given in (4.35).

In the infrared frequency range, the lattice vibration modes of solids lead to sharp absorption lines in the infrared, as discussed in Chap. 2. Associated with each absorption line is a resonance in the dieletric constant, as displayed in (2.27). The four wave mixing process described here allows one to probe the dielectric response of the material in the infrared, though an experiment carried out at visible frequencies. This can be most useful, when applied to materials opaque in the infrared, but nominally transparent in the visible. One measures, through this technique, the dispersion curves of electromagnetic radiation in the near vicinity of ω_0; these are displayed in Fig. 2.4. It is noted in Chap. 2 that such modes are referred to in the literature as polaritons, on frequent occasions. If the condition $c^2|\boldsymbol{k}_1 - \boldsymbol{k}_2|^2 \gg \varepsilon_\infty(\omega_1 - \omega_2)^2$ can be realized, then from Fig. 2.4, one sees the resonance in the four wave mixing process lies very near the frequency ω_0 of the vibration which produces the singularity in $\varepsilon(\omega)$. One thus can use the method as a form of vibrational spectroscopy. More generally, one can probe features of the dispersion curves of electromagnetic waves near a vibrational resonance by this means; see Fig. 2.4 again.

We may refer to the resonance just described as the polariton resonance. For it to be excited, one must arrange for the nonlinear dipole moment $P_\alpha^{(\mathrm{NL})}$ in (4.38 a) to have a component transverse to the wave vector $\boldsymbol{k}_1 - \boldsymbol{k}_2$. One may appreciate this by noting the presence of the factor $[\delta_{\mu\nu} - \hat{n}_\mu(12)\hat{n}_\nu(12)]$ in the second term of (4.39). There is a second resonance in (4.39), as $\omega_1 - \omega_2$ is scanned through frequency regime where $\varepsilon(\omega)$ itself has a singularity such as that displayed in (2.27). This comes from the third term in (4.39), in which $\varepsilon(\omega_1 - \omega_2)$ appears in the denominator. Just above ω_0, as one sees from (2.32), there is a frequency where the real part of the dielectric constant vanishes, and $\varepsilon_2(\omega)$ will be very small if the absorption line is sharp. If we ignore absorption, then this frequency, denoted below as ω_L, is found by setting $\varepsilon(\omega)$ in (2.27) to zero. One finds

$$\omega_L^2 = \omega_0^2 + \frac{\Omega_p^2}{\varepsilon_\infty} \equiv \left(\frac{\varepsilon_s}{\varepsilon_\infty}\right)\omega_0^2 , \tag{4.41}$$

where $\varepsilon_s = \varepsilon_\infty + \Omega_p^2/\varepsilon_\infty$ is the static dielectric constant, from (2.27). For this resonance to be excited, the nonlinear dipole moment at $(\omega_1 - \omega_2)$ must have a nonzero component parallel to $\boldsymbol{k}_1 - \boldsymbol{k}_2$, as one sees from the factors $\hat{n}_\mu(12)\hat{n}_\nu(12)$ in (4.39).

The resonance in the second term, excited by the transverse component of $\boldsymbol{P}^{(\mathrm{NL})}$ at the frequency $\omega_1 - \omega_2$, had its origin in the coincidence between the intermediate state wave at frequency $(\omega_1 - \omega_2)$ and wave vector $(\boldsymbol{k}_1 - \boldsymbol{k}_2)$, and the electromagnetic normal mode of the medium subjected to this driving force. It is also the case that the resonance at ω_L has its origin in a true normal mode of the system excited by the $\omega_1 - \omega_2$ wave. This is a longitudinal normal mode, whose character is understood from the following:

The normal modes of such a dielectric medium requires us to explore the full set of Maxwell equations including the statement, for a disturbance of frequency ω

$$\nabla \cdot \mathbf{D} = \varepsilon(\omega)\nabla \cdot \mathbf{E} = 0 . \tag{4.42}$$

For a transverse electromagnetic wave of wave vector $\boldsymbol{k}$, the electric field $\boldsymbol{E}$ is perpendicular to $\boldsymbol{k}$, so $\nabla \cdot \boldsymbol{E} = i\boldsymbol{k} \cdot \boldsymbol{E} \equiv 0$. Hence (4.42) is satisfied at any frequency.

Consider a longitudinal wave, in which $\boldsymbol{k}$ and $\boldsymbol{E}$ are parallel. Now $\nabla \cdot \boldsymbol{E} \neq 0$, and in general such a disturbance fails to satisfy (4.42). However, at the special frequency ω_L where $\varepsilon(\omega_L) \equiv 0$, this equation is satisfied, and the remaining Maxwell equations are as well; one has $\boldsymbol{k} \times \boldsymbol{E} = 0$, and this requires $\boldsymbol{B} = 0$. Thus, no magnetic field is set up by the longitudinal wave.

Thus, a dielectric medium described by the dielectric constant $\varepsilon(\omega)$ supports not only transverse electromagnetic modes whose dispersion relation is illustrated in Fig. 2.4, but in addition supports a longitudinal normal mode whose frequency ω_L is independent of wave vector (within our macroscopic theory, which is correct for wavelengths long compared to the underlying microscopic lengths). In ionic solids, such longitudinal modes are referred to as longitudinal optical phonons; the relation in (4.41) is used often and is referred to as the Lyddane-Sachs-Teller relation.

Four wave mixing, with attention to the resonances in the output induced by the second order process, has evolved into a powerful form of laser spectroscopy. The discussion presented here confines its attention to the case where the second order susceptibility $\chi^{(2)}_{\alpha\beta\gamma}$ is responsible for the appearance of the second order term. Thus, the phenomenon is confined to materials (crystals) which lack an inversion center. In Chap. 5, we discuss another nonlinearity that enters modern optics most importantly, the Raman nonlinearity. This term contributes to the optical response of materials which have inversion symmetry, including liquids and gasses. The Raman nonlinearity also provides a second order contribution to the effective third order susceptibility tensor, which exhibits a resonance rather similar to those in (4.39), as $(\omega_1 - \omega_2)$ is tuned through a vibrational resonance of the material. Thus, four wave mixing proves to be a versatile probe of matter, that can be applied to a wide class of materials. The technique is referred to by the acronym CARS, which is an abbreviation for Coherent Anti-Stokes Raman Spectroscopy. We shall appreciate the significance of the acronym when we turn our attention to four wave mixing spectroscopy once again in Chap. 5. An excellent review article on this topic has been written by *Levenson* [4.5].

Before we leave this topic, we call attention to the frequency dependence of the basic tensors $\chi^{(3)}_{\alpha\beta\gamma\delta}$ and $\chi^{(2)}_{\alpha\beta\gamma}$ in (4.39). If ω_1 and ω_2 lie in the visible frequency range, these response tensors are dominated by the contribution from the electronic degrees of freedom. As we have discussed earlier, there are resonances in these response functions when either input frequency lies close to that of an isolated electronic excited state, and also when the output frequency satisfies this condition. If infrared beams are used there can be contributions to the nonlinear susceptibility from nuclear motions excited by frequencies ω_1, ω_2 and $2\omega_1 - \omega_2$. We then encounter resonances in these quantities directly, when a wave frequency lies close to a vibrational resonance. This is clear from our examination of the response of the anharmonic oscillator in Chap. 3. From (3.22), we can see that the presence of anharmonicity leads to resonances in $\chi^{(3)}_{\alpha\beta\gamma\delta}$ when $\omega_1 - \omega_2$, or $2\omega_1$ lies near the vibrational frequency of the molecule. There have been few experimental studies of the resonances in nonlinear susceptibility tensors in the infrared, at the time of this writing.

The discussion so far has assumed the various interacting waves are simple plane waves, such as those emitted by a cw laser. The CARS technique proves most powerful when used in the pulsed mode of operation. One generates very short pulses with frequency and wave vector $(\omega_1\boldsymbol{k}_1)$ and $(\omega_2\boldsymbol{k}_2)$, with the experimental geometry arranged so that one is in resonance with a vibrational mode or polariton of frequency $(\omega_1 - \omega_2)$, and wave vector $(\boldsymbol{k}_1 - \boldsymbol{k}_2)$. The pulses are arranged to physically overlap in the material. Assume the pulse duration T is short compared to the natural lifetime τ of the mode excited; this condition can be realized through use of pulses in the picosecond range, if we have the vibrational and polariton modes of dense matter in mind. The mode thus "rings" for a time τ after the pulses disappear. One then sends in a pulse $(\omega_1\boldsymbol{k}_1)$ into the excitation region a time Δt later. This mixes with the excitation $(\omega_1 - \omega_2, \boldsymbol{k}_1 - \boldsymbol{k}_2)$, to produce output at $2\omega_1 - \omega_2$, $2\boldsymbol{k}_1 - \boldsymbol{k}_2$. The intensity of the signal is proportional to $\exp(-\Delta t/\tau)$, assuming exponential decay of the mode. By measuring the dependence of the relative intensity of the CARS signal on Δt, one has a direct, real time measurement of the lifetime of the mode. It can be the case that the decay is not a simple exponential. In this case, the method allows direct study of the real time decay profile.

This is an example of a class of experiments which have come to be known as "pulse-probe" experiments [4.6]. Measurements can be devised to monitor the real time evolution of excited state populations in diverse systems. Current laser technology allows one to generate pulses as short as 10 femtoseconds. Through use of such pulses, one can obtain "snapshots" of the time profile of phenomena that are characterized by time scales as short as 10^{-14} s.

4.4.2 Optical Phase Conjugation

We conclude this chapter with a brief discussion of a fascinating application of four wave mixing, referred to as the process of optical phase conjugation.

This is once again a four wave mixing process; this time each of the waves involved has precisely the same frequency ω in the simplest case. One has two counter-propagating input waves, one with wave vector $+\boldsymbol{k}$, and one wave vector $-\boldsymbol{k}$, see Fig. 4.4. There is a signal wave to be processed, and this has frequency ω also, but need not be a simple plane wave. Thus, in the medium, within the interaction volume, we have an electric field we may write in real notation as

$$E_\alpha(\boldsymbol{r},t) = E_\alpha^{(+)}\cos(\boldsymbol{k}\cdot\boldsymbol{r} - \omega t + \phi_+) + E_\alpha^{(-)}\cos(\boldsymbol{k}\cdot\boldsymbol{r} + \omega t - \phi_-)$$

$$+ \sum_i E_\alpha^{(s)}(i)\cos(\boldsymbol{k}_i\cdot\boldsymbol{r} - \omega t + \phi_i)\ . \tag{4.43}$$

The last term is the signal wave of frequency ω, which may have a complex wave form represented by superimposing a range of wave vectors $\{\boldsymbol{k}_i\}$. A plane wave which passes through a lense, or an inhomogenous distorting medium could be represented by such a form.

The interaction between the three waves provided by the third order susceptibility $\chi^{(3)}_{\alpha\beta\gamma\delta}$ is responsible for phase conjugation. We thus require the form of the function

$$F_\alpha(\boldsymbol{r},t) = \cos(\boldsymbol{k}\cdot\boldsymbol{r} - \omega t + \phi_+)\cos(\boldsymbol{k}\cdot\boldsymbol{r} + \omega t - \phi_-)$$

$$\times \sum_i E_\alpha^{(s)}(i)\cos(\boldsymbol{k}_i\cdot\boldsymbol{r} - \omega t + \phi_i) \tag{4.44}$$

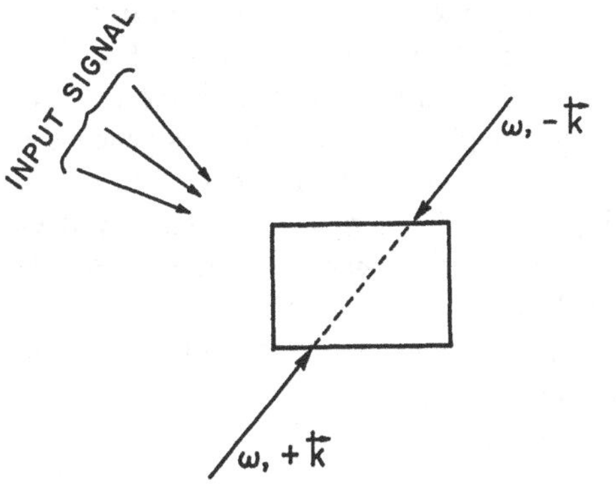

Fig. 4.4. The input waves in the optical phase conjugation process. We have two counter-propagating laser beams of frequency ω, and wave vectors $+\boldsymbol{k}$ and $-\boldsymbol{k}$. The input signal consists of an electromagnetic signal of frequency ω, and the nature of the output signal is discussed in the text. The rectangular region is the interaction volume

to understand the output. Simple trigonometric identities give, with $\Delta t = (\phi_+ + \phi_-)/\omega$

$$
\begin{aligned}
F_\alpha(\boldsymbol{r}, t) = {} & \frac{1}{4}\sum_i E_\alpha^{(s)}(i) \cos[\boldsymbol{k}_i \cdot \boldsymbol{r} + \omega(t - \Delta t) + \phi_i] \\
& + \frac{1}{4}\sum_i E_\alpha^{(s)}(i) \cos(\boldsymbol{k}_i \cdot \boldsymbol{r} - 3\omega t + \phi_i + \phi_- + \phi_+) \\
& + \frac{1}{4}\sum_i E_\alpha^{(s)}(i) \cos[(2\boldsymbol{k} + \boldsymbol{k}_i) \cdot \boldsymbol{r} - \omega t + \phi_- + \phi_+] \\
& + \frac{1}{4}\sum_i E_\alpha^{(s)}(i) \cos[(2\boldsymbol{k} - \boldsymbol{k}_i) \cdot \boldsymbol{r} + \omega t - \phi_i + \phi_- + \phi_+] \,. \qquad (4.45)
\end{aligned}
$$

The first term is the term of interest. Its form is precisely the same as the input signal in (4.43), save for the replacement $t \to -t$, and the introduction of an uninteresting time delay Δt. The four wave mixing process is thus a time reversal operation, which sends a signal back to the generating source, whose wave form is identical to that of the input signal.

The process that leads to the time reversal can be described as a nonlinear mixing of three waves of frequency ω, one with wave vector $+\boldsymbol{k}$, one with wave vector $-\boldsymbol{k}$, and one with wave vector $\boldsymbol{k}_i$, to produce output at frequency ω and wave vector $-\boldsymbol{k}_i$. This interaction is always phase matched, notice. Thus, if the interaction volume is large enough, or the counter-propagating beams sufficiently intense, one can actually amplify an incoming signal by this means. The remaining terms in (4.45) describe various nonlinear mixings between the input waves that in general are not phase matched, and which thus led to feeble output beams.

It is interesting to compare the effect of the phase conjugation process to that of an ordinary (passive) mirror, that also reflects a pulse back in the direction of the generating source, at least at normal incidence. We illustrate the two cases in Fig. 4.5. In Figs. 4.5 a,b, we show the influence of a lens on the wave-fronts of a perfect plane wave. A reflector is placed behind the lens. If the reflector is a phase conjugate reflector, its effect is the time reversal operation illustrated in Fig. 4.5 c. After the reflected pulse passes through the lens, a perfect plane wave emerges, as illustrated in Fig. 4.5 d. The lens can be replaced by any inhomogeneous or nonuniform medium, such as shower door glass, and the emergent beam will be identical in form to the incident beam, if phase conjugate reflection is employed. On the other hand, as we see in Figs. 4.5 e,f, reflection from an ordinary mirror provides a distorted wave form which is modified further by its second pass through the inhomogeneous medium. The property of phase conjugate reflection just described opens up very interesting possible applications [4.7].

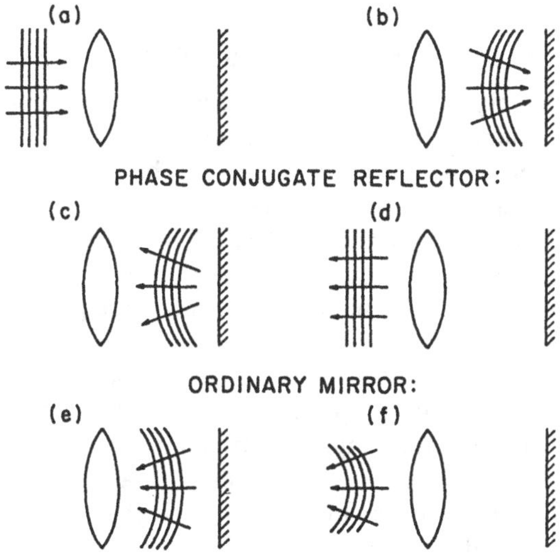

Fig. 4.5. A comparison between a phase conjugate reflector, and an ordinary mirror. In **a**, a plane wave is incident on a lens, which will distort its wave front. Behind the lens is a mirror. In **b**, the influence of the lens on the wave front is illustrated. If the mirror is a phase conjugate reflector, we show the emerging wave pattern **c** before and **d** after it has passed through the lens. In **e** and **f** we show the influence of an ordinary mirror

The two examples discussed above provide illustrations of the fascinating applications of optical techniques opened up by nonlinear optics. We shall see further examples as we proceed.

Problems

4.1 In our discussion of second harmonic generation, we have assumed the amplitude of the second harmonic electric field to vanish at $z = 0$, then integrated the equations forward in this variable. We should consider a true boundary between dielectric and vacuum, and solve the equations subject to the proper boundary conditions. We explore this question here, within the framework of the perturbation theory description of second harmonic generation.

(a) Consider a semi-infinite dielectric described by the dielectric tensor in (2.58). The optic axis, which coincides with the z axis, is perpendicular to the surface. The transmitted laser field gives rise to a nonlinear dipole moment everywhere in the material described by (4.4); the laser strikes the sample at normal incidence. Solve for the second harmonic fields everywhere, in a perturbation theory which uses the nonlinear dipole moment as a source term in Maxwell's equations. Do not make the slowly varying envelope approximation, and assume both $P_{\parallel}$ and $\boldsymbol{P}_{\perp}^{(\mathrm{NL})}$ are nonzero, in the langauge of (4.5). (*Proceed by finding a particular solution of the appropriate inhomogeneous Maxwell equations, then combine with the appropriate solution to the homogeneous equations, and arrange matters so that proper boundary conditions are obeyed at the surface.*)

(b) Examine and discuss the behavior of the reflected and transmitted second harmonic fields in the limit that perfect phase matching occurs.

4.2 The material GaAs is a cubic crystal, whose dielectric tensor $\varepsilon_{\alpha\beta}$ is isotropic. At the same time, $\chi^{(2)}_{\alpha\beta\gamma}$ is nonozero because the atomic arrangement in the crystal unit cell is such that there is no center of inversion. The Ga atoms form a face centered cubic lattice, and the coordinate system used standardly has the x, y and z axes aligned along the principal edges of the basic face centered cube. The only nonzero elements of $\chi^{(2)}_{\alpha\beta\gamma}$ in this system are $\chi^{(2)}_{xyz}$, and elements formed by permuting x, y and z. All nonzero elements are equal in magnitude.

Discuss the direction of the nonlinear dipole moment, and of the second harmonic radiation when (a) the incident wave propagates parallel to the x axis, with electric field in the yz plane, inclined at an angle of θ from the y direction, and (b) when the wave vector $\boldsymbol{k}_1$ makes equal angles with the x, y and z axes (this is propagation along the [111] direction of the crystal), and the electric field lies in the plane formed by $\boldsymbol{k}_1$ and the z axis.

4.3 Consider a four wave mixing experiment in GaAs, whose properties are outlined partly in Problem 4.2. The nonzero elements of $\chi^{(3)}_{\alpha\beta\gamma\delta}$ for this material are $\chi^{(3)}_{xxxx}$, $\chi^{(3)}_{xxyy}$, $\chi^{(3)}_{xyxy}$, and other elements obtained from these through use of cubic symmetry.

Write out the explicit form of the right-hand side of (4.39) for two cases:

(a) the beam of frequency ω_1 and that of frequency ω_2 each propagate parallel to $\hat{\mathbf{z}}$; the electric field of the ω_1 beam is parallel to $\hat{\mathbf{x}}$, and that of the ω_2 beam is parallel to $\hat{\mathbf{y}}$;

(b) The ω_1 beam propagates parallel to $\hat{\mathbf{x}}$, with electric field parallel to $\hat{\mathbf{z}}$, and the ω_2 beam propagates parallel to $\hat{\mathbf{y}}$, with electric field parallel to $\hat{\mathbf{x}}$.

5. Inelastic Scattering of Light from Matter: Stimulated Raman and Brillouin Scattering

Up until now, the discussion has assumed that the dipole moment per unit volume of a material is influenced only by application of an external electric field. The emphasis is then on the influence of the nonlinear terms in the expansion of the dipole moment per unit volume in powers of the electric field.

The subject is in fact richer than the above picture suggests. To see this, consider a molecule composed of n nuclei. If the nuclei are supposed to be clamped into a fixed position, then it is indeed the case that the electronic structure and hence the electric dipole moment are perturbed only through application of an external influence such as an electric field. In an actual molecule, the nuclei are engaged in vibrational motions. At any instant of time, to speak in classical language, the molecule is distorted a bit from its nominal, time averaged equilibrium state. As a consequence, the dipole moment induced by an external electric field will be a function not only of the external field, but of the instantaneous nuclear positions in addition. We may account for this dependence, at least for the rather small amplitude nuclear motions encountered in the analysis of thermal vibrations, by expanding the various polarizability coefficients in powers of the nuclear displacements.

When this is done, we encounter a new class of nonlinearities. One, which we shall call the Raman nonlinearity, has its origin in the dependence of the linear susceptibility on the nuclear positions. To approach this, we consider a single molecule, and let p_α be its electric dipole moment. This depends in general on the collection of nuclear positions, denoted by $(\boldsymbol{u}^{(i)})$, where $\boldsymbol{u}^{(i)}$ is the displacement of nucleus i away from its nominal equilibrium position. Since p_α is a function also of electric field, we may write

$$p_\alpha[(\boldsymbol{u}^{(i)}), \boldsymbol{E}] = p_\alpha^{(0)}[(\boldsymbol{u}^{(i)})] + \sum_\beta a_{\alpha\beta}[(\boldsymbol{u}^{(i)})]E_\beta + \cdots . \tag{5.1}$$

Here $p_\alpha^{(0)}[(u^{(i)})]$ is the electric dipole moment in zero electric field, which is possibly nonzero, $a_{\alpha\beta}[(u^{(i)})]$ is the first order polarizability, and so on. We use $a_{\alpha\beta}$ rather than $\chi_{\alpha\beta}^{(1)}$ to emphasize we are considering a single molecule, rather than the dipole moment per unit volume of a macroscopic sample of matter.

We now expand the various coefficients in (5.1) in powers of the nuclear displacements:

$$p_\alpha^{(0)}((\boldsymbol{u}^{(i)})) = p_\alpha^{(0)} + \sum_\gamma \sum_{i=1}^{n} \left[\frac{\partial p_\alpha^{(0)}}{\partial u_\gamma^{(i)}}\right]_0 u_\gamma^{(i)} + \cdots . \tag{5.2 a}$$

and if we refer to the second term in (5.1) as $p_\alpha^1[(u_\alpha^{(i)})]$, then

$$p_\alpha^{(1)}[(u^{(i)})] = \sum_\beta a_{\alpha\beta}^{(0)} E_\beta + \sum_{\beta\gamma} \sum_{i=1} a_{\alpha\beta\gamma}^{(\mathrm{R})} E_\beta u_\gamma^{(i)} + \cdots . \tag{5.2b}$$

The first term in (5.2 a) is the static dipole moment and is of little further interest for reasons discussed in our brief remarks on ferroelectrics in Chap. 1. Henceforth, it will be dropped. Upon noting $[\partial p_\alpha^{(0)}/\partial u_\gamma^{(i)}]_0$ has dimensions of electric charge, we rewrite the second term in (5.2 a) as

$$p_\alpha^{(0)} = \sum_\gamma \sum_{i=1}^{n} e_{\alpha\gamma}^{*}(i) u_\gamma^{(i)} , \tag{5.3a}$$

where $e_{\alpha\gamma}^{*}(i)$ is referred to as the dynamic effective charge tensor associated with nucleus i. In general, this quantity is unrelated to the static charge accumulation or deficit near the nucleus in question, since the electronic charge distribution of the molecule is perturbed by the nuclear motion. However, for highly ionic molecules, $e_{\alpha\gamma}^{*}(i)$ will be nearly diagonal, with value close to that of the nominal static charge of the species. Thus, the contribution to the dipole moment associated with motion of nucleus i will be simply $e_s(i)\boldsymbol{u}^{(i)}$ for this case, where $e_s(i)$ is the static charge of ion i.

The second term in (5.2 b) is the Raman nonlinearity, which will be the focus of our attention in what follows next, though we return to (5.3 a) shortly.

It is useful to rewrite the Raman term, as well as the expression in (5.3 a). The vibrational motions of molecules (and of atoms in crystals) are discussed most readily in terms of the normal modes of the system. Our molecule has a potential energy $V[(\boldsymbol{u}^{(i)})]$ that depends on the nuclear positions. We may expand this in powers of $u_\gamma^{(i)}$; the first order term must vanish, because the quantity $-[\partial V/\partial u_\gamma^{(i)}]_0$ is the force on nucleus i, in the Cartesian direction γ, evaluated at the equilibrium position. In the limit of small amplitude vibrations the leading term in the expansion of $V[(\boldsymbol{u}^{(i)})]$ is quadratic in the displacements. When this is combined with the kinetic energy, we have a Hamiltonian that describes harmonic vibrations of the array of $3n$ masses. The vibrational normal modes may be determined through the standard methods of classical mechanics [5.1]. There are $3n$ normal modes, which we label by an index ν, $\nu = 1, \ldots, 3n$. For a general molecule in free space, three of these are rigid body translations, and three are rigid body rotations. These six special modes have zero frequency, in the vibration analysis. There are thus $3n$-6 modes that have nonzero frequency.[1]

Each normal mode is described by a normal coordinate $q(\nu)$ that is a linear combination of the nuclear displacements $u_\gamma^{(i)}$; conversely, $u_\gamma^{(i)}$ can be related to

[1] Linear molecules are a special case, with three translation modes, but only two zero frequency rotational modes, each about an axis normal to the line of nuclei.

the $q(\nu)$ by means of a unitary transformation, constructed by methods described elsewhere [5.1]. Thus, we can write

$$u_\gamma^{(i)} = \sum_{\nu=1}^{3n} \Gamma(\gamma i;\nu) q(\nu) . \tag{5.4}$$

This allows us to cast (5.3 a) and the Raman nonlinearity in the form

$$p_\alpha = \sum_{\nu=1}^{3n} e_\alpha^*(\nu) q(\nu) \tag{5.5 a}$$

and

$$p_\alpha^{(\mathrm{R})} = \sum_\beta \sum_{\nu=1}^{3n} a_{\alpha\beta}^{(\mathrm{R})}(\nu) E_\beta q(\nu) , \tag{5.5 b}$$

where

$$e_\alpha^*(\nu) = \sum_\gamma \sum_{i=1}^{n} e_{\alpha\gamma}^*(i) \Gamma(\gamma i;\nu) . \tag{5.5 c}$$

and $a_{\alpha\beta}^{(\mathrm{R})}$ is defined similarly.

We comment next on the symmetry considerations that control the contributions to (5.5 a,b). Suppose our molecule has a structure with inversion symmetry, for simplicity. Then all of the normal modes of the system have either even parity or odd parity, a result easily established by group theoretic reasoning [5.2].

For an even parity mode, $q(\nu)$ is left unchanged by the inversion operation, and for an odd parity mode it changes sign, $q(\nu) \to -q(\nu)$. In Fig. 5.1 a we show an even parity mode for a simple linear triatomic molecule. For the even mode we have, ignoring normalization, $q(\nu) \sim u(A_\mathrm{R}) - u(A_\mathrm{L})$, with u the amplitude of the displacement of the right and left "A nucleus," respectively. The operation of inversion requires $u(A_\mathrm{R}) \to -u(A_\mathrm{L})$, and thus since the right and left nucleus are interchanged by inversion, $q(\nu)$ is left invariant. In Fig. 5.1 b, we illustrate an odd parity mode. In crystals, the very long wavelength vibrational modes probed directly by optical infrared spectroscopies can be classified as either even or odd parity, if an inversion center is present [5.3].

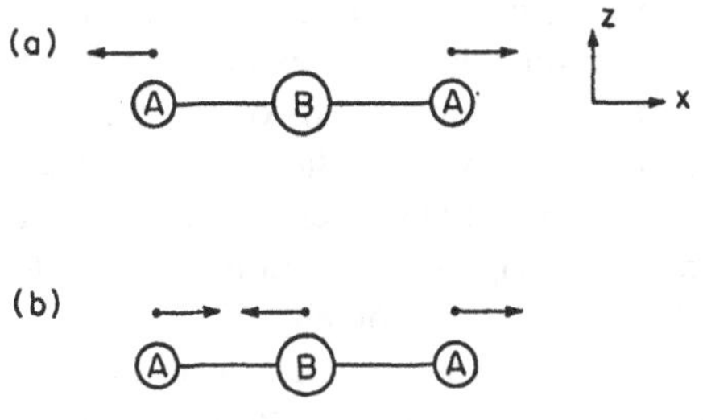

Fig. 5.1. For a linear triatomic molecule of the type A_2B, we illustrate **a** an even parity mode, and **b** an odd parity mode

If we examine (5.5 b), both the electric field and the dipole moment are odd under inversion. It follows that only even parity modes contribute to the Raman nonlinearity. Similarly, only odd parity modes contribute to (5.5 a). We call the modes which contribute to (5.5 b) Raman active modes.

The significance of these statements is the following: Consider first the modes which contribute to (5.5 a). Such modes, the odd parity modes, will be excited by a spatially uniform, time-dependent, electric field. One sees this by noting that a molecule placed in a time-dependent electric field will have in its potential energy the term $-\Sigma_\alpha p_\alpha E_\alpha = -\Sigma_\alpha \Sigma_\nu q(\nu) e_\alpha^*(\nu) E_\alpha$. The molecule exhibits a resonance if the frequency ω of the electric field matches that ω_ν of the molecule. There is strong resonant absorption when $\omega = \omega_\nu$; this is infrared absorption by the vibrational mode ν. One refers to the odd parity modes as infrared active modes. They give rise to sharp absorption lines such as those illustrated in Fig. 2.6, which have the label "lattice resonance absorption." The even parity, Raman active vibrations are "silent" in an infrared absorption experiment.

If the molecule (or crystal) lacks an inversion center, the modes cannot be classified as even and odd parity in character. In general, all of the modes are then simultaneously infrared and Raman active.

Suppose we consider the character of the Raman nonlinearity displayed in (5.5 b), within the framework of a classical discussion. The mode ν will be excited, if the molecule is in thermal equilibrium with its environment at some finite temperature. Thus, we may write

$$q(\nu) = Q_\nu \mathrm{e}^{-\mathrm{i}\omega_\nu t} + Q_\nu^* \mathrm{e}^{+\mathrm{i}\omega_\nu t} , \tag{5.6}$$

where Q_ν, Q_ν^* are the thermal vibration amplitudes. Now suppose the molecule is exposed to a time-dependent electric field with frequency ω_l. We write for the field at the site of the molecule

$$E_\beta(t) = \hat{e}_\beta(l)(E_l \mathrm{e}^{-\mathrm{i}\omega_l t} + E_l^* \mathrm{e}^{+\mathrm{i}\omega_l t}) . \tag{5.7}$$

The Raman nonlinearity then gives rise to a time dependent dipole moment with the form, confining our attention to only one mode ν,

$$p_\alpha^{(\mathrm{R})} = \sum_\beta a_{\alpha\beta}^{(\mathrm{R})}(\nu)\hat{e}_\beta(l)[E_l Q_\nu \mathrm{e}^{-\mathrm{i}(\omega_l+\omega_\nu)t} + E_l Q_\nu^* \mathrm{e}^{-\mathrm{i}(\omega_l-\omega_\nu)t}] + \mathrm{c.c.} . \tag{5.8}$$

We thus have a component of the oscillatory dipole moment of the molecule at the frequency $\omega_l - \omega_\nu$, and a component at the frequency $\omega_l + \omega_\nu$. The molecule will thus radiate electromagnetic energy at these frequencies, as a consequence of its vibrational motion. The molecule is, in essence, a point dipole whose size is very small compared to the wavelength of the emitted radiation. The radiation is then distributed over the whole solid angle, with an angular distribution controlled by the familiar $\sin^2(\theta)$ law of classical radiation theory [5.4]. Here θ is the angle between the propagation direction of the scattered radiation, and the direction of the Raman dipole moment $\boldsymbol{p}^{(\mathrm{R})}$.

Thus, if a laser beam of frequency ω_l propagates through a medium that has molecules with Raman active vibrational modes, or in a crystal with Raman active vibrations, one will observe radiation scattered out of the laser beam with frequencies $\omega_l \pm \omega_v$, shifted in frequency from that of the laser light. One refers to the phenomena as Raman scattering.

Detection of the frequency spectrum of the Raman radiation allows one to access modes that are "silent" in infrared studies, for molecules (or crystals) that possess an inversion center. Raman spectroscopy is thus complementary to infrared spectroscopy. A complete Raman study, combined with infrared data provides access to all of the vibrational modes of a molecular species.

The radiation of frequency $\omega_l - \omega_v$, downshifted from that of the exciting radiation, is referred to as Stokes radiation, while the upshifted component at $\omega_l + \omega_v$ is the anti-Stokes component of the scattered light. One may inquire into the origin of this terminology, because Raman scattering was discovered a number of decades after the completion of Stokes' career, and also in the current literature his name is associated primarily with a well-known theorem in vector calculus. In fact, Stokes was an experimentalist with an active laboratory in Cambridge. He studied the luminescence of materials, and discovered that excitation of fluorescence with ultraviolet radiation led to subsequent emission of radiation with frequency downshifted into the visible, an effect we now understand has its origin in the internal relaxation that occurs after a molecule or substance is placed in an excited state by absorbing a photon. This shift is referred to as a Stokes shift, and the terminology was applied to the downshifted component of the Raman radiation by *Woods* [5.5] at an early stage in the development of the field, though the physical origin of the downshift in frequency is different in the two cases.[2]

We now return to a detailed description of Raman scattering, using the above discussion as the base.

5.1 Quantum Theory of Raman Scattering

The quantum mechanical description of Raman scattering proves essential if the account is to be complete, as we discuss shortly.

The picture of Raman scattering in quantum theory begins by noting that the vibrational levels of the molecule are quantized, given by $(n_v + \frac{1}{2})\hbar\omega_v$, with n_v an integer. An incoming photon of frequency ω_l encounters the molecule in the quantum level n_v, and excites it to the level $n_v + 1$ as it scatters, thus leaving with energy $\hbar\omega_s = \hbar(\omega_l - \omega_v)$. This is Stokes scattering. In the anti-Stokes event, the molecule is de-excited from n_v to $n_v - 1$, a process clearly impossible if the molecule is in the vibrational ground state $n_v = 0$, and the photon thus emerges with energy $\hbar\omega_s = \hbar(\omega_l + \omega_v)$.

[2] The author is grateful to Professor A.K. Ramdas for a historical account of the introduction of this terminology.

An account of these processes is obtained as follows: We have a quantization volume V, within which our molecule is placed. The molecule is located at the origin of the coordinate system. The Hamiltonian consists of that of the radiation field [5.6]

$$H_{\mathrm{R}} = \sum_{k\lambda} \hbar\omega_k \left(a_{k\lambda}^{+} a_{k\lambda} + \frac{1}{2} \right) \tag{5.9}$$

with $a_{k\lambda}$, $a_{k\lambda}^{+}$ the annihilation and creation operator of a photon of wave vector $\boldsymbol{k}$ and polarization λ. Then we have that of the vibrating molecule,

$$H_{\nu} = \sum_{\nu} \hbar\omega_{\nu} \left(b_{\nu}^{+} b_{\nu} + \frac{1}{2} \right) \tag{5.10}$$

with b_{ν}, b_{ν}^{+} the lowering and raising operators of the quantum theory of the harmonic oscillator.

The Raman nonlinearity provides the coupling between the radiation field and the molecule, which one may write, using the classical expression for the energy of a polarized dipole in an electric field [5.7],

$$H_I = -\int_0^{E} \boldsymbol{p} \cdot d\boldsymbol{E}' = -\frac{1}{2} \sum_{\alpha\beta} \sum_{\nu} a_{\alpha\beta}^{(\mathrm{R})} E_{\alpha} E_{\beta} q(\nu) \,. \tag{5.11}$$

In (5.11), we use the quantum theoretic operator for the normal mode displacements $q(\nu)$, and the electric field. For a quantum oscillator of frequency ω_{ν} and effective mass M, we have

$$q(\nu) = \left(\frac{\hbar}{2M\omega_{\nu}} \right)^{1/2} (b_{\nu} + b_{\nu}^{+}) \tag{5.12a}$$

and the electric field at the origin, expressed in terms of photon annihilation and creation operators is [5.6]

$$E_{\alpha} = \mathrm{i} \sum_{k\lambda} \hat{e}_{\alpha}(\boldsymbol{k}\lambda) \left(\frac{2\pi\hbar\omega_k}{V} \right)^{1/2} (a_{k\lambda} - a_{k\lambda}^{+}) \,, \tag{5.12b}$$

where $\hat{\mathbf{e}}(\boldsymbol{k}\lambda)$ is unit vector directed along the electric field associated with the mode $\boldsymbol{k}\lambda$.

When the expressions are inserted into (5.11), we have terms proportional to $a_{k'\lambda'}^{+} a_{k\lambda}$ that describe scattering of a photon from $\boldsymbol{k}\lambda$ to $\boldsymbol{k}'\lambda'$, accompanied by excitation or de-excitation of the oscillator. We consider scattering of a photon from a particular state I to a final state S, using these symbols everywhere rather than the more cumbersome notation $\boldsymbol{k}\lambda$, $\boldsymbol{k}'\lambda'$. When these terms are collected together, and those not of interest discarded, we have

$$H_I = \frac{2\pi\hbar}{V} \left(\frac{\hbar}{2M\omega_{\nu}} \right)^{1/2} (\omega_S \omega_I)^{1/2} \tilde{a}(IS; \nu) a_S^{+} a_I (b_{\nu} + b_{\nu}^{+}) \,, \tag{5.13}$$

where we consider scattering with excitation or de-excitation of a particular vibrational mode ν. We have defined

$$\tilde{a}(IS;\nu) = \sum_{\alpha\beta} a_{\alpha\beta}(\nu)\hat{e}_{\alpha}(I)\hat{e}_{\beta}(S) . \tag{5.14}$$

We now calculate the transition rate using the Fermi golden rule, from a state with (n_I, n_S) photons in the initial and scattered photon states, to one with $(n_I - 1, n_S + 1)$ photons in the two states. The molecule is in state n_ν initially, and is excited to $n_\nu + 1$, so we are considering a Stokes event. The transition rate is also dn_S/dt, the increase in number of photons of frequency ω_S with time. We calculate the rate for one molecule, and multiply by N, the total number of molecules in V, to obtain a result appropriate to a sample with density $n = N/V$ molecules per unit volume. Upon recalling the Dirac delta function identity $\delta(ax) = \delta(x)/|a|$, the result is

$$\frac{dn_S}{dt} = \frac{8\pi^3 n\omega_S\omega_I}{V}\left(\frac{\hbar}{2M\omega_\nu}\right) |\tilde{a}(IS;\nu)|^2 n_I(1+n_S)(1+n_\nu)\delta(\omega_s - \omega_I + \omega_\nu) . \tag{5.15}$$

The scattering rate for the anti-Stokes process is found by replacing $1 + n_\nu$ by n_ν in (5.15), then changing the sign of ω_ν in the energy conserving delta function.

It is useful to rearrange (5.15) a bit, for some purposes. It provides the rate for scattering into a *single* final photon state S. We can calculate the total number of photons scattered per unit time into solid angle $d\Omega_s$, $d^2N_s/d\Omega_s dt$. This is done by noting that the total number of photon states (of given polarization) in solid angle $d\Omega_s$, with frequency between ω_s and $\omega_s + d\omega_s$ is $[V/(2\pi)^3](\omega_s^2 d\omega_s d\Omega_s/c^3)$. We multiple (5.15) by this density of states and integrate on ω_s to find

$$\frac{d^2N_s}{d\Omega_S dt} = \frac{n\omega_S^3\omega_I}{c^3}\left(\frac{\hbar}{2M\omega_\nu}\right)|\tilde{a}(IS;\nu)|^2 n_I(1+n_S)(1+n_\nu) . \tag{5.16}$$

One speaks commonly in this field of the scattering efficiency per unit length, per unit solid angle. This is the fraction of photons scattered out of the beam, per unit length of travel in the medium. We find the scattering efficiency S by dividing the number of photons scattered per unit time by n_I, and convert to a scattering rate per unit length by dividing by the velocity of light, which we are taking to be c, in our gaseous medium. Hence, setting $n_S = 0$, appropriate for spectroscopic studies with rather weak exciting radiation,

$$\frac{d^2S}{d\Omega_s dz} = \frac{n\omega_s^3\omega_I}{c^4}\left(\frac{\hbar}{2M\omega_\nu}\right)|\tilde{a}(Is;\nu)|^2(1+n_\nu) . \tag{5.17}$$

If the molecules are in thermal equilibrium, one averages over the quantum number n_ν, to find it replaced by $n_\nu = [\exp(\hbar\omega_\nu/k_B T) - 1]^{-1}$, with k_B Boltzmann's constant and T the temperature. For anti-Stokes scattering, $1 + n_v$ is replaced by n_v itself. For visible frequency radiation, $\omega_S^3\omega_I \cong \omega_S^4$. Then the ratio of the Stokes intensity to that of the anti-Stokes intensity in a Raman spectrum is $\exp(\hbar\omega_v/k_BT)$. The measurement of this ratio thus provides the sample temperature, a most useful result. For example, in a spectroscopic study, by monitoring the Stokes/anti-Stokes ratio, one may learn whether the sample has been heated by the laser beam, or one can infer the temperature of a remote object.

The frequency spectrum of the Raman radiation is a delta function, according to (5.15). This is, of course, an idealization. In actual matter, the excited vibration levels of the molecule are lifetime broadened, and in this circumstance the delta function is replaced by a suitably normalized Lorentzian, whose width is controlled by that of the excited vibrational levels. We use γ_v as a measure of this width, and replace (5.15) by

$$\frac{dn_S}{dt} = \frac{8\pi^2 n\omega_S\omega_I}{V}\left(\frac{\hbar}{2M\omega_v}\right)|\tilde{a}(IS;v)|^2\frac{n_I(1+n_S)(1+n_v)\gamma_v}{\gamma_v^2+(\omega_S-\omega_I+\omega_v)^2}. \tag{5.18}$$

It is very useful to pause, and consider a semi-classical treatment of the Raman process.

In classical electrodynamics, one calculates the power radiated per unit solid angle by a point dipole to find [5.4], supposing the radiated power to have frequency ω_S,

$$\frac{dP}{d\Omega_s} = \frac{\omega_S^4}{8\pi c^3}|\hat{\mathbf{n}}_S \times (\hat{\mathbf{n}}_S \times \boldsymbol{p})|^2 , \tag{5.19}$$

where $\hat{\mathbf{n}}_S$ is a unit vector in the emission direction. The right-hand side is to be time averaged. Noting $\hat{\mathbf{n}}_S \times (\hat{\mathbf{n}}_S \times \boldsymbol{p}) = \boldsymbol{p} - \hat{\mathbf{n}}_S(\hat{\mathbf{n}}_S \cdot \boldsymbol{p})$, one sees the intensity is proportional to the square of the projection of $\boldsymbol{p}$ onto the plane perpendicular to $\hat{\mathbf{n}}_S$. This projection is in fact parallel to the radiated electric field, which we suppose is parallel to the unit vector $\hat{\mathbf{e}}(S)$. Thus, (5.19) may also be written

$$\frac{dP}{d\Omega_S} = \frac{\omega_S^4}{8\pi c^3}|\hat{\mathbf{e}}(S)\cdot\boldsymbol{p}|^2 \tag{5.20}$$

We apply this to the Raman process, using for $\boldsymbol{p}$ the prefactor of $\exp[-i(\omega_I - \omega_v)t]$ in (5.8). We are thus examining the Stokes radiation. We then have, noting our earlier conventions, $|\hat{\mathbf{e}}(s)\cdot\boldsymbol{p}|^2 \rightarrow 2|\tilde{a}(Is;\ v)|^2|E_I|^2|Q_v|^2$. Also, $(dP/d\Omega_s) = \hbar\omega_s[d^2N_s/(d\Omega_s dt)]$ where N_s is the number of scattered photons. Hence, the classical treatment gives, for the radiation from one molecule,

$$\frac{d^2N_s}{d\Omega_s dt} = \frac{\omega_S^3}{8\pi c^3} |\tilde{a}(Is;\nu)|^2 |E_I|^2 |Q_\nu|^2 . \tag{5.21}$$

If there are n_I photons in the incident beam, distributed over the volume V, then $n_I\hbar\omega_I c/V$ is the energy per unit area per unit time in the incident beam. This is to be identified with the time average of the Poynting vector, $c|E_I|^2/2\pi$.

In classical theory, $|Q_\nu|^2$ should be replaced, for a molecule in contact with a thermal reservoir at temperature T, by the mean square displacement, or $k_B T/M\omega_\nu^2$. If we treat the vibrating molecule as a quantum system, we invoke the correspondence principle to replace $k_B T/\hbar\omega_\nu$ by $(1 + n_\nu)$ for a Stokes process.

When the comments of the two previous paragraphs are accounted for, (5.21) becomes, for N molecules in our volume V,

$$\frac{d^2N_s}{d\Omega_s dt} = \frac{n\omega_s^3\omega_I}{c^3}\left(\frac{\hbar}{2M\omega_\nu}\right)|\tilde{a}(IS;\nu)|^2 n_I(1 + n_\nu) . \tag{5.22}$$

This is the result for the number of photons per unit time per unit solid angle provided by a theory which treats the electromagnetic field classically, acknowledging the quantum character of the molecular vibrations. The factor of $(1 + n_s)$ which enters the full quantum treatment, however, is absent.

The factor of $(1 + n_s)$ is in fact required by another consideration. Consider our volume V, with its N molecules, which are in thermal equilibrium at temperature T. Black-body radiation is also necessarily present. In thermal equilibrium, the average value of the vibrational quantum number $\bar{n}_\nu = [\exp(\hbar\omega_\nu/k_B T) - 1]^{-1}$ cannot change in time. Also, the Planck function describes the population of the various photon states, and these states must have a time independent average occupancy. The Raman process of Stokes character scatters photons from a state I to a final state S, at the rate given by (5.15), with n_I, n_S and n_ν replaced by their thermal equilibrium values $\bar{n}_I$, $\bar{n}_S$ and $\bar{n}_\nu$. This removes photons from I, and increases the number in S. For the system to remain in thermal equilibrium, this process must be exactly counterbalanced by an inverse process. This is, in fact, the anti-Stokes process, wherein S is regarded as the initial state, and I the final state. This rate is given by interchanging the sign of I and S in the algebraic prefactor in (5.15) (the prefactor is left unchanged by this), and replacing $n_I(1 + n_S)(1 + n_\nu)$ by $n_S(1 + n_I)n_\nu$. One sees easily that for the Planck distribution function, $1 + \bar{n}_S = \exp(\hbar\omega_S/k_B T)\bar{n}_S$, we then have

$$\bar{n}_I(1 + \bar{n}_S)(1 + \bar{n}_\nu) = \exp[\hbar(\omega_S - \omega_I + \omega_\nu)](1 + \bar{n}_I)\bar{n}_S\bar{n}_\nu . \tag{5.23}$$

With due account taken of energy conservation, we see the two processes balance exactly, and on the time average neither $\bar{n}_I$, $\bar{n}_S$ nor $\bar{n}_\nu$ are changed by the presence of the Raman process. This is known as the principle of detailed balance in statistical mechanics.

The presence of the factor $(1 + n_S)$ is essential for detailed balancing to be achieved; the system of black-body radiation in contact with vibrating molecules could not remain in thermal equilibrium, if we accept the semi-classical formula in (5.22). The reader will recognize this argument follows closely that given by Einstein, who was led to introduce a similar factor in the expressions for the rate of emissions of photos by an atom in its excited state [5.8].

The factor of $(1 + n_S)$ controls a most important phenomenon, the stimulated Raman effect. Initially, one may have a state for which $n_S \cong 0$, and initiate Raman scattering which produces photons in the state S. The presence of the photons in the final state then increases the scattering rate, through the factor $(1 + n_S)$. The process is unstable, and produces an avalanche of Raman photons, until the supply (the incident beam) is exhausted. We turn to a discussion of this in the next section.

5.2 Stimulated Raman Effect

At high power levels, the Raman output can be appreciable, particularly at the center of a line associated with a long-lived vibrational mode, and in a medium where the scattered photons are absorbed weakly. Then the parameter γ_ν in (5.18) is small, and at the center of the line where $\omega_s = \omega_I - \omega_\nu$, the scattering rate scales inversely with γ_ν. The population n_S of the final state can then build up, so the role of the stimulated emission term $(1 + n_S)$ enters crucially. This is the regime of stimulated Raman scattering.

If we ignore the role of absorption, then we can obtain a simple description of the phenomenon. Suppose the laser beam enters a medium at $z = 0$, and propagates along the z axis, with the medium in the upper half space $z > 0$. We assume the molecules responsible for the Raman process reside in the ground state, so $n_\nu = 0$. Then only Stokes scattering can occur. The number of Stokes photons n_S will be a function of z, and $n_S(z)$ will vanish at $z = 0$, where the beam enters the material. As we move along the beam, the Stokes intensity will increase. The rate of increase dn_S/dz is found from (5.18), after dividing by the propagation velocity c in the Raman active medium. Thus,

$$\frac{dn_S}{dz} = Gn_I(1 + n_S) \, , \tag{5.24}$$

where

$$G = \frac{8\pi^2 n \omega_S \omega_I}{cV\gamma_\nu} \left(\frac{\hbar}{2M\omega_\nu} \right) |\tilde{a}(IS;\nu)|^2 \, . \tag{5.25}$$

Each Raman event depletes the incident beam, so we also have

$$\frac{dn_I}{dz} = -Gn_I(1 + n_S) , \tag{5.26}$$

which means that

$$n_I(z) + n_S(z) = n_I(0) , \tag{5.27}$$

when $n_S(0)$ vanishes. We can then easily eliminate n_I from (5.24), then solve the result for $n_S(z)$. The result is

$$n_S(z) = n_I(0)\frac{\exp\{G[1 + n_I(0)]z\} - 1}{n_I(0) + \exp\{G[1 + n_I(0)]z\}} . \tag{5.28}$$

For small values of z, the number of Stokes photons increases linearly,

$$n_S(z) \cong G[1 + n_I(0)]z \tag{5.29 a}$$

while

$$\lim_{z\to\infty} n_S(z) = n_I(0) . \tag{5.29 b}$$

In this simple picture, if the path length in the medium is sufficiently long or the incident power sufficiently high, then all of the incident radiation is converted to Raman radiation. We have another example of a nonlinear effect which, though the basic nonlinearity may be very small, in the sense that the magnitude of the Raman dipole moment $p_\alpha^{(\mathrm{R})}$ is small compared to the first term of (5.2 b), the influence of the small term can be dramatic indeed.

Stimulated Raman emission can be realized in the presence of absorption. A full solution of the problem is not simple; if the Stokes photon is absorbed, (5.24) is modified to read

$$\frac{dn_S}{dz} = Gn_I(1 + n_S) - \alpha_S n_S$$

$$= Gn_I + (Gn_I - \alpha_S)n_S . \tag{5.30}$$

Near $z = 0$, we may replace $n_I(z)$ by $n_I(0)$ in this equation. Then if $Gn_I(0) > \alpha_S$, the number of Stokes photons will grow with increasing z. Now, however, a power threshold must be exceeded before growth is realized.

The simple theory outlined above contains the essential physics of stimulated Raman scattering, but does not account for the observations in a quantitative manner. First of all, at the high power levels used in such studies, self-focusing effects lower the threshold, and increase the gain very substantially over that expected from the simple picture. Self-focusing occurs because at high power levels, the index of refraction of a medium is modified by the presence of the beam itself. Nonlinearities similar to those discussed in Chap. 3 cause the effective index to depend locally on field strength, through terms in the nonlinear dipole moment $\chi^{(3)}_{\alpha\beta\gamma\delta}$ with frequency equal to that of the pump beam itself. Heating effects provide an additional source of power dependence. A laser beam will have maximum intensity near the beam center. If the effective index of refraction increases with power, the medium in which the beam propagates will act as a lens that focuses the beam down. We shall discuss this further in Chap. 7.

Also, as the Stokes beam increases in intensity, new nonlinearities enter importantly. For example, one then has the pump beam at the frequency ω_l, and the Stokes beam at $\omega_l - \omega_v$. These may be mixed, possibly through the Raman nonlinearity itself, to produced output at the frequency $2\omega_l - \omega_v > \omega_l$. This is referred to in the literature as the anti-Stokes radiation associated with the stimulated Raman process. Note that the terminology differs from that used in low power Raman spectroscopy, wherein the Stokes radiation appears at $\omega_l - \omega_v$, but the anti-Stokes at $\omega_l + \omega_v$. The term anti-Stokes is employed in the present context, because the $2\omega_l - \omega_v$ radiation can be viewed as the anti-Stokes scattering of a pump photon at frequency ω_l, with an excitation in the medium (the Stokes wave) at the frequency $\omega_l - \omega_v$. We shall discuss the generation of such radiation through the Raman nonlinearity in the next subsection. From this discussion, the reason why the $2\omega_l - \omega_v$ output is so intense will become apparent. In the text by *Shen* [5.9], the reader will find a detailed discussion of both the influence of self-focusing on the stimulated Raman effect, along with generation of intense anti-Stokes radiation.

5.3 Contribution to Four Wave Mixing from the Raman Nonlinearity

In Chap. 4, we discussed the four wave mixing process, in which beams of frequency ω_1 and ω_2 are mixed through action of the third order susceptibility $\chi^{(3)}_{\alpha\beta\gamma\delta}$, to produce output at the frequency $2\omega_1 - \omega_2$. In addition, we saw there was a contribution from the second order susceptibility $\chi^{(2)}_{\alpha\beta\gamma}$, treated to second order in perturbation theory. In the end, the second piece exhibits a resonance when $\omega_1 - \omega_2$, along with $\boldsymbol{k}_1 - \boldsymbol{k}_2$ are "tuned" to an internal mode of the system. The interference between these two terms provides the basis for the very powerful four wave mixing spectroscopy. However, to judge from the discussion in Chap. 4, the technique of four wave mixing can be used only to probe noncentro-symmetric crystals, since if an inversion center is present, $\chi^{(2)}_{\alpha\beta\gamma}$ vanishes.

In the presence of the Raman nonlinearity, there is a new contribution to the output of the four wave mixing experiment which exhibits a resonance whenever $(\omega_l - \omega_S)$ matches the frequency ω_v of a Raman active vibrational mode. As we have seen, a material with an inversion center can contain molecules or atomic groups with Raman active vibrations. Thus, the spectroscopy in fact can be applied to a very wide class of materials, including liquids and gasses.

One begins by noting, returning once again to a single, isolated molecule, that the Raman nonlinearity leads to a term in the potential energy given in (5.11). If we then place the molecule in an external electric field, this introduces a term in the equation of motion of the normal mode coordinate $q(\nu)$ proportional to $-\partial V/\partial q(\nu)$. Hence, the equation of motion reads

$$\ddot{q}(\nu) + \omega_\nu^2 q(\nu) = \frac{1}{2M} \sum_{\alpha\beta} a_{\alpha\beta}^{(\mathrm{R})}(\nu) E_\alpha E_\beta \,. \tag{5.31}$$

Now suppose that at the site of the molecule, we superimpose an electromagnetic wave of frequency ω_1 with an electromagnetic wave of frequency ω_2. Then we have

$$E_\alpha(t) = E_\alpha(\omega_1)\mathrm{e}^{-\mathrm{i}\omega_1 t} + E_\alpha(\omega_2)\mathrm{e}^{-\mathrm{i}\omega_2 t} + \mathrm{c.c.} \,. \tag{5.32}$$

If we insert this form into (5.31), and retain only the term that leads to the resonance of interest,[4] one has

$$\ddot{q}(\nu) + \omega_\nu^2 q(\nu) = \frac{1}{M} \sum_{\alpha\beta} a_{\alpha\beta}^{(\mathrm{R})}(\nu) E_\alpha(\omega_1) E_\beta^*(\omega_2) \mathrm{e}^{-\mathrm{i}(\omega_1 - \omega_2)t} + \mathrm{c.c.} \,. \tag{5.33}$$

The oscillator is thus driven into motion at the frequency $(\omega_1 - \omega_2)$. The amplitude of vibration $q(\nu)$ is given by

$$q(\nu) = \frac{1}{M} \sum_{\alpha\beta} \frac{a_{\alpha\beta}^{(\mathrm{R})}(\nu) E_\alpha(\omega_1) E_\beta^*(\omega_2) \mathrm{e}^{-\mathrm{i}(\omega_1 - \omega_2)t}}{\omega_\nu^2 - (\omega_1 - \omega_2)^2} \,. \tag{5.34}$$

If we insert this expression into the Raman dipole moment in (5.5 b), realizing the electric field in (5.32) is present simultaneously, the molecule acquires a contribution to its electric dipole moment from the Raman nonlinearity which has the frequency $2\omega_1 - \omega_2$:

$$p_\alpha^{(\mathrm{R})} = \frac{1}{M} \sum_{\beta} \sum_{\nu=1}^{3n} \sum_{\gamma\delta} \frac{a_{\alpha\beta}^{(\mathrm{R})}(\nu) a_{\gamma\delta}^{(\mathrm{R})}(\nu) E_\beta(\omega_1) E_\gamma(\omega_1) E_\delta^*(\omega_2) \mathrm{e}^{-\mathrm{i}(2\omega_1 - \omega_2)t}}{\omega_\nu^2 - (\omega_1 - \omega_2)^2} \,. \tag{5.35}$$

If there are n molecules per unit volume, then the dipole moment per unit volume is found by multiplying (5.35) by n. Upon comparing with (4.39), we see the system is described by an effective third order susceptibility given by

$$(\chi_{\alpha\beta\gamma\delta}^{(3)})_{\mathrm{eff}} = \chi_{\alpha\beta\gamma\delta}^{(3)} + \frac{1}{3M} \frac{a_{\alpha\beta}^{(\mathrm{R})}(\nu) a_{\gamma\delta}^{(\mathrm{R})}(\nu)}{\omega_\nu^2 - (\omega_1 - \omega_2)^2} \,, \tag{5.36}$$

which exhibits the resonance when $\omega_1 - \omega_2$ equals ω_ν.

---

[4] The coefficient $a_{\alpha\beta}^{(\mathrm{R})}(\nu)$ depends on the frequencies of the fields E_α and E_β in (5.31), in a manner familiar from our earlier discussion of the nonlinear electric susceptibilities. We should then write this coefficient as $a_{\alpha\beta}^{(\mathrm{R})}(\omega_i, \omega_j; \nu)$ with ω_i the frequency of the field component E_β. The coefficient $a_{\alpha\beta}^{(\mathrm{R})}(\omega_i, \omega_j; \nu)$ must be invariant under the interchange of the pair (α, ω_i) and (β, ω_j). One uses this to obtain (5.33).

As remarked in Chap. 4, four wave mixing experiments such as that just described, where one detects ω_v by tuning $\omega_1 - \omega_2$ through the resonance, are described by the acronym CARS, which is an abbreviation for coherent anti-Stokes Raman spectroscopy. We now can appreciate that the resonance is detected here by means of an anti-Stokes scattering of an ω_1 photon from the vibrational motion of the molecule at the frequency $\omega_1 - \omega_2$. This motion is a response to the Fourier component at the frequency $\omega_1 - \omega_2$ which results from the superposition of the two laser beams, each of which is coherent.

In Fig. 5.2, we show a CARS spectrum of liquid benzene, from the work of *Levinson* and *Bloembergen* [5.10]. The feature is introduced into the intensity of the output at the frequency $2\omega_1 - \omega_2$ by a Raman active vibration at 992 cm^{-1}. The peculiar non-Lorentzian line shape has its origin in the interference between $\chi^{(3)}_{\alpha\beta\gamma\delta}$, and the resonant term in (5.36). To obtain a complete description of the line shape, it must be recognized that the vibrational motion is damped. If a term in $\gamma_v \dot{q}(v)$ is added to the left-hand side of the equation of motion given in (5.31), the energy denominator in (5.36) is replaced by $[\omega_v^2 - (\omega_1 - \omega_2)^2 - i(\omega_1 - \omega_2)\gamma_v]^{-1}$.

In our discussion of stimulated Raman scattering, it was noted that in addition to the strong Stokes radiation at the frequency $\omega_l - \omega_v$, one has intense anti-Stokes radiation at $2\omega_l - \omega_v$. As we discussed, this may be viewed as the anti-Stokes scattering of a photon of frequency $\omega_1 = \omega_l$, from the Stokes photon at $\omega_2 = \omega_l - \omega_v$, which is generated by the Raman nonlinearity. The process, considered in total, is an example of four wave mixing. For these photons, ω_1

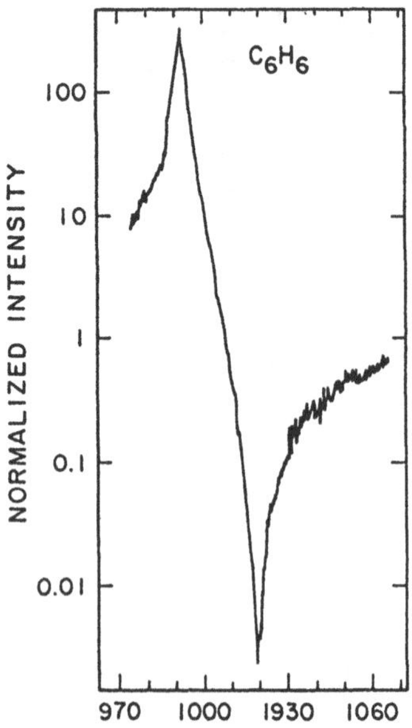

Fig. 5.2. A CARS spectrum of liquid benzene, illustrating the feature in the output produced by the 992 cm^{-1} Raman active vibration. The figure is reproduced from the work of *Levinson* and *Bloembergen* [5.10].

$- \omega_2 = \omega_v$, so one is operating right on the resonance in (5.36). Thus, when the Stokes beam builds up in intensity, the anti-Stokes radiation at $2\omega_l - \omega_v$ is generated with very substantial efficiency.

5.4 Brillouin Scattering of Light

The physical origin of Raman scattering is the modulation of the dielectric tensor of a material by the nuclear motions associated with the internal vibrations of a molecule, or perhaps the ions associated with a unit cell of a crystal.

The dielectric constant is influenced by the fluctuations in other properties of a material. These fluctuations, of thermodynamic origin, also produce inelastic scattering of light.

For example, the dielectric constant depends on the density ρ of the material and in any medium there are necessarily thermodynamic fluctuations in density. For simplicity, consider an isotropic dielectric. If there is a local fluctuation in density, $\delta\rho(\boldsymbol{r}, t)$, the dielectric constant changes by an amount

$$\delta\varepsilon(\boldsymbol{r}, t) = \left(\frac{\partial \varepsilon}{\partial \rho}\right) \delta\rho(\boldsymbol{r}, t) . \tag{5.37}$$

Suppose that we follow the position of a particular point in the material as it engages in thermodynamic fluctuations. Let $\boldsymbol{u}(\boldsymbol{r}, t)$ be its displacement. The fractional change in volume of a small piece of material, in response to the thermodynamic fluctuations is easily shown to be $\Sigma_\alpha \, \partial u_\alpha / \partial x_\alpha$, so that $\delta\rho(\boldsymbol{r}, t)/\rho = \Sigma_\alpha \, \partial u_\alpha / \partial x_\alpha$. Hence,

$$\delta\varepsilon(\boldsymbol{r}, t) = \rho \left(\frac{\partial \varepsilon}{\partial \rho}\right) \sum_\alpha \frac{\partial u_\alpha}{\partial x_\alpha} . \tag{5.38}$$

The time-dependent fluctuations in density produce inelastic scattering of light, in just the same manner as the time-dependent modulations of the dipole moment per unit volume provided by the internal vibrations of molecules. Before we proceed with the discussion, however, we need to extend the above remarks.

Thermodynamic fluctuations in particle positions produce not only changes in density, but in addition lead to more complex alternations in local dielectric properties. Consider an isotropic dielectric, and an infinitesimal volume element dV that is cubic in shape on the time average. As the particles within dV engage in thermodynamic fluctuations, not only does the volume and hence the density change, but the shape changes as well, as illustrated in Fig. 5.3. In the original quiescent state, the volume element dV is cubic, and as a consequence its dielectric constant remains isotropic. However, in the presence of the fluctuation, dV changes shape, and the new low symmetry form is described

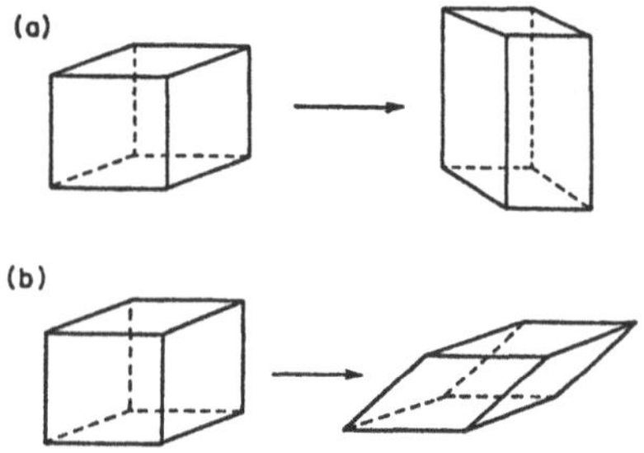

Fig. 5.3. Shape changes in an elemental volume induced by thermodynamic fluctuations in particle positions are schematically illustrated

by a dielectric response tensor. Hence, even if the material is initially isotropic, we should speak of $\delta\varepsilon$ as a tensor quantity.

We may expand $\delta\varepsilon$ in powers of the displacement gradients $(\partial u_\alpha / \partial x_\beta)$. It is usual to consider combinations which are, respectively, symmetric and antisymmetric under interchange of α and β. We have

$$e_{\alpha\beta} = \frac{1}{2}\left(\frac{\partial u_\alpha}{\partial x_\beta} + \frac{\partial u_\beta}{\partial x_\alpha}\right) \tag{5.39 a}$$

and

$$\omega_{\alpha\beta} = \frac{1}{2}\left(\frac{\partial u_\alpha}{\partial x_\beta} - \frac{\partial u_\beta}{\partial x_\alpha}\right) . \tag{5.39 b}$$

One may readily show that a rigid body rotation of the volume element dV may be described in terms of $\omega_{\alpha\beta}$. If a small volume element is rotated through the infinitesimal angle $d\theta$ about the direction $\hat{\boldsymbol{n}}$, and we let $d\boldsymbol{\theta} = \hat{\boldsymbol{n}} d\theta$, then $d\theta_\alpha = \sum_\beta \omega_{\alpha\beta} \hat{n}_\beta$.

The symmetric combination $e_{\alpha\beta}$ is the strain tensor of elasticity theory; changes in shape of an infinitesimal volume are described with this tensor [5.11]. We may write

$$\delta\varepsilon_{\alpha\beta}(\boldsymbol{r}, t) = \sum_{\gamma\delta} k_{\alpha\beta\gamma\delta} e_{\gamma\beta}(\boldsymbol{r}, t) , \tag{5.40}$$

where the tensor $k_{\alpha\beta\gamma\delta}$ is referred to as the elasto optic tensor. The tensor $k_{\alpha\beta\gamma\delta}$ is invariant under interchange of γ and δ.

A rigid body rotation of the element dV leaves the dielectric tensor of the material unchanged, if the material is initially isotropic. In an anisotropic material, however, a rotation of an element dV does lead to a change in the dielectric tensor; the principal axes have rotated relative to a fiducial set of coordinate axes in the laboratory frame, thus modulating the dielectric tensor. Thus, in an anisotropic material, we have an additional contribution to $\delta\varepsilon_{\alpha\beta}(\boldsymbol{r}, t)$:

$$\delta\varepsilon_{\alpha\beta}(\boldsymbol{r}, t) = \sum_{\gamma\delta} \lambda_{\alpha\beta\gamma\delta} \omega_{\alpha\delta}(\boldsymbol{r}, t) , \tag{5.41}$$

where now $\lambda_{\alpha\beta\gamma\delta}$ is antisymmetric with respect to exchange of γ and δ. Clearly, through use of geometrical reasoning, $\lambda_{\alpha\beta\gamma\delta}$ may be related to the elements of the dielectric tensor of the unrotated material. This has been done by *Nelson* and *Lax* [5.12], who first pointed out that the contribution to $\delta\varepsilon_{\alpha\beta}$ from rigid rotations of volume elements had been omitted from earlier discussions of elasto optic effects in crystals.

Suppose we have a crystal with volume V. We may make a Fourier decomposition of the displacements induced by thermal fluctuations at any instant of time:

$$u_\alpha(\boldsymbol{r}, t) = \frac{1}{\sqrt{V}} \sum_{\boldsymbol{q}} u_\alpha(\boldsymbol{q}, t) \mathrm{e}^{\mathrm{i}\boldsymbol{q}\cdot\boldsymbol{r}} . \tag{5.42}$$

We shall discuss the time dependence of $u_\alpha(\boldsymbol{q}, t)$ shortly. For the moment we suppose that we simply have a single frequency $\omega(\boldsymbol{q})$ that characterizes the time variation of the Fourier component with wave vector $\boldsymbol{q}$. Then

$$u_\alpha(\boldsymbol{r}, t) = \frac{1}{\sqrt{V}} \sum_{\boldsymbol{q}} u_\alpha(\boldsymbol{q}) \mathrm{e}^{\mathrm{i}[\boldsymbol{q}\cdot\boldsymbol{r} - \omega(\boldsymbol{q})t]} . \tag{5.43}$$

When this statement is combined with (5.40,41) where applicable, we see that the thermal fluctuations of wave vector $\boldsymbol{q}$ lead to the presence of a diffraction grating, that moves through the medium with phase velocity $\omega(q)/q$. Light is diffracted from the diffraction grating, and suffers a Doppler shift in the process. This inelastic light scattering is referred to as Brillouin scattering.

Clearly, the wave vectors $\boldsymbol{q}$ probed by Brillouin scattering are the order of the wave vector of the probing radiation; a grating whose spacing is more than half of that of the probe radiation can never satisfy the Bragg condition. The technique thus probes rather long wavelength fluctuations. The time evolution of the displacement field $u_\alpha(\boldsymbol{r}, t)$ can then be described by the macroscopic theory of elasticity, at least for solid media. The equations of motion of the displacement field then have the form [5.13]

$$\ddot{u}_\alpha = \frac{1}{\rho} \sum_{\beta\gamma\delta} c_{\alpha\beta\gamma\delta} \frac{\partial^2 u_\beta}{\partial x_\gamma \partial x_\delta} \tag{5.44}$$

with $c_{\alpha\beta\gamma\delta}$ the elastic constant tensor of the material, and ρ its density. If we insert the space and time variation of (5.43) into this equation of motion, the allowed frequencies associated with wave vector $\boldsymbol{q}$ are found by setting to zero the determinant of the matrix

$$D_{\alpha\beta}(\boldsymbol{q}) = \delta_{\alpha\beta}\omega^2(\boldsymbol{q}) - \frac{1}{\rho} \sum_{\gamma\delta} c_{\alpha\beta\gamma\delta} q_\gamma q_\delta . \tag{5.45}$$

The modes described by this procedure are acoustic waves, whose sound velocity depends in general on the propagation direction, and whose magnitude is controlled by the elastic constants. We may write $q_\gamma = q\hat{n}_\gamma$, with $\hat{\mathbf{n}}$ a unit vector in the direction of $\boldsymbol{q}$. Then the frequencies which satisfy (5.44) have the form $\omega^2(q) = c_s^2(\hat{\mathbf{n}})q^2$, where the sound velocity $c_s(\hat{\mathbf{n}})$ is found from

$$\det\left|\delta_{\alpha\beta}c_s^2(\hat{n}) - \frac{1}{\rho}\sum_{\gamma\delta} c_{\alpha\beta\gamma\delta}\hat{n}_\gamma\hat{n}_\delta\right| = 0 . \tag{5.46}$$

The task of finding the sound velocities associated with a given propagation direction involves examining the roots of a 3 × 3 matrix. Thus, for each choice of $\hat{\mathbf{n}}$, there are three roots $c_s(\hat{\mathbf{n}})$. The discussion that follows (5.43) is thus oversimplified, in that for each choice of $\boldsymbol{q}$, the disturbance in the medium is a superposition of that associated with the three acoustic waves associated with this propagation direction. In the isotropic elastic medium, for each choice of $\boldsymbol{q}$ we have a longitudinal acoustic wave with displacement $\boldsymbol{u}$ parallel to $\hat{\mathbf{n}}$, and two transverse waves each with the same propagation velocity [5.13].[5]

In solids, the acoustic vibrations are quantized. They are the acoustical phonons of solid state theory [5.14]. We may then view the Brillouin scattering event from the vantage point of quantum theory, recognizing that the energy of the phonons is quantized in units of $\hbar\omega(\boldsymbol{q})$. A photon of frequency ω_I may emerge with frequency $\hbar\omega_S = \hbar\omega_I \pm \hbar\omega(\boldsymbol{q})$. For the purpose of describing the kinematics of the scattering process, one may suppose phonons carry an effective momentum $\hbar\boldsymbol{q}$. Thus, $\boldsymbol{k}_S = \boldsymbol{k}_I \pm \boldsymbol{q}$, where $\boldsymbol{k}_S$ and $\boldsymbol{k}_I$ are the wave vectors of the scattering and incident photons.[6] From the discussion of phase matching in Chap. 4, we recognize the energy and wave vector conservation condition are also the phase matching requirements for a three wave interaction.

A description of Brillouin scattering may be obtained by following our earlier discussion of Raman scattering. We have a contribution to the energy density of the material from the presence of $\delta\varepsilon_{\alpha\beta}$ which may be written

$$U_B = -\frac{1}{8\pi}\sum_{\alpha\beta}\sum_{\gamma\delta}[k_{\alpha\beta\gamma\delta}e_{\gamma\delta}(\boldsymbol{r}, t) + \lambda_{\alpha\beta\gamma\delta}\omega_{\gamma\delta}(\boldsymbol{r}, t)]E_\alpha(\boldsymbol{r}, t)E_\beta(\boldsymbol{r}, t) . \tag{5.47}$$

When integrated over the sample volume V, this provides us with a potential energy which contains a description of the inelastic scattering of photons from the phonon quanta. The scattering cross section may be obtained, as in our discussion of Raman scattering, through use of (5.12 b) to describe the electric field operator, and through use of similar expressions which express the displacement $u_\alpha(\boldsymbol{r})$ in terms of the phonon annihilation and creation operators. We omit the details.

It follows that the Brillouin cross section for scattering from initial state I to a final state S is proportional to $(1 + n_S)$, with n_S the number of photons in the final state. With sufficient intensity in the incident beam, one may realize

[5] In a true crystalline medium, the character of the waves is more complex [5.13].

[6] The mechanical momentum carried by a phonon in a crystal with wave vector **q** is, in fact, identically zero, as discussed in [5.14].

intense stimulated Brillouin scattering, in a manner very similar to the stimulated Raman effect.

From a conceptual point of view, Brillouin scattering and Raman scattering are very similar. In the laboratory, very different experimental methods must be used to study the two phenomena. Raman spectra explore vibrational frequencies roughly in the range of 100–2500 cm^{-1}. Grating spectrometers may be used to detect the frequency shift. The acoustic modes probed in Brillouin scattering have wave vectors comparable to those of light, so $q \sim 10^5\ \text{cm}^{-1}$. Sound velocities in dense matter lie in the range of 3×10^5 cm/s, so the frequency shift in the Brillouin event lies in the range of a few Gigahertz. A Fabry Perot interferometer is needed to scan the frequency spectrum of the scattered light, to detect such small shifts. While we have confined our attention to frequency shifts associated with vibrational motions of atoms incorporated in molecules or solids, which are the first phenomena to be studied by these methods, in fact, thermal fluctuations of diverse origin may "modulate" the dielectric constant of a material. Other examples include thermal fluctuations in the magnitude and orientation of the magnetization vector in a ferromagnet, those of the electric polarization of a ferroelectric, and density fluctuations in the current carrying electrons in metals and semiconductors.[7] The scattering event is referred to as Raman scattering or Brillouin scattering, depending on the detection scheme employed in the experiment.

Problems

5.1 In Fig. 2.4, we show the dispersion curves of electromagnetic waves in an isotropic dielectric medium, for frequencies ω near that of a resonance in the dielectric constant at ω_0. As discussed in Chap. 2, this picture applies near the frequencies of sharp lattice vibrational absorption lines in crystals, in the infrared. Typical values of ω_0 are then a few hundred wave numbers. The electromagnetic modes in such a spectral region are frequently called polaritons, as we noted.

These modes may be explored with Raman scattering, with visible radiation as a probe. The incident photon has wave vector $\boldsymbol{k}_I$, the scattered photon $\boldsymbol{k}_S$, and the wave vector of the polariton created in the Raman event is $\boldsymbol{k}$.

(a) Make rough numerical estimates of the angle θ between the incident and scattered photon, if one wishes to explore the interesting regime sketched in Fig. 2.4, where $ck/\sqrt{\varepsilon_s} \approx \omega_0$ for the polariton. You should conclude a very near forward scattering geometry is required.

[7] An excellent collection of papers which explore diverse aspects of the inelastic scattering of light in solid materials can be found in the proceedings of a conference held early in the development of the field of laser light scattering in solids, as many of the papers address basic issues [5.15].

(b) Now consider the kinematics of the scattering event in more detail. The wave vector of the polariton is small compared to that of the incident photon, and the frequency shift is small as well. The dielectric constant [or index of refraction $n(\omega)$] depends on frequency in the visible, and thus assumes a different value for the incident and scattered photon. If θ is the angle between the wave vector of the incident photon and the polariton, show

$$\cos\theta = \frac{v_p(\text{pol.})}{v_p(I)}\left[1 + \frac{\omega_I}{n(\omega_I)}\left(\frac{\partial n}{\partial \omega}\right)_{\omega_I}\right].$$

Here v_p(pol.) and $v_p(I)$ are the phase velocity of the polariton and the incident photon. Explain why this relation forbids one from "seeing" polaritons on the upper branch in Fig. 2.4.

5.2 Develop a physical argument that will provide you with a rough order of magnitude estimate of the parameter $\tilde{a}(IS; \nu)$ in (5.17). Then, for visible radiation, make an estimate of the order of magnitude of the Raman scattering efficiency per unit length, per unit solid angle, from the N_2 vibration in air at standard temperature and pressure.

5.3 Consider a Brillouin scattering event produced by a density variation $\delta\rho \sim \cos(\boldsymbol{q}\cdot\boldsymbol{x})$. Show that the wave vector conservation condition $\boldsymbol{k}_S = \boldsymbol{k}_I \pm \boldsymbol{q}$ can be derived by considering the Bragg diffraction of light by a suitable grating whose spacing $d = 2\pi/|\boldsymbol{q}|$. In this discussion, suppose the grating is stationary, (the velocity of sound is very small compared to the velocity of light), and ignore the influence of the small frequency shift of the light upon the kinematics of the scattering process. (This problem can be addressed by tracing path differences between various light rays that strike the grating.)

6. Interaction of Atoms with Nearly Resonant Fields: Self-Induced Transparency

The discussion we have presented so far has supposed throughout that the primary quantity of interest, the electric dipole moment per unit volume $\boldsymbol{P}$, can be expanded in powers of the electric field $\boldsymbol{E}$. The justification for this procedure is the notion that the largest applied fields encountered in practice are very small compared to those encountered by an electron within an atom or a molecule. It follows that the electronic arrangement in the atom or molecule is perturbed only very slightly from that arrangement found in zero applied external field. In the concluding remarks of Chap. 3, we referred to this as the regime of weak nonlinearity. We now appreciate, however, that even though the nonlinearities are weak in the sense just described, their influence can be substantial, as illustrated by our model discussions of second harmonic generation, and the stimulated Raman effect. A bucket full of water may have only a small leak, but if one waits long enough, the bucket will be quite empty.

We also see that if the frequency ω of an oscillating electric field lies very close to that ω_{m0} of a transition between the ground state and an isolated excited state of the system, both the linear susceptibility and the various nonlinear susceptibilities can be enhanced enormously, over values appropriate to the nonresonant case. Indeed, the perturbation theoretic expressions produce a divergence as $\omega \to \omega_{m0}$, if no account is taken of the influence of the finite lifetime of the excited state. For frequencies very close to resonance, and for very intense applied fields, the perturbation theory breaks down. We then enter the regime of strong nonlinearities. This chapter is devoted to the analysis of the response of the system under these conditions, and the array of rather fascinating phenomena one encounters in this regime. For instance, when $\omega = \omega_{m0}$, elementary theory asserts that radiation will be absorbed by the system. We shall encounter the remarkable phenomenon of self-induced transparency later in the chapter; when $\omega = \omega_{m0}$ and suitable conditions are satisfied, the medium is quite transparent!

Our first task is to describe the response of the system to near resonant radiation. For simplicity, we shall confine our attention to the response of an isolated atom, with a single nondegenerate excited level. We shall exploit the spherical symmetry of the atom in what follows:

We begin by examining (A.22) in Appendix A, which provides the perturbation theoretic description of an atom subject to the perturbation described in (A.21). We rewrite (A.22) to read, with $\omega \cong \omega_{m0}$ in mind,

$$|\psi\rangle = e^{-iE_0/\hbar t}\left[|\psi_0\rangle - \frac{1}{\hbar}\frac{\langle\psi_m|v|\psi_0\rangle}{\omega_{m0} - \omega - i\eta}|\psi_m\rangle e^{-i(\omega+i\eta)t}\right.$$
$$- \frac{1}{\hbar}\sum_{n\neq m}\frac{\langle\psi_n|v|\psi_0\rangle}{\omega_{n0} - \omega - i\eta}|\psi_n\rangle e^{-i(\omega+i\eta)t}$$
$$\left. - \frac{1}{\hbar}\sum_{n}\frac{\langle\psi_n|v^+|\psi_0\rangle}{\omega_{n0} + \omega + i\eta}|\psi_n\rangle e^{i(\omega-i\eta)t} + \cdots\right]. \tag{6.1}$$

A weak perturbation is one for which matrix elements such as $\langle\psi_n|v|\psi_0\rangle$ are very small compared to the level splittings of the system. Our interest is clearly in this limit. Then the third and fourth terms on the right-hand side of (6.1) represent small corrections to the ground state wave function $|\psi_0\rangle$. If ω is very close to ω_{m0}, however, the second term can be very large. While perturbation theory clearly breaks down in this case, quite clearly in this circumstance the full wave function must have the form

$$|\psi(t)\rangle = a_0(t)|\psi_0\rangle + a_m(t)|\psi_m\rangle + \text{small terms}\,. \tag{6.2}$$

The treatment that follows proceeds by approximating the wave function by the first two terms only:

$$|\psi(t)\rangle = a_0(t)|\psi_0\rangle + a_m(t)|\psi_m\rangle\,. \tag{6.3}$$

This approximation, surely accurate when ω is close to ω_{m0}, allows one to obtain a simple and appealing description of the response of the atom.

6.1 Description of the Wave Function under Near Resonant Conditions

It will be instructive to obtain an explicit description of the coupling of the atom to an applied electric field, which we take to have the form $\boldsymbol{E}(t) = \hat{\mathbf{n}}E_0 \cos(\omega t) \equiv \hat{\mathbf{n}}E(t)$, with $\hat{\mathbf{n}}$ a unit vector. The Hamiltonian is then

$$H = H_0 - (\hat{\mathbf{n}}\cdot\boldsymbol{p})E(t)\,, \tag{6.4}$$

where $\mathbf{p} = e\sum_{i=1}^{n}\boldsymbol{r}_j$ is the electric dipole moment operator of the atom. The sum is over the n electrons in the atom. One may treat the coupling of an electric field to a localized object such as an atom either through use of the vector potential, as described in Appendix A, or through use of the dipole moment operator as in (6.4). One may readily verify that application of (6.4) to the calculation of the frequency dependent susceptibility from an atom leads to an expression equivalent to (A.41) for its contribution to the dielectric constant. One must suppose the wave vector $\boldsymbol{k} = 0$ on the right-hand side of this expression.

We now wish to solve the Schrödinger equation

$$i\hbar\frac{\partial|\psi\rangle}{\partial t} = H|\psi\rangle \tag{6.5}$$

with $|\psi\rangle$ given in (6.3). On the subspace spanned by the two states $|\psi_0\rangle$ and $|\psi_m\rangle$, the Hamiltonian may be represented by a 2×2 matrix. With ε_0 and ε_m the energies of the ground and excited state, we have

$$H = \begin{pmatrix} \varepsilon_0 & -E\gamma e^{-i\phi} \\ -E\gamma e^{+i\phi} & \varepsilon_m \end{pmatrix}, \tag{6.6}$$

where the matrix element $\langle\psi_0|(\hat{\mathbf{n}}\cdot\boldsymbol{p})|\psi_m\rangle$ is written $\gamma e^{-i\phi}$. One now lets

$$|\psi\rangle = e^{-i/\hbar(\varepsilon_m+\varepsilon_0)t}|\Phi\rangle \tag{6.7}$$

to find

$$i\hbar\frac{\partial}{\partial t}|\Phi\rangle = H|\Phi\rangle\,, \tag{6.8}$$

where, with $\omega_{m0} = (\varepsilon_m - \varepsilon_0)/\hbar$,

$$H = \begin{pmatrix} -\frac{1}{2}\hbar\omega_{m0} & -\gamma E(t)e^{-i\phi} \\ -\gamma E(t)e^{i\phi} & \frac{1}{2}\hbar\omega_{m0} \end{pmatrix}. \tag{6.9}$$

The transformation in (6.7) just shifts the zero of energy of the problem, to the midpoint between level ε_0 and ε_m. For any operator A, notice that $\langle\psi|A|\psi\rangle = \langle\Phi|A|\Phi\rangle$. Since the eigenstates of the atom necessarily have well defined parity, $\langle\psi_0|(\hat{\mathbf{n}}\cdot\boldsymbol{p})|\psi_0\rangle = \langle\psi_m|(\hat{\mathbf{n}}\cdot\boldsymbol{p})|\psi_m\rangle = 0$.

The Hamiltonian in (6.9) may be written in terms of the three Pauli spin $\frac{1}{2}$ matrices, which we write

$$s_1 = \frac{1}{2}\begin{pmatrix} 0 & 1 \\ 1 & 0 \end{pmatrix}, \tag{6.10a}$$

$$s_2 = \frac{1}{2}\begin{pmatrix} 0 & -i \\ i & 0 \end{pmatrix}, \tag{6.10b}$$

and

$$s_3 = \frac{1}{2}\begin{pmatrix} 1 & 0 \\ 0 & -1 \end{pmatrix}. \tag{6.10c}$$

We suppose s_1, s_2 and s_3 to be components of a spin vector in a fictitious space with axes parallel to the three orthogonal unit vectors $\hat{\mathbf{x}}_1$, $\hat{\mathbf{x}}_2$ and $\hat{\mathbf{x}}_3$. There is no relationship between this coordinate system, and the x, y and z axes of the laboratory frame. The Hamiltonian in (6.9) is then, with $\gamma_1 = \gamma\cos\phi$ and $\gamma_2 = \gamma\sin\phi$

$$H = -\hbar\omega_{m0}s_3 - \frac{2E(t)}{\hbar}(\gamma_1 s_1 + \gamma_2 s_2)\,. \tag{6.11}$$

This Hamiltonian is identical in form to that of a spin $\frac{1}{2}$ particle placed in a static magnetic field directed along $\hat{\mathbf{x}}_3$, with time-dependent, linearly polarized magnetic field in the $\hat{\mathbf{x}}_1 - \hat{\mathbf{x}}_2$ plane. The orientation of the magnetic field is such that it makes an angle ϕ with respect to the $\hat{\mathbf{x}}_1$ axis. The parameter $\hbar\omega_{m0}$ measures the strength of the effective static magnetic field, and the time dependence of the oscillating magnetic field is proportional to $E(t) = E_0\cos(\omega t)$.

All possible information about the wave function is contained in the expectation values

$$\langle s_1\rangle = \langle \Phi|s_1|\Phi\rangle = \frac{1}{2}(a_0^* a_m + a_m^* a_0)\,, \tag{6.12a}$$

$$\langle s_2\rangle = \frac{i}{2}(a_m^* a_0 - a_0^* a_m)\,, \tag{6.12b}$$

and

$$\langle s_3\rangle = \frac{1}{2}(|a_0|^2 - |a_m|^2)\,. \tag{6.12c}$$

The constants a_0 and a_m are both complex, so four real numbers are required to specify the wave function fully. But normalization requires $|a_0|^2 + |a_m|^2 = 1$ at all times, so only three numbers suffice. We choose $\langle s_1\rangle$, $\langle s_2\rangle$ and $\langle s_3\rangle$ to be these three.

One may calculate the time dependence of the expectation values, noting $|\Phi\rangle$ obeys the Schrödinger equation:

$$\begin{aligned} i\hbar\frac{d}{dt}\langle s_i\rangle &= i\hbar\frac{d}{dt}[\langle\Phi(t)|s_i|\Phi(t)\rangle] \\ &\equiv \langle\Phi(t)|[s_i, H]|\Phi(t)\rangle\,. \end{aligned} \tag{6.13}$$

A short calculation which makes use of the well-known commutation relations for spin operators gives

$$\frac{d}{dt}\langle \mathbf{s}\rangle = \mathbf{\Lambda}(t) \times \langle \mathbf{s}\rangle\,, \tag{6.14}$$

where

$$\mathbf{\Lambda}(t) = \hat{\mathbf{x}}_3\omega_{m0} + \frac{2E(t)}{\hbar}(\gamma_1\hat{\mathbf{x}}_1 + \gamma_2\hat{\mathbf{x}}_2)\,. \tag{6.15}$$

The equation of motion (6.14) is identical in form to that of a classical magnetic moment placed in an effective magnetic field described by (6.15). The behavior of the atom may be visualized by exploiting this analogy. First notice that

$$\frac{d}{dt}\langle \mathbf{s}\rangle^2 = \langle \mathbf{s}\rangle\cdot\frac{d}{dt}\langle \mathbf{s}\rangle = 0\,, \tag{6.16}$$

so that the vector $\langle \boldsymbol{s} \rangle$ is constant in length, with its tip always lying on the sphere in our three-dimensional space with radius $\frac{1}{2}$. When $\langle \boldsymbol{s} \rangle$ is directed upward in the $\hat{\mathbf{x}}_3$ direction, $\langle \boldsymbol{s} \rangle = \hat{\mathbf{x}}_3/2$, and the atom is in the ground state. Conversely, when $\langle \boldsymbol{s} \rangle = -\hat{\mathbf{x}}_3/2$, the atom is in the excited state. More generally, the projection of $\langle s \rangle$ onto the $\hat{\mathbf{x}}_3$ axis provides a measure of the degree of excitation of the atom.

The electric dipole moment of the atom, $\langle \Phi | \boldsymbol{p} | \Phi \rangle$, is necessarily parallel to $\hat{\mathbf{n}}$, since the atom is spherically symmetric. Hence a short calculation gives

$$\begin{aligned} \langle \Phi | \boldsymbol{p} | \Phi \rangle &= \hat{\mathbf{n}} \langle \Phi | (\hat{\mathbf{n}} \cdot \boldsymbol{p}) | \Phi \rangle \\ &= 2\hat{\mathbf{n}}\gamma [\langle s_1 \rangle \cos\phi + \langle s_2 \rangle \sin\phi] \, . \end{aligned} \tag{6.17}$$

Thus, for the atom to possess a nonzero dipole moment, either $\langle s_1 \rangle$ or $\langle s_2 \rangle$ must be non-zero.

We could solve (6.14) directly; when written out, we have three coupled first order differential equations. However, there is a simple and elegant means of extracting information about the nature of the solution. We make a transformation to a coordinate system that rotates about the $\hat{\mathbf{x}}_3$ axis with angular velocity ω. We thus transform to new variables given by

$$\langle s_1' \rangle = \langle s_1 \rangle \cos(\omega t) + \langle s_2 \rangle \sin(\omega t) \, , \tag{6.18 a}$$

$$\langle s_2' \rangle = -\langle s_1 \rangle \sin(\omega t) + \langle s_2 \rangle \cos(\omega t) \, , \tag{6.18 b}$$

$$\langle s_3' \rangle = \langle s_3 \rangle \, . \tag{6.18 c}$$

One then finds, using primes to denote vectors seen by an observer stationary with respect to the rotating system

$$\frac{d}{dt} \langle \boldsymbol{s}' \rangle = \boldsymbol{\Lambda}' \times \langle \boldsymbol{s}' \rangle - \omega \hat{\mathbf{x}}_3' \times \langle \boldsymbol{s}' \rangle \, . \tag{6.19}$$

The vector Λ' has the components, recalling $E(t) = E_0 \cos(\omega t)$,

$$\Lambda_1' = \frac{E_0 \gamma_1}{\hbar} + \frac{E_0}{\hbar} [\gamma_1 \cos(2\omega t) + \gamma_2 \sin(2\omega t)] \, , \tag{6.20 a}$$

$$\Lambda_2' = \frac{E_0 \gamma_2}{\hbar} + \frac{E_0}{\hbar} [-\gamma_1 \sin(2\omega t) + \gamma_2 \cos(2\omega t)] \, , \tag{6.20 b}$$

$$\Lambda_3' = \omega_{m0} \, . \tag{6.20 c}$$

If we ignore the time-dependent terms in (6.20), then the motion of $\langle \boldsymbol{s}' \rangle$ is a simple Larmor precession around the static effective magnetic field

$$\boldsymbol{\Lambda}_{\text{eff}} = (\omega_{m0} - \omega) \hat{\mathbf{x}}_3' + \frac{E_0}{\hbar} (\gamma_1 \hat{\mathbf{x}}_1' + \gamma_2 \hat{\mathbf{x}}_2') \, . \tag{6.21}$$

One can argue that the influence of the terms in $\cos(2\omega t)$ and $\sin(2\omega t)$ are small, under the conditions of interest. If one treats E_0 as small, and uses the present formalism to construct a description of the wave function of the atom to first order in E_0, one can demonstrate that the time-independent terms in (6.20 a,b) lead to the second term in (6.1) with the energy denominator $(\omega_{m0}-\omega)$, while the terms in $\sin(2\omega t)$ and $\cos(2\omega t)$ lead to a term proportional to $(\omega_{m0}+\omega)$, which is one of the contributions to the last term displayed in (6.1). By ignoring the time-dependent terms in (6.20 a,b), we thus overlook a correction to the wave function that is small even right on resonance. Indeed, if we were to retain these terms, we should then include the contribution from excited states other than $|\psi_m\rangle$, since all the nonresonant terms contained on the right-hand side of (6.1) are comparable in magnitude. It would thus be inconsistent to retain the time-dependent terms in (6.20), while at the same time confining our attention to the approximate description of the wave function given in (6.2).

In the rotating frame, the motion of the vector $\langle s'\rangle$ is now simply visualized. It is a simple precession around the vector $\boldsymbol{\Lambda}_{\text{eff}}$ given in (6.21), with the angular velocity

$$\omega_r = [(\omega_{m0}-\omega)^2 + (E_0\gamma/\hbar)^2]^{1/2} . \tag{6.22}$$

The vector $\langle s'\rangle$ sweeps out a conical figure oriented so that the projection of $\langle s'\rangle$ on $\boldsymbol{\Lambda}_{\text{eff}}$ is a constant. Thus, the cone is oriented so $\boldsymbol{\Lambda}_{\text{eff}}$ lies along its principal axis. The angle of the apex of the cone is controlled by the initial conditions. The precession frequency ω_r is referred to often as the Rabi frequency.

Suppose the atom is in its ground state at $t=0$, so the boundary conditions at $t=0$ read $\langle s_3'\rangle = 1/2$, $\langle s_1'\rangle = \langle s_2'\rangle = 0$. The explicit solution of the differential equations then reads

$$\langle s_1'\rangle = \frac{\gamma_2 E_0}{2\hbar\omega_r}\sin(\omega_r t) + \frac{\gamma_1 \Delta\omega E_0}{2\hbar\omega_r^2}[1-\cos(\omega_r t)] , \tag{6.23 a}$$

$$\langle s_2'\rangle = -\frac{\gamma_1 E_0}{2\hbar\omega_r}\sin(\omega_r t) + \frac{\gamma_2 \Delta\omega E_0}{2\hbar\omega_r^2}[1-\cos(\omega_r t)] , \tag{6.23 b}$$

and

$$\langle s_3'\rangle = \frac{1}{2}\cos(\omega_r t) + \frac{(\Delta\omega)^2}{2\omega_r^2}[1-\cos(\omega_r t)] , \tag{6.23 c}$$

where $\Delta\omega = \omega_{m0} - \omega$.

Of particular interest is the case of exact resonance, $\omega = \omega_{m0}$, or $\Delta\omega = 0$. Then we have

$$\langle s_3\rangle = \langle s_3'\rangle = \frac{1}{2}\cos\left(\frac{\gamma E_0}{\hbar}t\right) . \tag{6.24}$$

The system is in the ground state initially, and after the time $\Delta t = \pi\hbar/\gamma E_0$, it resides in the first excited state. But it has returned to the ground state once again in the time interval $2\Delta T$.

Clearly, on the time average, the atom absorbs no net energy from the electromagnetic field, since it returns to the ground state at the times $2\pi n\hbar/\gamma E_0$, for each integer n.

When $\omega = \omega_{m0}$, one ordinarily assumes the system absorbs energy from the field, so we have reached a conclusion very different than found in elementary texts. A complete microscopic treatment of absorption must in fact take account of the role of collisions. Suppose our atom is in a gas or liquid, and suffers collisions with τ the mean time between the collisions. If an atom in the ground state is exposed to a resonant field, it will begin to be excited, as described by (6.24). If $\tau \ll 2\pi\hbar/\gamma E_0$, it will suffer many collisions before it may advance to the excited state. The collisions will return it to the ground state, and the collision partner will carry off the excess kinetic energy, which then appears as heat. When $\tau \gg 2\pi\hbar/\gamma E_0$, the atom will complete many excursions to the excited state and back before suffering a collision. The usual elementary theory of resonant absorption found in quantum mechanics texts assumes implicitly that $\tau \ll 2\pi\hbar/\gamma E_0$. The atom is never highly excited, and at all times of interest the influence of the applied electromagnetic field may be treated by perturbation theory. Any physical system is characterized by some characteristic internal energy transfer or collision time. Thus, at sufficiently low intensity (small E_0), the condition $\tau \ll 2\pi\hbar/\gamma E_0$ will be satisfied.

On resonance, when $\omega = \omega_{m0}$, one can find the time dependence of the dipole moment of the atom from (6.17), recognizing that the expression in (6.23) applies to the rotating frame. One finds

$$\langle \Phi | p | \Phi \rangle = \hat{\mathbf{n}}\gamma \sin\left(\frac{\gamma E_0}{\hbar} t\right) \sin(\omega t) \,. \tag{6.25}$$

The largest value that can be assumed by the dipole moment is then γ. This is achieved at times for which $|a_0| = |a_m| = 1/\sqrt{2}$, and the wave function of the atom is an equal admixture of ground and excited state character. Recall the applied electric field is $E(t) = E_0 \cos(\omega t)$, and the $\sin(\omega t)$ factor is thus 90° out of phase with $\cos(\omega t)$, as the above argument suggests.

The periodic cycling of the atom between its ground and excited state is referred to often as "Rabi flopping". One sees that the dipole moment in (6.25) oscillates with frequency ω, but there is a modulation of the amplitude at frequency $\gamma E_0/\hbar$. The emission intensity is thus modulated. One refers to these modulations as "quantum beats".

Just after (6.24), it was argued that in the absence of collisions, or equivalently other processes that depopulate the excited state, the atom fails to absorb energy from an electric field of frequency exactly in resonance with the atom, $\omega = \omega_{m0}$. It is interesting to follow this reasoning through, and show that one arrives at a result equivalent to that derived in texts on quantum theory for the absorption rate, in the collision dominated regime.

We shall assume in what follows that the excited state has a lifetime τ, possibly as a consequence of collisions with other atoms in the gas phase, radiative decay, or any other mechanism that destroys the dipole moment of the atom. Assume a de-excitation occurs at time $t = 0$. The subsequent time development of the dipole moment is described by the above relations, until the next de-excitation event. If the frequency ω of the electromagnetic field is off resonance, (6.25) is replaced by

$$\langle\Phi|\boldsymbol{p}|\Phi\rangle = \hat{\mathbf{n}}\,\frac{\gamma^2 E_0}{\hbar\omega_r}\left\{\sin(\omega_r t)\sin(\omega t) + \frac{\gamma^2\Delta\omega}{\omega_r}[1-\cos(\omega_r t)]\cos(\omega t)\right\}. \tag{6.26}$$

The rate at which the atomic dipole absorbs energy from the electromagnetic field is

$$\frac{dU}{dt} = \boldsymbol{E}(t)\cdot\frac{d}{dt}\langle\Phi|\boldsymbol{p}|\Phi\rangle , \tag{6.27}$$

with $\boldsymbol{E}(t) = E_0\cos(\omega t)$. Of interest is the rate of energy absorption, averaged over a cycle of the electromagnetic field, $\langle dU/dt\rangle$. When this average is calculated, treating $\sin(\omega_r t)$ and $\cos(\omega_r t)$ as slowly varying functions of time, but replacing $\langle\cos^2\omega t\rangle = 1/2$, $\langle\cos(\omega t)\sin(\omega t)\rangle = 0$, one finds, recalling $\Delta\omega = \omega_{m0} - \omega$,

$$\left\langle\frac{dU}{dt}\right\rangle = \frac{\gamma^2 E_0^2}{2\hbar\omega_r}\sin(\omega_r t) . \tag{6.28}$$

Now the probability of finding the atom in the time interval dt with time evolution still described by (6.26) is $(dt/\tau)\exp(-t/\tau)$, where τ is the mean time between events that de-excite the atom. The average energy ΔW absorbed by the atom between de-excitation events is thus

$$\Delta W = \frac{\gamma^2 E_0^2\omega_{m0}}{2\hbar\omega_r}\int_0^\infty dt\,\sin(\omega_r t)e^{-t/\tau} , \tag{6.29}$$

which equals

$$\Delta W = \frac{\pi\gamma^2\omega_{m0}E_0^2}{2\hbar}\,\frac{1}{\pi}\,\frac{1}{(\omega-\omega_{m0})^2 + (1/\tau)^2} , \tag{6.30}$$

where we have replaced ω_r^2 by its weak field limit $(\omega - \omega_{m0})^2$.

In the time interval τ, the total amount of energy per unit area that has passed by the atom is just the time average of the Poynting vector, multiplied by τ, or $\Phi_0 = cE_0^2\tau/8\pi$. The absorption cross section of the atom, $\sigma(\omega)$ is just the ratio of ΔW to Φ_0. Hence,

$$\sigma(\omega) = \frac{4\pi^2\gamma^2\omega_{m0}}{\hbar c}\left\{\frac{1}{\pi}\,\frac{1/\tau}{(\omega-\omega_{m0})^2 + (1/\tau)^2}\right\} , \tag{6.31}$$

a result which agrees with the expression quoted in quantum mechanics texts[1], provided the unphysical conserving Dirac delta function found in such expressions is replaced by a lifetime broadened Lorentzian.

We can appreciate from this discussion that radiation with frequency in resonance with an atomic transition can be absorbed only if the excited state has a finite lifetime, by virtue of collisions in the gas or liquid phase, or other de-excitation processes. If the lifetime of the excited state τ is long compared to $2\pi\hbar/\gamma E_0$, then the description of the interaction of the atom with a resonant electric field must proceed along the lines given in this section, and the behavior of the atomic system is very different.

6.2 Bloch Equations: Power Broadening and Saturation Effects in Absorption Spectra

In the previous subsection, we considered the interaction between an isolated atom, and an electric field whose frequency ω nearly matches that ω_{m0} associated with a transition from the ground state to a (nondegenerate) excited state. At the end of our discussion, we examined the influence of the finite lifetime of the excited state on the response of the atom, in a crude fashion. This section is devoted to a more careful examination of the nature of the interaction of such an atom with its environment. We shall be led to a well-known set of equations introduced in the literature on nuclear magnetic resonance [6.2]. These are known as the Bloch equations, and in various forms have been applied extensively in the analysis of the response of two level systems of diverse nature.

We begin by examining not just the response of one single atom to a nearby resonant field, but instead to that of an array of N such atoms confined to a very small volume ΔV, within which the applied electric field may be regarded as constant in magnitude. We wish to examine the time evolution of the average dipole moment per atom of this collection, rather than that of one particular atom. Thus, we wish to explore the behavior of

$$\boldsymbol{S} = \frac{1}{N}\sum_{i=1}^{N} \langle \boldsymbol{s}(i)\rangle \,. \qquad (6.32)$$

The electric dipole moment per unit volume that enters our macroscopic electromagnetic theory is then

$$\boldsymbol{P} = 2n\gamma[S_1 \cos\phi + S_2 \sin\phi]\hat{\boldsymbol{n}} \,. \qquad (6.33)$$

where n is the number of atoms per unit volume.

[1] One may compare our (6.31) with (13.51) of *Baym* [6.1]. Also, one recognizes differences in convention, and needs to use our (A.35) to convert matrix elements of the current operator to that of the electric dipole moment operator. Our current density is the electrical current, and includes the charge e, while Baym's refers to the number current.

For each individual atom, $\langle \boldsymbol{s}(i) \rangle$ evolves in time as prescribed by (6.14). Each atom interacts with its environment, including nearby resonant atoms. If we excite the atoms coherently with a pulse, then switch off the field, we induce a nonzero value of $\boldsymbol{S}$, which will decay to zero as a consequence of the interactions. Thus, $\boldsymbol{S}$ obeys an equation of motion that we may write as

$$\frac{d\boldsymbol{S}}{dt} = \boldsymbol{\Lambda}(t) \times \boldsymbol{S} + \left(\frac{d\boldsymbol{S}}{dt}\right)_{\text{int}}, \tag{6.34}$$

where the new term describes the influence on $\boldsymbol{S}$ of the surrounding environment.

It is a complex matter to construct the form of $(d\boldsymbol{S}/dt)_{\text{int}}$ from first principles. We thus resort to the phenomenological arguments which lead us to the Bloch equations.

In the absence of any applied field, all atoms will reside in the ground state. Then $\boldsymbol{S} \equiv \boldsymbol{S}_0 = \hat{\mathbf{x}}_3/2$. If the array of atoms is perturbed by an applied field, then of course we will have $\boldsymbol{S} \neq \boldsymbol{S}_0$. If the applied field is then shut off, the atoms will return to their ground state, in an exponential manner characterized by some relaxation time.

The Bloch equations recognize the precence of two distinct means of returning $\boldsymbol{S}$ to its equilibrium value $\boldsymbol{S}_0$. Consider first the behavior of S_1 and S_2 for an array such as envisioned in (6.32). Suppose the atoms are imbedded in a crystal lattice. In general, one may expect the transition frequency ω_{m0} to differ from atom to atom, since the local environment of two atoms may differ. We may write $\omega_{m0}(i) = \bar{\omega}_{m0} + \Delta\omega_{m0}(i)$, where $\Delta\omega_{m0}(i)$ varies from site to site, possibly because a given atom may lie close to a defect which perturbs its energy, or be near one or more impurities, etc. The precession frequencies of the various atoms will then differ, and as time evolves, $\langle s_1(i) \rangle$ and $\langle s_2(i) \rangle$ will precess at different rates for different atoms. In the absence of an external field, $\langle s_3(i) \rangle$ is constant and left unchanged, so S_3 is constant in time in the absence of the external field. The fluctuations in the value of $\Delta\omega_{m0}(i)$ will cause $\langle \boldsymbol{s}_{\parallel}(i) \rangle = \hat{\mathbf{x}}_1 \langle s_1(i) \rangle + \hat{\mathbf{x}}_2 \langle s_2(i) \rangle$ to point in random directions after sufficient time has passed, with the consequence that S_1 and S_2 willl decay to zero in a characteristic time T_2 called the transverse relaxation time, or the dephasing time.

The discussion just given assumes that the phenomenon of dephasing has its origin in differences in the static local environment of the various atoms. Even if the local environment of each atom is identical on the time average, fluctuations in the local environment always occur because of the presence of thermal motions. If all the atoms are placed on perfectly equivalent sites in a crystal lattice, at any instant of time thermal motions of neighboring ions will lead to differences in the local geometry experienced by the various atoms. We may regard $\Delta\omega_{m0}(i)$ to be a randomly fluctuating function of time in this instance, and at any particular fixed time $\Delta\omega_{m0}(i)$ will differ as we sample different members of our ensemble. In a gas or a liquid, collisions or time dependent potentials which result from thermal motions of nearby atoms will produce dephasing in a similar manner.

The dephasing processes just discussed leave S_3 constant in time, in the absence of an applied field. Clearly, if the atom is in the excited state, interaction with its environment will lead to transitions through which it returns to the ground state. Even an isolated atom in free space returns to its ground state, by radiating one or more photons. In the gas phase, inelastic collisions with other atoms provide a de-excitation channel, and in solids or liquids a variety of mechanisms can lead to decay to the ground state from an excited state. Such processes cause S_3 to relax to its equilibrium value $\hat{\mathbf{x}}_3/2$ in a characteristic time T_1 referred to as the longitudinal relaxation time, or population relaxation time.

The dephasing processes examined earlier contribute to the relaxation of S_1 and S_2, but not to that of S_3, for the reasons given. However, a T_1 process also contributes to the relaxation of S_1 and S_2, since the return of an atom to the ground state destroys its dipole moment. Thus, one finds $T_2 < T_1$, quite generally. It is not uncommon for T_2 to be shorter than T_1 by one or more orders of magnitude.

The above discussion leads one to the Bloch equation, which has the form

$$\frac{d\mathbf{S}}{dt} = \mathbf{\Lambda}(t) \times \mathbf{S} - \frac{1}{T_2}(S_1\hat{\mathbf{x}}_1 + S_2\hat{\mathbf{x}}_2) - \frac{1}{T_1}\left(S_3 - \frac{1}{2}\right)\hat{\mathbf{x}}_3 \,. \tag{6.35 a}$$

We turn to a most important application of the Bloch equation. This is the description of the absorption of radiation by a nearly resonant system, with attention to its dependence on the applied power.

When the system is exposed to an applied field of frequency ω, and we transform (6.35 a) to the rotating frame. Upon ignoring the terms in $\cos(2\omega t)$ and $\sin(2\omega t)$ once again in (6.20), we have

$$\frac{d\mathbf{S}'}{dt} = \mathbf{\Lambda}_{\text{eff}} \times \mathbf{S}' - \frac{1}{T_2}(S_1'\hat{\mathbf{x}}_1' + S_2'\mathbf{x}_2') - \frac{1}{T_1}\left(S_3' - \frac{1}{2}\right)\hat{\mathbf{x}}_3' \,, \tag{6.35 b}$$

with $\mathbf{\Lambda}_{\text{eff}}$ given by (6.21).

In the rotating frame, the vector $\mathbf{S}'$ senses the effective field $\mathbf{\Lambda}_{\text{eff}}$, as we have discussed. When we ignore relaxation by setting T_1 and T_2 to infinity, the vector $\mathbf{S}'$ precesses about $\mathbf{\Lambda}_{\text{eff}}$ with the Rabi frequency given in (6.22). In the presence of relaxation, the precessional motion is damped out, and $\mathbf{S}'$ assumes the time-independent orientation for which $d\mathbf{S}'/dt$ vanishes, after sufficient time. Thus, we will have, solving for the components of $\mathbf{S}'$ when this state is realized,

$$S_1' = \frac{1}{2}\frac{E_0}{\hbar}\frac{\gamma_1(\omega_{m0} - \omega) + \gamma_2/T_2}{(\omega_{m0} - \omega)^2 + \dfrac{1}{T_2^2}\left(1 + \dfrac{\gamma^2 E_0^2}{\hbar^2}T_1T_2\right)} \,, \tag{6.36 a}$$

$$S_2' = \frac{1}{2}\frac{E_0}{\hbar}\frac{\gamma_2(\omega_{m0} - \omega) - \gamma_1/T_2}{(\omega_{m0} - \omega)^2 + \dfrac{1}{T_2^2}\left(1 + \dfrac{\gamma^2 E_0^2}{\hbar^2}T_1T_2\right)} \tag{6.36 b}$$

and

$$S_3' = \frac{1}{2} - T_1 \frac{E_0}{\hbar}(\gamma_2 S_1' - \gamma_1 S_2') \,. \tag{6.36c}$$

With these expressions, we can obtain the rate at which energy is absorbed by the array of atoms. With n atoms per unit volume, the induced dipole moment per unit volume $\boldsymbol{P}$ is just $\boldsymbol{P} = 2n\gamma[S_1 \cos\phi + S_2 \sin\phi]\hat{\mathbf{n}}$, or when we express this in terms of the vector $\boldsymbol{S}'$ in the rotating frame,

$$\boldsymbol{P} = 2n[\cos(\omega t)(\gamma_1 S_1' + \gamma_2 S_2') + \sin(\omega t)(\gamma_2 S_1' - \gamma_1 S_2')] \,. \tag{6.37}$$

The average rate at which energy is absorbed by the array is, per unit volume,

$$\frac{dU}{dt} = \left\langle \boldsymbol{E}(t) \cdot \frac{d\boldsymbol{P}}{dt} \right\rangle ,$$

where the angular brackets denote a time average.

One finds

$$\frac{dU}{dt} = \frac{n\omega\gamma^2 E_0^2}{2\hbar} \frac{1/T_2}{(\omega_{m0} - \omega)^2 + \dfrac{1}{T_2^2}\left(1 + \dfrac{\gamma^2 E_0^2}{\hbar^2} T_1 T_2\right)} \,. \tag{6.39}$$

We may convert this to an absorption cross section per atom, by deleting the factor of n, and dividing by $cE_0^2/8\pi$, the time averaged energy per unit area per unit time which flows by a given atom. Thus, we have

$$\sigma(\omega) = \frac{4\pi\gamma^2\omega}{\hbar c} \frac{1/T_2}{(\omega_{m0} - \omega)^2 + \dfrac{1}{T_2^2}\left(1 + \dfrac{\gamma^2 E_0^2}{\hbar^2} T_1 T_2\right)} \,. \tag{6.40}$$

We may compare (6.40) with (6.31) in the limit of low power, $\gamma^2 E_0^2 T_1 T_2/\hbar \ll 1$. The two expressions agree in form, save for the replacement of ω_{m0} by ω, in the prefactor of (6.40). The crude argument that leads to (6.31) in fact produces the incorrect prefactor, though close to resonance the two formulae agree. Clearly, as $\omega \to 0$, $\sigma(\omega)$ must vanish.

In the low power limit, the absorption profile is Lorenzian in shape, with width controlled by the dephasing rate $1/T_2$. In our discussion of relaxation processes, we described two classes of processes which contribute to the dephasing phenomenon. One, for atoms imbedded in a solid material, has origin in the differences in the static, or time averaged, environment felt by the various atoms. Then $\Delta\omega_{m0}(i)$ differs from atom to atom. Also, dynamic processes lead to dephasing. As we have noted, examples are time-dependent modulations of $\Delta\omega_{m0}(i)$ produced by thermal motions of nearby ions, and interactions which de-excite the atom (thus giving contributions to the depopulation rate $1/T_1$ also). When dephasing is dominated by differences in the static environment, one says the absorption line is inhomogeneously broadened. When $1/T_2$ receives its dominant contributions from dephasing processes of dynamical origin, one says the line is homogeneously broadened.

The absorption rate, and the cross section both exhibit a dependence on power. Most particularly, when $\gamma^2 E_0^2 T_1 T_2/\hbar^2 \gg 1$, the cross section is reduced greatly from the value appropriate to the linear response, or low power regime. The absorption line also broadens. As $E_0^2 \to \infty$, $\sigma(\omega) \to 0$, and the rate of power absorption saturates at the rate

$$\lim_{E_0^2\to\infty} \frac{dU}{dt} = \frac{1}{2} n\hbar\omega \frac{1}{T_1} . \tag{6.41}$$

Notice also that

$$\lim_{E_0^2\to\infty} S_3 = \lim_{E_0^2\to\infty} S_3' = 0 . \tag{6.42}$$

One refers to this as saturation of the absorption line. The limits in (6.42) show that at high powers, the vector $\boldsymbol{S}$ lies in the xy plane. Thus, $|a_0|^2 = |a_m|^2$; the probability of finding the atom in the excited state is the same as in the ground state. In our ensemble of atoms, one has half the atoms in the ground state, and half in the excited state.

Within quantum mechanical transition rate theory, one may understand this as follows: The rate for absorbing a photon of frequency ω_{0m}, and promoting atoms to the excited state may be written as $MN_0 n_\omega$, with n_ω the number of photons, N_0 the number of atoms in the ground state, and M all the various other factors in the transition rate formula. The rate of emission by an atom in the excited state is $MN_m(1 + n_\omega) \cong MN_m n_\omega$ in the limit discussed here, where the applied electric field may be described classically (this requires $n_\omega \gg 1$). As the applied field raises atoms from the ground to excited state, $(N_0 - N_m)$ decreases, and absorption drops to zero when $N_m = N_0$, or when $S_3 = 0$, in the language of the present discussion. As atoms are placed in the excited state, they radiate energy back into the electromagnetic field via stimulated emission.

The remarks of the previous paragraph suggest that at high powers, the absorption rate should drop to zero while (6.41) shows it saturates. The previous paragraph overlooks the energy leak to the surrounding environment described by the finiteness of T_1. Atoms in an excited state may lose energy by suffering deexcitation by the mechanisms described earlier. For example, in the gas phase, inelastic collisions which de-excite the atom lead to an increase in the kinetic energy of thermal motions; the gas is heated as a consequence. In the presence of "T_1 processes," the rate at which energy leaks out of our system of resonant atoms subjected to a high intensity field is controlled by the rate of such energy transfers, which occur at the rate of $1/T_1$ per atom. In the high power limit, $|a_m|^2 = \frac{1}{2}$, so the right-hand side of (6.41) is also $n\hbar\omega|a_m|^2/T_1$.

For laser action to be initiated by an array of atoms, one must establish a population inversion; there must be more atoms in the excited state than in the ground state [6.3]. We see that the best one can achieve by pumping atoms directly into an excited state is $N_m = N_0$.

The phenomenon of saturation is a feature of a two-level quantum system, which has been exposed to an intense, nearly resonant field. The harmonic oscillator, with its infinite ladder of equally spaced levels, will not exhibit the effect. If the oscillator absorbs a photon, and is promoted to the first excited state, it may absorb a second photon, and be promoted to the $n = 2$ level. If it is exposed to a resonant field of constant amplitude, it may continue to climb up the ladder of oscillator levels. The total energy absorbed thus increases linearly with time. In a two-level quantum system, the particle can only return to the ground state by photon emission, and be excited no further once it is in the excited state. We can generalize the opening statement of this paragraph that any quantum system with a finite number of equally spaced levels will exhibit saturation, but the harmonic oscillator will fail to do so. An example of a system with a finite ladder is an object with angular momentum $\boldsymbol{J}$ in a static magnetic field. Here one has $(2J + 1)$ equally spaced levels. The power required to saturate the system increases with J, and approaches infinity in the limit $J \rightarrow \infty$.

While the width of the absorption line at low powers is given by $1/T_2$, notice the power required to saturate the resonance is controlled in part by T_1. Thus, the power dependence of the absorption profile allows one access to both relaxation rates, T_1 and T_2.

There are a variety of spectroscopic methods that exploit the phenomenon of saturation. Suppose we have an inhomogeneously broadened line, with width $1/T_2$. The sample is illuminated by intense radiation with spectral width $\Delta\nu \ll 1/T_2$. The radiation will saturate the transitions of that fraction of the atoms in resonance with the applied field, while those whose resonance frequency lies outside the linewidth of the exciting source will remain unaffected. Now suppose the absorption line profile is scanned with a weak probe beam. There will be a dip in the absorption profile detected by the probe, centered about the frequency of the pump, as illustrated in Fig. 6.1. One refers to this phenom-

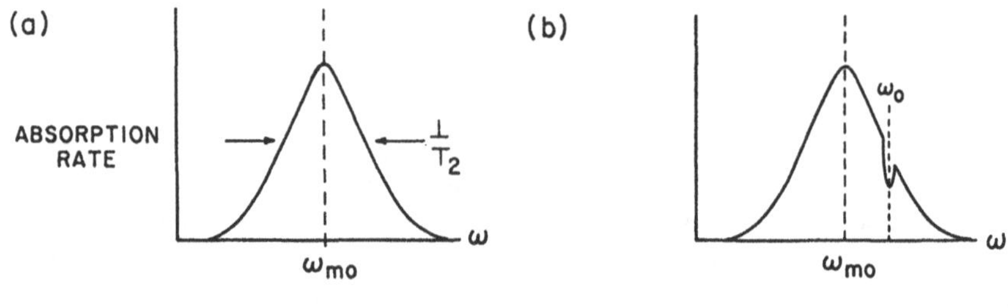

Fig. 6.1. An illustration of the absorption profile in a hole burning experiment. In **a** we show the Lorentzian line with width $1/T_2$ seen by a weak probe beam, and in **b** we illustrate the absorption profile seen by a weak probe beam, if the sample is illuminated by an intense pump beam of frequency ω_0, with spectral width small compared to $1/T_2$. The figure applies to the case where the line is inhomogeneously broadened

enon as hole burning. If the pump is shut off, the probe beam may then be used to follow the recovery of the system in real time, if the relaxation times are long enough. Transfer of the excitation to atoms which lie outside the profile of the pump beam will cause the hole to broaden with time, a process referred to as spectral diffusion. One can study the dynamics of energy transport within an array of excited atoms or molecules by this means.

6.3 Self-Induced Transparency

As remarked earlier, the Bloch equations apply to spin systems, and have been employed extensively in the description of both electron and nuclear resonance. Indeed, the formalism was developed many years ago with these applications in mind. Thus, the phenomena discussed above, such as the saturation of absorption lines at high power density, were well known and exploited by researchers in these fields, well before that advent of modern laser spectroscopy. Nuclear resonance involves application of radio frequency radiation, and electron spin resonance studies are carried out with microwaves. Under usual conditions, the wavelength of the radiation is large compared to the sample size, and one can assume the driving field is spatially uniform, as we have done above. A unique aspect of the optical frequency probes is the fact that radiation wavelength is small compared to the sample size in nearly all experiments. Thus, we must combine our description of the response of a collection of atoms to a driving field, with the full Maxwell equations which describe propagation of an electromagnetic disturbance in the medium. We do this in this section; we shall then encounter the remarkable phenomenon of self-induced transparency discovered some years ago by *McCall* and *Hahn* [6.4].

We have an electric field which we write as $\boldsymbol{E} = \hat{\boldsymbol{n}}E(z,t)$, and this has associated with it a dipole moment density $\boldsymbol{P}(z,t) = \hat{\boldsymbol{n}}P(z,t)$, where $P(z,t) = 2n[\gamma_1 S_1(z,t) + \gamma_2 S_2(z,t)]$. The Maxwell equation then reads

$$\left(\frac{\partial^2}{\partial z^2} - \frac{1}{\bar{c}^2}\frac{\partial^2}{\partial t^2}\right)E(z,t) = \frac{8\pi n}{c^2}\frac{\partial^2}{\partial t^2}[\gamma_1 S_1(z,t) + \gamma_2 S_2(z,t)] \,. \tag{6.43}$$

Here $\bar{c} = c/\sqrt{\varepsilon}$, where ε is the dielectric constant of the medium in which the resonant atoms are embedded.

We are to calculate S_1 and S_2 from the Bloch equation, (6.35). We do this by describing the electric field as follows:

$$E(z,t) = e(z,t)\cos[\omega t - kz + \psi(z,t)] \,, \tag{6.44}$$

where the amplitude $e(z,t)$ and the phase $\psi(z,t)$ are assumed to be slowly varying functions of both position z and the time t.

We now turn to the Bloch equations. We make a transformation to a local rotating frame, in a manner quite similar to (6.18). We have

$$S_1' = S_1 \cos(\omega t - kz + \psi) + S_2 \sin(\omega t - kz + \psi)\ , \tag{6.45a}$$

$$S_2' = -S_1 \sin(\omega t - kz + \psi) + S_2 \cos(\omega t - kz + \psi)\ , \tag{6.45b}$$

and

$$S_3' = S_3\ . \tag{6.45c}$$

One finds, upon ignoring terms in $(\partial\psi/\partial t)$ in transforming from the laboratory to the rotating frame,

$$\frac{dS'}{dt} = \Lambda'_{\text{eff}}\, x\, S' - \frac{1}{T_1}\hat{\mathbf{x}}_3'\left(S_3' - \frac{1}{2}\right) - \frac{1}{T_2}(\hat{\mathbf{x}}_1'S_1' + \hat{\mathbf{x}}_2'S_2')\ , \tag{6.46}$$

where

$$\Lambda'_{\text{eff}} = (\omega_{m0} - \omega)\hat{\mathbf{x}}_3' + \frac{1}{\hbar}e(z,t)(\gamma_1\hat{\mathbf{x}}_1' + \gamma_2\hat{\mathbf{x}}_2') \tag{6.47}$$

is a function of position and time, because $e(z,t)$ varies.

In what follows, we let both T_1 and T_2 become infinite. We also introduce new variables

$$S_\parallel' = S_1' \cos(\phi) + S_2' \sin(\phi) \tag{6.48a}$$

and

$$S_\perp' = -S_1' \sin(\phi) + S_2' \cos(\phi) \tag{6.48b}$$

where the phase angle ϕ is defined just after (6.6). The component of the effective field in the rotating frame proportional to $e(z,t)$ lies in the $\hat{\mathbf{x}}_1' - \hat{\mathbf{x}}_2'$ plane, directed at an angle ϕ relative to the $\hat{\mathbf{x}}_1'$ axis. The variables $S_\parallel'$ and $S_\perp'$ are the components of $S_1'\hat{\mathbf{x}}_1' + S_2'\hat{\mathbf{x}}_2'$ parallel and perpendicular to this effective field, respectively.

One finds the following equations of motion, when these variables are introduced:

$$\frac{\partial}{\partial t}S_\parallel' = -(\omega_{m0} - \omega)S_\perp'\ , \tag{6.49a}$$

$$\frac{\partial}{\partial t}S_\perp' = (\omega_{m0} - \omega)S_\parallel' - \frac{\gamma}{\hbar}e(z,t)S_3'\ , \tag{6.49b}$$

and

$$\frac{\partial}{\partial t}S_3' = \frac{\gamma}{\hbar}e(z,t)S_\perp'\ . \tag{6.49c}$$

One has also

$$\gamma_1 S_1(z,t) + \gamma_2 S_2(z,t) = \gamma[S_\parallel'(z,t)\cos(\omega t - kz + \psi) - S_\perp'\sin(\omega t - kz + \psi)]\ . \tag{6.50}$$

We next apply the slowly varying envelope approximation to the left-hand side of (6.43). After requiring that $k = \omega/\bar{c}$, one finds

$$\left(\frac{\partial^2}{\partial z^2} - \frac{1}{\bar{c}^2}\frac{\partial^2}{\partial t^2}\right)E \cong \frac{2\omega}{\bar{c}}\left[\left(\frac{\partial e}{\partial z}\right) + \frac{1}{\bar{c}}\left(\frac{\partial e}{\partial t}\right)\right]\sin(\omega t - kz + \psi)$$

$$+ \frac{2\omega}{\bar{c}} e\left[\left(\frac{\partial \psi}{\partial z}\right) + \frac{1}{\bar{c}}\left(\frac{\partial \psi}{\partial t}\right)\right]\cos(\omega t - kz + \psi) \tag{6.51}$$

When (6.51,50) are employed in (6.43), we obtain two first order differential equations which link the amplitude and phase of the electric field to the variables $S'_{\parallel}$ and $S'_{\perp}$. We have

$$\frac{\partial e}{\partial z} + \frac{1}{\bar{c}}\frac{\partial e}{\partial t} = \frac{4\pi n\omega\gamma}{c\sqrt{\varepsilon}} S'_{\perp} \tag{6.52 a}$$

and

$$e\left(\frac{\partial \psi}{\partial z} + \frac{1}{\bar{c}}\frac{\partial \psi}{\partial t}\right) = -\frac{4\pi n\omega\gamma}{c\sqrt{\varepsilon}} S'_{\parallel} . \tag{6.52 b}$$

The propagation of electromagnetic disturbances through the medium can be analyzed by solving (6.49) in combination with (6.52).

We shall consider a pulse which encounters the atoms in the ground state, so as $t \to -\infty$, we have $S'_3 = +\frac{1}{2}$, $S'_{\parallel} = S'_{\perp} = 0$. We shall suppose also that the carrier frequency of the pulse is in precise resonance with the atomic transition. Thus, we set $\omega = \omega_{m0}$. Then (6.49 a) in combination with the boundary condition requires that $S'_{\parallel}$ vanish at all times. The vector $\boldsymbol{S}'$ then precesses in the plane perpendicular to the effective field $e(\gamma_1\hat{\mathbf{x}}'_1 + \gamma_2\hat{\mathbf{x}}'_2)$. It then follows that $\psi(z,t)$ vanishes identically at all times, from (6.52 b). Since the vector $\boldsymbol{S}'$ has constant length, we may write

$$S'_3(z,t) = \frac{1}{2}\cos[\theta(z,t)] \tag{6.53 a}$$

and

$$S'_{\perp}(z,t) = \frac{1}{2}\sin[\theta(z,t)] , \tag{6.53 b}$$

where as $t \to -\infty$, $\theta(z,t) \to 0$. Equations (6.49 b,c) then collapse into a single statement which describes the time evolution of $\theta(z,t)$,

$$\frac{\partial \theta(z,t)}{\partial t} = -\frac{\gamma}{\hbar} e(z,t) \tag{6.54}$$

and we may then eliminate e from (6.52 a), to find

$$\frac{\partial^2 \theta}{\partial t^2} + \bar{c}\frac{\partial^2 \theta}{\partial t \partial z} = -\frac{2\pi n\omega\gamma^2}{\hbar\varepsilon}\sin(\theta) . \tag{6.55}$$

As we shall see later, (6.55) is very close in form to a classical equation of nonlinear physics, the sine-Gordon equation. Indeed, an elementary transformation will convert this equation into precisely the sine-Gordon equation, as we shall see later. There are a rich and fascinating array of solutions to this equation. For the moment, we will be content to examine one special, but very important solution.

We seek solutions in the form of a pulse, which propagates through the system with some velocity $v < \bar{c}$. Thus, all quantities are assumed to be functions of the variable $\tau = t - z/v$. Equation (6.55) then reduces to

$$\frac{\partial^2 \theta}{\partial \tau^2} = \Lambda^2 \sin \theta , \tag{6.56}$$

where $\Lambda^2 = 2\pi n \omega \gamma^2 v / \hbar \varepsilon(\bar{c} - v)$.

One may integrate (6.56) exactly, for our boundary conditions. Upon multiplying each side by $\partial\theta/\partial\tau$, we obtain the statement

$$\frac{\partial}{\partial \tau}\left[\frac{1}{2}\left(\frac{\partial \theta}{\partial \tau}\right)^2 + \Lambda^2 \cos \theta\right] = 0 \tag{6.57}$$

which, with the requirement $\theta \to 0$ as $\tau \to -\infty$ requires

$$\left(\frac{\partial \theta}{\partial \tau}\right)^2 + 2\Lambda^2[\cos \theta - 1] = 0 , \tag{6.58}$$

or

$$\frac{\partial \theta}{\partial \tau} = \pm 2\Lambda \sin \frac{\theta}{2} . \tag{6.59}$$

One more integration yields two solutions:

$$\theta_{\pm}(\tau) = 4 \tan^{-1}(e^{\pm \Lambda \tau}) . \tag{6.60}$$

Consider first the solution $\theta_{-}(\tau)$. For any fixed z, as $t \to -\infty$, $\theta_{-}(\tau) \to 0$. As $t \to +\infty$, $\theta_{-}(\tau) \to 2\pi$, so after the pulse has passed over a given set of atoms, the atoms are returned to their ground state. Halfway through the pulse, when $\tau = 0$, $\theta_{-} = \pi$, all atoms reside in the excited state. One refers to such a pulse as a 2π pulse. The vector $\boldsymbol{S}'$ rotates precisely through the angle 2π, in the clockwise direction as one looks at the precession down the vector $\hat{\mathbf{x}}_1' \cos\phi + \hat{\mathbf{x}}_2' \sin\phi$. The second solution, $\theta_{+}(\tau)$ is also at 2π pulse, in which the vector $\boldsymbol{S}'$ precesses in the counterclockwise direction around the direction of the electric field in the rotating frame.

It is a straightforward matter to find the expression for the envelope function $e(\tau)$ that describes the electric field $e(z, t)$:

$$e(\tau) = \pm \frac{2\hbar\Lambda}{\gamma} \frac{1}{\cosh(\Lambda\tau)} . \tag{6.61}$$

There are two remarkable features of this pulse. First of all, the carrier frequency ω precisely matches the transition frequency ω_{m0} of our atoms. Under these circumstances, elementary considerations suggest the electromagnetic radiation should be absorbed, as we have discussed. In fact, the medium is perfectly transparent to this resonant radiation! The pulse propagates forever!

Also, for frequencies ω very close to the resonance frequency ω_{m0}, the dielectric constant $\varepsilon(\omega)$ of the medium exhibits the resonant behavior discussed in Chap. 2. If we construct an electromagnetic wave packet localized in space, we must synthesize the wave packet from plane waves of different frequency and wave vector. The different wavelengths propagate at different phase velocities, with the consequence that the wave packet distorts and spreads as it propagates through the medium, if its propagation is described by linear dielectric theory [6.5]. We shall address this question explicitly in Chap. 7. The rate of spreading is particularly severe for frequencies near resonance. The wave packet described by (6.61) not only has infinite mean free path, but in addition propagates through the medium perfectly undistorted in shape!

We have clearly encountered new solutions that describe propagation of radiation through the medium with characteristics that differ qualitatively from those encountered in the theory of wave propagation in the small amplitude, linear response limit (the limit of ordinary dielectric theory). This is a fascinating aspect of the physics of highly nonlinear systems. We have just encountered an object referred to as a soliton. The solution $\theta_-(\tau)$ is a soliton pulse, and $\theta_+(\tau)$ is an anti soliton. We shall comment further on the properties of these objects later in this chapter, when we discuss general properties of the sine-Gordon equation.

One refers to the phenomenon just described as self-induced transparency. The atoms absorb energy from the leading edge of the pulse, but subsequently re-radiate it back into the tail in such a way as to produce a rigid, stable, propagating pulse. This remarkable phenomenon was discovered experimentally by *McCall* and *Hahn* [6.4], who also presented in their original papers a full and complete description of the theory. To address a number of questions not explored here, the reader will do well to read through their paper for a complete analysis.

Our soliton solution contains a parameter, the soliton propagation velocity v. The parameter Λ provides a measure of the width of the soliton, and we have from the definition of Λ,

$$\frac{1}{v} = \frac{1}{\bar{c}} + \frac{2\pi n\omega\gamma^2}{\hbar\varepsilon\Lambda^2\bar{c}} \tag{6.62}$$

The propagation velocity is thus always less than $\bar{c}$.

There remains the question of how the soliton pulse evolves from an input pulse whose shape may differ greatly from the soliton form. This is a complex question that in general must be addressed by a numerical solution of the full equations [6.4]. However, a most important theorem proved by McCall and

Hahn, the area theorem, tells one a great deal about this question. We turn next to a discussion of the area theorem.

6.4 Area Theorem

In general, if we consider an arbitrary pulse and not the special soliton of the previous section, both $e(z,t)$ and $\theta(z,t)$ depend on z and t, and not just on a special combination such as $t - z/v$. The pulse will change shape as it propagates. One defines the area of the pulse $A(z)$ by the statement

$$A(z) = \frac{\gamma}{\hbar} \int_{-\infty}^{+\infty} e(z,t)dt \tag{6.63}$$

which for later purposes we write as

$$A(z) = \frac{\gamma}{\hbar} \lim_{T\to\infty} \int_{-\infty}^{T} e(z,t)dt\,. \tag{6.64}$$

We do suppose that the pulse is localized in extent so that for fixed z, $e(z,t)$ vanishes both as $t \to +\infty$, and as $t \to -\infty$.

We begin by considering set of atoms that are a bit off resonance with the field, so $\Delta\omega = \omega_{m0} - \omega \neq 0$. One integrates both sides of (6.52 a) on time, taking note of (6.49 a) to find

$$\frac{dA}{dz} = -\frac{4\pi n\omega\gamma^2}{\hbar c\sqrt{\varepsilon}} \frac{1}{\Delta\omega} \lim_{T\to\infty} S'_{\parallel}(z,T;\Delta\omega)\,, \tag{6.65}$$

where the notation on the right-hand side makes explicit the fact that we are considering a set of dipoles off resonance.

Let T be very large, and introduce a time $t_0 < T$ with the property that $e(z,t) = 0$ for $t > t_0$. The combination of (6.49 a,b) then allows us to write, for $t_0 \leq t$

$$\begin{aligned} S'_{\parallel}(z,t;\Delta\omega) = {} & S'_{\parallel}(z,t_0;\Delta\omega)\cos[\Delta\omega(t-t_0)] \\ & - S'_{\perp}(z,t_0;\Delta\omega)\sin[\Delta\omega(t-t_0)]\,. \end{aligned} \tag{6.66}$$

Now any real ensemble of atoms will not have identical transition frequencies ω_{m0}, but inhomogeneous broadening is necessarily present. We account for the influence of inhomogeneous broadening by averaging the right hand side of (6.65) over $\Delta\omega$. We do this after inserting (6.66) on the right-hand side. Let $P(\Delta\omega)d(\Delta\omega)$ be the fraction of atoms in our ensemble characterized by a transition frequency such that $\Delta\omega$ lies in the range $d(\Delta\omega)$. Then

$$\frac{dA}{dz} = -\frac{4\pi n\omega\gamma^2}{\hbar c\sqrt{\varepsilon}} \lim_{T\to\infty} \left\{ \int d(\Delta\omega) P(\Delta\omega) \frac{\cos[\Delta\omega(T - t_0)]}{\Delta\omega} S'_{\parallel}(z, t_0; \Delta\omega) \right.$$
$$\left. - \int d(\Delta\omega) P(\Delta\omega) \frac{\sin[\Delta\omega(T - t_0)]}{\Delta\omega} S'_{\perp}(z, t_0; \Delta\omega) \right\} . \tag{6.67}$$

The function $P(\Delta\omega)$ is assumed to be an even function of $\Delta\omega$. Then for all times T, the first integral on the right-hand side of (6.67) vanishes. If this condition is not satisfied, the structure of the integrand is such that as $T \to \infty$, the region near $\Delta\omega = 0$ is all that matters, and the integral, viewed as a principal part integral, will vanish in the limit. Note the identity

$$\lim_{T\to\infty} \frac{\sin[\Delta\omega(T - t_0)]}{\Delta\omega} = \pi\delta(\Delta\omega) \tag{6.68}$$

so we have

$$\frac{dA}{dz} = \frac{4\pi^2 n\omega\gamma^2}{\hbar c\sqrt{\varepsilon}} P(0) S'_{\perp}(z, t_0; 0) . \tag{6.69}$$

The right-hand side of (6.69) is controlled by those spins precisely in resonance with the pulse. For these spins, we have, from our earlier analysis,

$$S'_{\perp}(z, t_0) = \frac{1}{2} \sin[\theta(z, t_0)] \tag{6.70}$$

and

$$\theta(z, t_0) = -\frac{\gamma}{\hbar} \int_{-\infty}^{t_0} dt e(z, t) \equiv -A(z) . \tag{6.71}$$

We thus have

$$\frac{dA}{dz} = -\alpha \sin(A) , \tag{6.72}$$

where

$$\alpha = \frac{2\pi^2 n\omega\gamma^2}{\hbar c\sqrt{\varepsilon}} P(0) . \tag{6.73}$$

The solution of (6.72) is

$$A(z) = 2 \tan^{-1}\left(\tan\left[\frac{A(z_0)}{2}\right] e^{-\alpha z} \right) . \tag{6.74}$$

We can now make general statements about the behavior of the pulse. Suppose $0 < A(z_0) < \pi$. Then as $z \to \infty$, $A(z) \to 0$. The energy in the initial pulse is simply absorbed by the atoms. For large z, $A(z) \sim \mathrm{e}^{-\alpha z}$, and one appreciates that α is just the standard expression for the absorption constant provided by elementary theory. If we compare (6.73) with (6.31) applied to resonant absorption with $\omega = \omega_{m0}$, we must multiply the cross section by the density of atoms n to obtain the absorption constant in cm^{-1}. This then describes the rate of decay of the intensity of the wave, while α describes that of the electric field itself. The intensity is proportional to the square of the electric field. Hence the factor of 4 in (6.31), while (6.73) has a 2. We are to identify $P(0)$ in (6.73) with the Lorentzian in curly brackets in (6.31), evaluated at $\omega = \omega_{m0}$.

Now suppose $\pi < A(z_0) < 3\pi$. Then from (6.74), $\lim_{z\to\infty} A(z) = 2\pi$. The initial pulse then evolves into the soliton pulse of the previous section. If $3\pi < A(z_0) < 5\pi$, one has $\lim_{z\to\infty} A(z) = 4\pi$. Numerical studies, and data reported by McCall and Hahn show the initial pulse breaks up into two solitons in this regime. From (6.60), if $(2n-1)\pi < A(z_0) < (2n+1)\pi$, then $\lim_{z\to\infty} A(z) = 2n\pi$, and the input pulse breaks up into a sequence of n solitons.

Thus, while a full analytic discussion of the evolution of an input pulse of arbitrary shape is not possible, the area theorem allows one to obtain insight into its final state. Most particularly, the threshold condition $A(z_0) > \pi$ must be satisfied to realize self-induced transparency.

6.5 Sine-Gordon Equation

During our discussion of self-induced transparency, when we considered an array of atoms with transition frequency ω_{m0} in resonance with the carrier frequency ω of the pulse, we arrived at (6.55) as the nonlinear differential equation which describes the evolution of the "tipping angle" $\theta(z,t)$. It was remarked that this equation is closely related to a classical equation of nonlinear physics, the sine-Gordon equation. This connection may be made explicit by transforming from our coordinates z and t, to a fictitious time τ, and fictitious spatial coordinate x, given by

$$\tau = t - \frac{1}{\bar{c}}(1-\alpha)z\,, \tag{6.75a}$$

$$x = (1+\alpha)z - \bar{c}t\,, \tag{6.75b}$$

where α is any nonzero number. After this change of variables, (6.55) becomes

$$\frac{\partial^2\theta}{\partial\tau^2} - \bar{c}^2\frac{\partial^2\theta}{\partial x^2} = -\mu^2\sin\theta\,, \tag{6.76}$$

where $\mu^2 = 2\pi n\omega\gamma^2/\alpha\hbar\varepsilon$.

The differential equation in (6.76) is the sine-Gordon equation, encountered usually when θ is the amplitude of some field of interest, while x and τ are in fact the true space and time variables of the system of interest. An example of a physical situation where the equation is encountered is provided by the theory of long wave length vibrations of a line of pendula in a gravitational field, each in the form of a rigid rod attached to a backbone that can be twisted by motion of the pendula. Also, the equation emerges again in the discussions of spin motions, in magnetic materials whose structure is such that one-dimensional lines of spins thread through the crystal [6.6].

The equation receives its name from the fact that when one examines small amplitude vibrations, where $\sin\theta \cong \theta$, the equation becomes identical to the Klein-Gordon equation of quantum mechanics, which was introduced to describe the relativistic motions of bosons [6.7]. If, in the limit of small amplitude vibrations, we seek solutions of the sine-Gordon equation with the space and time variation $\exp[i(Qx - \Omega t)]$, then one has

$$\Omega^2 = \mu^2 + \bar{c}^2 Q^2 . \tag{6.77}$$

If one applies the Klein-Gordon equation to the description of a relativistic particle, then $\hbar\Omega$ and $\hbar Q$ are its energy and momentum, respectively. Equation (6.77) is then the relation between the energy and momentum of a relativistic particle whose rest mass is proportional to μ.

One may easily extract the soliton solution from (6.76), as we did in Sect. 6.3. One can calculate the energy stored in this object by forming the Hamiltonian for the continuous field $\theta(x, t)$ from which (6.76) is derived, then inserting the soliton solution into the Hamiltonian to evaluate the energy. One finds (see Appendix B) that if $E(v)$ is the energy of a soliton with velocity v, then $E(v) = E(0)/(1 - v^2/\bar{c}^2)^{1/2}$. This is precisely the relation between the total energy and velocity of a relativistic particle moving in a medium whose maximum propagation velocity is $\bar{c}$. One is thus tempted to regard the soliton as an elementary particle, synthesized from interactions between the bosons which emerge as the solution of the linearized version of the sine-Gordon equation. This connection may be pursued further by a most remarkable mathematical property of the sine-Gordon equation: one can construct solutions which describe an arbitrary number N of solitons on the one-dimensional line, moving with various velocities [6.8]. As time progresses, two solitons necessarily "collide." They pass through each other quite unchanged in form! In Appendix B, we consider the properties of a particular solution that describes two separated solitons, as $t \to -\infty$, which are approaching each other. They collide, and as $t \to +\infty$, we again have two solitons. Their interaction does lead to a time delay in that if they collide at $t = 0$, at large times t their separation is $v(t - \Delta t)$, with v their relative velocity and Δt the time delay produced by their interaction.

The sine-Gordon equation thus describes states of N noninteracting "particles," or solitons. One can envision a primitive theory of elementary particles built around the sine-Gordon equation, with interactions between the basic "particles" provided by weak nonlinearities which supplement the sin θ term in (6.76). For reasons such as this, the basic mathematical properties of nonlinear field theories is an active topic of research in elementary particle physics, and in statistical mechanics. We should remark that in mathematical physics, the term soliton is reserved to describe localized objects which emerge from nonlinear differential equations, which also have many "particle" solutions within which the particles do not interact. Notice that such localized solutions have no counterpart in the linearized small amplitude limit, which in general describes propagating waves, with a characteristic dispersion relation such as that given in (6.77).

Other more complex solutions of the sine-Gordon equation can be found, and are discussed often in the literature. An example is the breather, which is a bound state of a soliton – anti-soliton pair. The terminology arises because the envelope function, in addition to being localized, also exhibits oscillations in time, in a reference frame where the object is at rest. We refer the reader elsewhere for a complete discussion [6.8] of the properties of this remarkable equation, along with methods for analyzing the time evolution of the original pulse.

Clearly, an array of atoms in contact with a resonant field can exhibit rich behavior, as we have seen. We conclude the present discussion with comments on the nature of the linearized waves which emerge from the present discussion. For this purpose, it is more useful to explore (6.55) directly, rather than its transformed version. If we seek small amplitude solutions for $\theta(z,t)$ proportional to $\exp[\mathrm{i}(\pm Qz \mp \Omega t)]$, then one sees easily that

$$\Omega^2 - \bar{c}Q\Omega - \frac{2\pi n\omega\gamma^2}{\hbar\varepsilon} = 0 \tag{6.78}$$

which has two solutions:

$$\Omega_\pm = \frac{1}{2}\bar{c}Q \pm \left[\left(\frac{2\pi n\omega\gamma^2}{\hbar\varepsilon}\right)^2 + \left(\frac{\bar{c}Q}{2}\right)^2\right]^{1/2} . \tag{6.79}$$

These solutions can be interpreted straightforwardly. The envelope function $e(z,t)$ has the same space and time variation as $\theta(z,t)$, from (6.54). Thus, we have side bands to the basic carrier wave $\cos(kz - \omega t)$ given in (6.44). Recall that when the atoms are all in resonance with the carrier wave, $\omega = \omega_{m0}$, we have $\psi(z,t) \equiv 0$. Also, it is the case that in our development of the slowly varying envelope approximation, we have assumed $\omega = ck/\sqrt{\varepsilon}$, where ε is the "background" dielectric constant.

Consider the dispersion relation of waves with wave vector $k \pm Q$ and frequency $\omega_{m0} \pm \Omega$ in a medium described by a frequency dependent dielectric constant $\varepsilon(\omega)$. From Chap. 2, we have (2.36)

$$\frac{c^2(k \pm Q)^2}{(\omega_{m0} \pm \Omega)^2} = \varepsilon(\omega_{m0} \pm \Omega) . \tag{6.80}$$

In the notation of the present chapter, and with use of (A.41) to describe the resonant term in $\varepsilon(\omega)$, we have

$$\varepsilon(\omega) = \varepsilon + \frac{8\pi n \gamma^2 \omega_{m0}}{\hbar} \frac{1}{\omega_{m0}^2 - \omega^2} . \tag{6.81}$$

For Ω very small, one has

$$\varepsilon(\omega_{m0} \pm \Omega) = \varepsilon \mp \frac{4\pi n \gamma^2}{\hbar} \frac{1}{\Omega} \tag{6.82 a}$$

while when Ω and Q are both small

$$\frac{c^2(k \pm Q)^2}{(\omega_{m0} \pm \Omega)^2} \cong \frac{c^2 k^2}{\omega_{m0}^2} \left(1 \pm 2\frac{Q}{k} \mp 2\frac{\Omega}{\omega_{m0}}\right) = \varepsilon \pm \frac{2\varepsilon}{\omega_{m0}} (\bar{c}Q - \Omega) , \tag{6.82 b}$$

where in the last step we realize k is chosen so that $ck/\omega_{m0} = \varepsilon^{1/2}$. When (6.82 a) and (6.82 b) are combined,

$$\Omega^2 - \bar{c}Q\Omega - \frac{2\pi n \gamma^2 \omega}{\hbar \varepsilon} = 0 , \tag{6.83}$$

which agrees with (6.78), where we realize $\omega \cong \omega_{m0}$ very near resonance.

Thus, the linearized version of (6.55) reproduces the dispersion relation discussed earlier in Chap. 2, with the results phrased in the language of the slowly varying envelope approximation. For each choice of Q, we have two frequencies which emerge from (6.78). This is a reflection to the two branch dispersion relation illustrated in Fig. 2.4. The present background dielectric constant ε plays the role of ε_∞ in the discussion of Chap. 2, ω_{m0} plays the role of ω_0, and (6.79) describes the two branches in the near vicinity of the intersection of the curve $\omega = ck/\sqrt{\varepsilon_\infty}$ with the line $\omega = \omega_0$ in Fig. 2.4.

Of course, the soliton pulse discussed in Sect. 6.3 has no counterpart in dielectric theory, and enters only within the framework of the complete nonlinear theory.

General aspects of the sine-Gordon equation are discussed in Appendix B.

Problems

6.1 Just after (6.21), it is asserted that retention of the time-independent terms in (6.20) generate the resonant terms in the original perturbation expression, (6.1), while the time-dependent terms in fact generate off resonant contributions. Verify this explicitly by solving (6.19) to first order in E_0, using the full form of (6.20). Then construct the wave function for the model, use it to calculate $\langle s' \rangle$, and compare with the result provided by (6.19). Assume that at time $t = -\infty$, the system is in the ground state, and that the perturbation is turned on adiabatically from time $t = -\infty$.

6.2 A two level system is described by the effective Hamiltonian in (6.11). At time $t = 0$, it resides in the ground state. A dc electric field of strength E_0 is switched on at time $t = 0$, then switched off at a later time T. Treat the atom as a classical dipole, which radiates energy at a rate controlled by the time dependence of the expectation value of the electric dipole moment, $\langle\Phi|p|\Phi\rangle$. Discuss the time dependence of the radiation rate for times $t > 0$.

6.3 An array of pendula, each in the form of a light but rigid rod with a mass M attached to the end, are attached to a "backbone" that will twist as the pendula swing. If θ_n and θ_{n+1} are the deflection angles of two neighboring pendula, the twisting of the backbone introduces a term in the potential energy of the array of the form $(\tau/2)(\theta_n - \theta_{n+1})^2$. Gravity acts downward.

Find the equation of motion of the nth pendulum, and show that in the limit that θ_n varies slowly with n, it has the form of the sine-Gordon equation. Find also the dispersion relation (relation between frequency and wave vector) of small amplitude waves in the system.

6.4 Consider a one dimensional field $\theta(x, t)$ that obeys the sine-Gordon equation, (6.76). A single soliton, described by $\theta_0(x)$, is at rest in the system, centered about the origin.

Examine the nature of small amplitude waves in the system, in the presence of the soliton. This is done by writing $\theta(x, t) = \theta_0(x) + \delta\theta(x, t)$, then linearizing (6.76) with respect to $\delta\theta$. Show there are solutions of the form

$$\delta\theta(x, t) = \exp(ikx)[1 + \alpha \tanh(\beta x)]e^{-i\omega t},$$

with appropriate choice of α and β.

The dispersion relation of these waves is the same as when the soliton is absent, notice. Also, the wave passes over the soliton, without generating a reflected wave!

7. Self-Interaction Effects in One-Dimensional Wave Propagation: Solitons in Optical Fibers and in Periodic Structures

In Chap. 6, we discussed phenomena which, in the language introduced at the end of Chap. 3, fall into the regime of strong nonlinearities. The electronic structure of the physical system exposed to the electromagnetic field is perturbed far from equilibrium in the course of its evolution. In the soliton pulses encountered in the theory of self-induced transparency, for example, the resonant atoms are moved from their ground state, to the first excited state, at the midpoint of the pulse. We now return to the regime of weak nonlinearities, where the nonlinear polarization $\boldsymbol{P}^{(\mathrm{NL})}$ may be expanded in powers of the electric field.

We consider a perfectly isotropic medium, where the second order susceptibility $\chi^{(2)}_{\alpha\beta\gamma}$ vanishes. We expose the medium to an electromagnetic pulse with carrier frequency ω, which propagates in the $\hat{\boldsymbol{z}}$ direction:

$$\boldsymbol{E}(z,t) = \hat{\boldsymbol{n}}[E_\omega(z,t)\mathrm{e}^{\mathrm{i}kz}\mathrm{e}^{-\mathrm{i}\omega t} + E^*_\omega(z,t)\mathrm{e}^{-\mathrm{i}kz}\mathrm{e}^{\mathrm{i}\omega t}]\,. \tag{7.1}$$

Necessarily, the third order susceptibility $\chi^{(3)}_{\alpha\beta\gamma\delta}$ is nonzero. This gives rise to a nonlinear polarization $\boldsymbol{P}^{(\mathrm{NL})}$, which must be parallel to $\hat{\boldsymbol{n}}$ from symmetry considerations. As we have discussed, a third harmonic will be generated in the presence of the nonlinearity. The nonlinear dipole moment will be written

$$\boldsymbol{P}^{(\mathrm{NL})}(z,t) = \hat{\boldsymbol{n}}P^{(\mathrm{NL})}(z,t)\,,$$

$$P^{(\mathrm{NL})}(z,t) = \bar{\chi}^{(3)}[E_\omega(z,t)\mathrm{e}^{\mathrm{i}kz}\mathrm{e}^{-\mathrm{i}\omega t} + E^*_\omega(z,t)\mathrm{e}^{-\mathrm{i}kz}\mathrm{e}^{+\mathrm{i}\omega t}]^3 + \cdots\,, \tag{7.2}$$

where the omitted terms describe the interaction of the third harmonic with the primary wave, and the various other nonlinear mixings which occur from the presence of $\chi^{(3)}_{\alpha\beta\gamma\delta}$. We have introduced $\bar{\chi}^{(3)} = \Sigma_{\alpha\beta\gamma\delta}\chi^{(3)}_{\alpha\beta\gamma\delta}\hat{n}_\alpha\hat{n}_\beta\hat{n}_\gamma\hat{n}_\delta$.

Expanding the right-hand side of (7.2) gives

$$P^{(\mathrm{NL})}(z,t) = \bar{\chi}^{(3)}[E^3_\omega(z,t)\mathrm{e}^{\mathrm{i}3kz}\mathrm{e}^{-\mathrm{i}3\omega t}$$
$$+ 3|E(z,t)|^2E_\omega(z,t)\mathrm{e}^{\mathrm{i}kz}\mathrm{e}^{-\mathrm{i}\omega t} + \cdots\,. \tag{7.3}$$

The first term in (7.3) could be used for a perturbation theoretic description of third harmonic generation, following our discussion of second harmonic generation in Chap. 4. However, if the third order harmonic generation process is not phase matched, the third harmonic will be very weak, because $\bar{\chi}^{(3)}$ is small. We thus ignore this term, assuming phase matching is not achieved, and focus our attention on the second and third term (not shown) in (7.3). These describe intensity dependent influences of the wave on itself. They are thus an example of a self-interaction effect, wherein a pulse modifies the medium within which it propagates.

In this chapter, we retain only the self-interaction terms just described, and analyze their effect on the nature of wave propagation in the medium. We shall see that we again encounter pulses that may be regarded as solitons.

We wish to confine our attention to the simple case of one-dimensional wave propagation, under the influence of the nonlinearity just described. However, we must pause to discuss how such a simple form of propagation can be realized.

If the laser pulse is incident on a sample of infinite extent whose surface coincides with the *xy* plane, and if the wave fronts are infinite planes parallel to the *xy* plane, then the simple one-dimensional picture can be applied to the subsequent evolution of the pulse. However, any real laser beam has a finite diameter, with intensity larger at the beam center than at the edges. Its profile thus has an intensity dependent on distance from the beam center.

The terms of interest then destabilize the transverse profile of the pulse. The second term of (7.3) enters the wave equation, and can be combined with the $(\varepsilon/c^2)\partial^2 E/\partial t^2$ term, with the consequence that the dielectric constant ε (assumed independent of frequency for the moment) can be replaced by $\varepsilon + 12\pi\bar{\chi}^{(3)}|E_\omega|^2$. Thus, if $\bar{\chi}^{(3)} > 0$, the medium acquires an effective index of refraction at the beam center larger than at the beam edges. As illustrated in Fig. 7.1 a, this causes the beam to decrease in diameter. The reason is that the effective propagation velocity at the beam center is slower than at the beam edges. Since the energy per unit time carried by the beam must be independent of position (the medium is assumed transparent), the intensity of the beam must increase as a consequence of the narrowing of its waist. This enhances the focusing process, and the beam collapses further. The phenomenon is referred

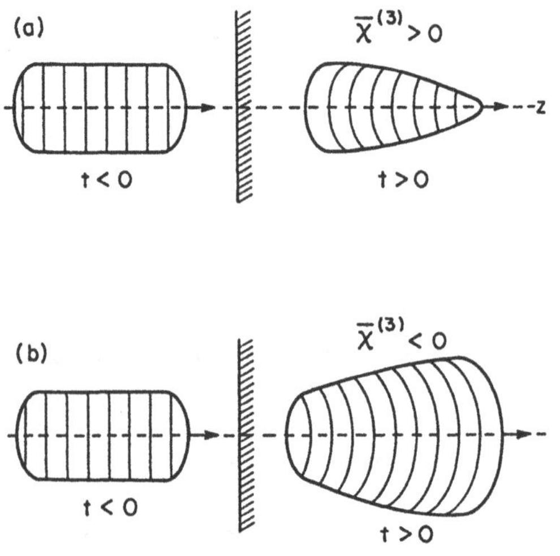

Fig. 7.1. A schematic illustration of the influence of the effective intensity dependence of the index of refraction on the shape of a pulse of finite diameter incident on a nonlinear medium. It is assumed the pulse enters the medium at the time $t \approx 0$. We have **a** collapse in diameter of the pulse if $\bar{\chi}^{(3)} > 0$, and the effective index at the pulse center is larger than at the edge, and **b** expansion of the diameter if $\bar{\chi}^{(3)} < 0$, and the effective index at the pulse center is smaller than at the edges

to as self-focusing. If $\bar{\chi}^{(3)} < 0$, as illustrated in Fig. 7.1 b, the beam diameter increases with distance. In either case, a discussion of the beam propagation characteristics requires consideration of the variations in electric field transverse as well as parallel to the propagation direction. A one-dimensional theory can thus not be used. We refer the reader elsewhere for a discussion of this problem [7.1].

However, in confined geometries, such as optical fibers, we can realize pulse propagation characteristics that are, in effect, quite one dimensional. In a sense described in the next subsection, by confining the wave in the two directions normal to the propagation direction, the transverse profile of the electric field acquires a certain rigidity.

7.1 Normal Modes of Optical Fibers

The simplest picture of an optical fiber is to regard it as a dielectric cylinder of radius R, with uniform dielectric constant ε. The cylinder is of infinite length, and perfectly straight, with axis parallel to the z direction.

Such a structure can serve as a guide of eletromagnetic waves, which propagate parallel to the axis of the cylinder, and are "bound" to the structure in a sense we shall appreciate shortly.

To see this, we examine the Maxwell equations for solutions in which all components of the electric and magnetic field are proportional to $\exp[\mathrm{i}(k_z z - \omega t)]$. It is convenient to work in cylindrical coordinates in the xy plane. A full discussion of the modes bound to the cylinder is complex, so we confine attention to the simplest case where all field components are independent of ϕ, and thus depend on only ρ.

One condition that must be satisfied is $\nabla \cdot \boldsymbol{E} = 0$, and also $\nabla \cdot \boldsymbol{B} = 0$ everywhere. In cylindrical coordinates, and in the absence of ϕ dependence of all field amplitudes, this gives

$$\frac{1}{\rho}\frac{\partial}{\partial \rho}(\rho E_\rho) + \mathrm{i}k_z E_z = 0 . \tag{7.4}$$

We have also the wave equations, which in the dielectric cylinder $0 < \rho < R$ read

$$\frac{1}{\rho}\frac{\partial}{\partial \rho}\left(\rho\frac{\partial E_\rho}{\partial \rho}\right) + \left(\frac{\omega^2}{c^2}\varepsilon - k_z^2 - \frac{1}{\rho^2}\right)E_\rho = 0 , \tag{7.5a}$$

$$\frac{1}{\rho}\frac{\partial}{\partial \rho}\left(\rho\frac{\partial E_\phi}{\partial \rho}\right) + \left(\frac{\omega^2}{c^2}\varepsilon - k_z^2 - \frac{1}{\rho^2}\right)E_\phi = 0, \tag{7.5b}$$

and

$$\frac{1}{\rho}\frac{\partial}{\partial \rho}\left(\rho\frac{\partial E_z}{\partial \rho}\right) + \left(\frac{\omega^2}{c^2}\varepsilon - k_z^2\right)E_z = 0 . \tag{7.5c}$$

In the vacuum, $R < \rho < \infty$, we use the same equations but with $\varepsilon = 1$.

Given the various components of the electric field, one generates the magnetic field from Faraday's Law, $\nabla \times \boldsymbol{E} = i\omega \boldsymbol{B}/c$. Hence, in cylindrical coordinates,

$$B_\rho = -\frac{ck_z}{\omega} E_\phi , \tag{7.6a}$$

$$B_\phi = \frac{ck_z}{\omega} E_\rho + i\frac{c}{\omega}\frac{\partial E_z}{\partial \rho} , \tag{7.6b}$$

and

$$B_z = -\frac{ic}{\omega\rho}\frac{\partial}{\partial \rho}(\rho E_\phi). \tag{7.6c}$$

There are two classes of solutions to the above set of equations. The first has E_ρ, E_z nonzero, but $E_\phi \equiv 0$. Then we see from (7.6) that B_ϕ is the only nonzero component of the magnetic field. The magnetic field thus lies in the plane perpendicular to the direction of propagation. Such waves are referred to as transverse magnetic, or TM waves. We may also have $E_\rho = E_z \equiv 0$, but $E_\phi \not\equiv 0$. Then both B_ρ and B_z are nonzero, linked by the requirement $\nabla \cdot \boldsymbol{B} = 0$. These are the transverse electric, or TE waves. Notice there are no solutions in which both $\boldsymbol{E}$ and $\boldsymbol{B}$ lie in the xy plane. The physical reason for this will be apparent later.

In what follows, we confine our attention to the TM waves. Consider first the vacuum region $\rho > R$, where E_z then obeys

$$\frac{1}{\rho}\frac{\partial}{\partial \rho}\left(\rho \frac{\partial E_z}{\partial \rho}\right) + \left(\frac{\omega^2}{c^2} - k_z^2\right)E_z = 0 . \tag{7.7}$$

There are two distinct regimes which may be examined. The first is $k_z > \omega/c$, where (7.7) reads

$$\frac{1}{\rho}\frac{\partial}{\partial \rho}\left(\rho \frac{\partial E_z}{\partial \rho}\right) - \kappa_0^2 E_z = 0 , \tag{7.8}$$

where $\kappa_0^2 = k_z^2 - \omega^2/c^2 > 0$. This equation admits solutions where the electric field decays to zero exponentially with increasing distance from the cylinder. Most particularly, if $K_0(\xi)$ is the modified Bessel function of order zero we have [7.2]

$$E_z(\rho) = A^> K_0(\kappa_0 \rho) \tag{7.9a}$$

and from (7.4) upon noting the identity $\xi K_0(\xi) = -d(\xi K_1)/d\xi$, we have

$$E_\rho(\rho) = i\frac{k_z}{\kappa_0} A^> K_1(\kappa_0 \rho) . \tag{7.9b}$$

Here $A^>$ is an arbitrary multiplicative constant.

The particular modified Bessel functions in (7.9) have the asymptotic behavior, for any order r,

$$\lim_{\rho\to\infty} K_\nu(\kappa_0\rho) = \left(\frac{\pi}{2\kappa_0\rho}\right)^{1/2} e^{-\kappa_0\rho}\,, \tag{7.10}$$

so, as remarked earlier, in the regime $k_z > \omega/c$, we have solutions in which the electromagnetic fields are "bound" to the dielectric cylinder. In the regime $k_z < \omega/c$, (7.5) also admit solutions in which E_z is a linear combination of the regular Bessel functions J_0 and N_0, while E_ρ is a linear combination of J_1 and N_1 [7.3]. If $k^2 = \omega^2/c^2 - k_z^2 > 0$, these have amplitude that falls off simply as $(k\rho)^{-1/2}$, multiplied by the oscillatory function $\cos(k\rho + \psi)$ where the phase angle ψ is controlled by the relative admixture of J_ν and N_ν in a given solution. These solutions describe cylindrically symmetric radiating waves which emanate from the cylinder, as opposed to waves "bound" to the cylinder. In the cylindrical geometry, radiation fields have amplitudes which fall off with distance from the source as $\rho^{-1/2}$.

Our interest here is thus in the regime $k_z > \omega/c$. We need to consider the behavior of the fields within the cylinder $0 < \rho < R$. We suppose for the moment that we have

$$\frac{\omega}{c} < k_z < \frac{\omega}{c}\,\varepsilon^{1/2}\,. \tag{7.11}$$

Then inside the cylinder, one sees that the solutions regular at the origin are, with $\kappa^2 = \omega^2\varepsilon/c^2 - k_z^2$,

$$E_z(\rho) = A^< J_0(\kappa\rho) \tag{7.12a}$$

and

$$E_\rho(\rho) = -\mathrm{i}\,\frac{k_z}{\kappa}\,A^< J_1(\kappa\rho). \tag{7.12b}$$

To obtain a solution, the boundary conditions at the surface of the cylinder $\rho = R$ must be satisfied. We use conservation of tangential components of $\boldsymbol{E}(E_z)$, and normal components of $\boldsymbol{D}(D_\rho)$, to find

$$A^> K_0(\kappa_0 R) = A^< J_0(\kappa R) \tag{7.13a}$$

and

$$\frac{A^>}{\kappa_0}\,K_\phi(\kappa_0 R) = -\frac{\varepsilon}{\kappa}\,A^< J_1(\kappa R). \tag{7.13b}$$

Since these are two homogeneous equations, we have a nontrivial solution only if

$$\frac{J_1(\kappa R)}{J_0(\kappa R)} = -\frac{\kappa}{\varepsilon\kappa_0}\,\frac{K_1(\kappa_0 R)}{K_0(\kappa_0 R)}\,. \tag{7.14}$$

The constraint in (7.14) may be viewed as an eigenvalue equation for the quantity $\kappa^2 = \omega^2\varepsilon/c^2 - k_z^2$. Notice that we have $\kappa_0^2 = \omega^2(\varepsilon - 1)/c^2 - \kappa^2$, so in fact κ_0 may be viewed as a function of κ^2, which must lie in the range $0 < \kappa < (\omega/c)(\varepsilon - 1)^{1/2}$.

Analysis of the consequences of (7.14) is assisted by letting $\kappa R = \Lambda \sin \theta$, where $\Lambda = (\omega/c)(\varepsilon - 1)^{1/2}R$. Then $\kappa_0 R = \Lambda \cos \theta$, so (7.14) becomes

$$\frac{J_1(\Lambda \sin \theta)}{J_0(\Lambda \sin \theta)} = -\frac{1}{\varepsilon} \tan \theta \frac{K_1(\Lambda \cos \theta)}{K_0(\Lambda \cos \theta)}. \tag{7.15}$$

Here θ lies in the range $0 < \theta < \pi/2$. The ratio K_1/K_0 is positive always, approaching unity for large values of its argument. One sees easily that the right-hand side diverges as $\theta \to \pi/2$, which is where κ_0 approaches zero, or where k_z approaches ω/c from above.

The behavior of the left-hand side, and the number of solutions that emerge from (7.15), depend on the value of Λ. The first zero of $J_0(x)$ occurs at $x = 2.4048$, and that of $J_1(x)$ at $x = 3.8317$. Both $J_0(x)$ and $J_1(x)$ are positive for $0 < x < 2.4048$. Hence, we have no solutions whatsoever if $0 < \Lambda < 2.4048$. This means that, for fixed R, there are no solutions of the wave equations which describe modes bound to the cylinder, if the frequency ω lies below the cutoff frequency ω_c, given by

$$\omega_c = 2.4048 \frac{c}{R(\varepsilon - 1)^{1/2}}. \tag{7.16}$$

The solutions of (7.15) are illustrated graphically, in Fig. 7.2. If $J_0[\Lambda \sin\theta]$, in the interval $0 < \theta < \pi/2$, has n zeros, there will be n solutions of (7.15), and thus n distinct modes are guided by the cylinder, at the frequency ω. For the ith such solution, for which $\Lambda \sin \theta$ assumes the value $\Lambda \sin\theta_i = \gamma_i\omega$, we

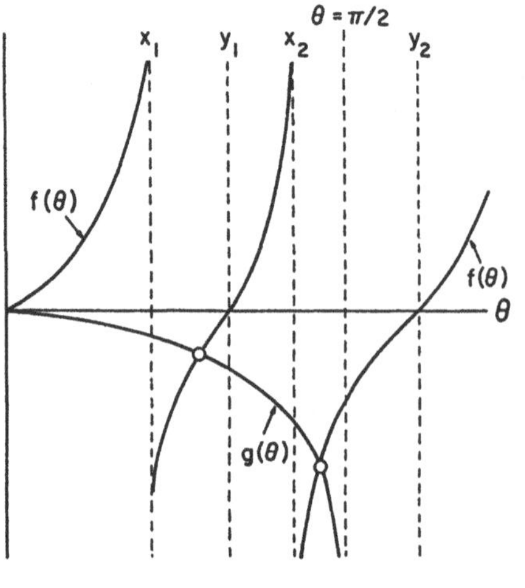

Fig. 7.2. A sketch of the graphical solution of (7.15). One has $g(\theta) = -\tan \theta K_1(\Lambda \cos \theta)/K_0(\Lambda \cos \theta)$, and $f(\theta) = J_1(\Lambda \sin \theta)/J_0(\Lambda \sin \theta)$. It is assumed the value of Λ is such that, as θ is scanned from 0 to $\pi/2$, two zeros of $J_0(\Lambda \sin \theta)$, and one zero of $J_1(\Lambda \sin \theta)$, are located in the interval of interest, as θ varies from 0 to $\pi/2$. The zeros of $J_0(x)$ are denoted by $x_1, x_2, \ldots$, and those of $J_1(x)$ by $y_1, y_2, \ldots$

have $\kappa_i = \gamma_i(\omega)/R$. The dispersion relation of the i-th mode may then be written

$$\omega^2 = \frac{c^2}{\varepsilon}\left[\left(\frac{\gamma_i(\omega)}{R}\right)^2 + k_z^2\right], \tag{7.17}$$

where the constraint $k_z > \omega/c$ must be recalled. As $k_z \to \omega/c$, $\gamma_i(\omega)$ approaches x_i, the ith zero of the Bessel function (for then $\kappa_0 \to 0$, and the solution is such that $\theta_i \to \pi/2$). As $k_z \to \infty$, and we follow the i-th branch, $\omega \to \infty$, and also Λ becomes large. The right-hand side of (7.15) becomes independent of frequency, since K_1/K_0 approaches unity, and γ_i approaches a frequency independent value, that zero of $J_1(x)$ which lies just above x_i. (As $k_z \to \infty$, $\omega \to \infty$, and so does Λ. Use as the variable in (7.17) $y = \Lambda \sin\theta$, and note the right-hand side vanishes for large Λ.) The dispersion relations are sketched in Fig. 7.3.

We have confined our attention to the region where $k_z < \omega\sqrt{\varepsilon}/c$. One may show that in the regime where $k_z > \omega\sqrt{\varepsilon}/c$, there are no solutions which satisfy the boundary condition, are regular at the origin $\rho \simeq 0$, and "bound" to the cylinder, in the sense described above. Also, one may construct a discussion of the TE modes, by following reasoning very similar to that given above. One finds dispersion relations of the various modes of the form given in (7.17); there is a cutoff frequency ω_c below which no propagating modes exist. The cutoff frequency for the TE modes is in fact identical to that for the TM modes, given in (7.16). One may also extend the discussion to the consideration of modes with azimuthal variation, where the various field components vary with the angle ϕ as $\exp(im\phi)$. When the azimuthal "quantum number" $m \neq 0$, the modes are no longer TE and TM in character, but rather have all components of $\boldsymbol{E}$ and $\boldsymbol{B}$ nonvanishing.

We can obtain a physical understanding of the origin of the condition $k_z > \omega/c$ required in our discussion of the guided modes, from the following argument: Consider a dielectric cylinder with radius R large compared to the wavelength of a light beam which is incident on one end of the cylinder, as illustrated in Fig. 7.4. In this limit, we can use ray optics to trace out the trajectory of the beam. We assume the beam lies in the xz plane illustrated,

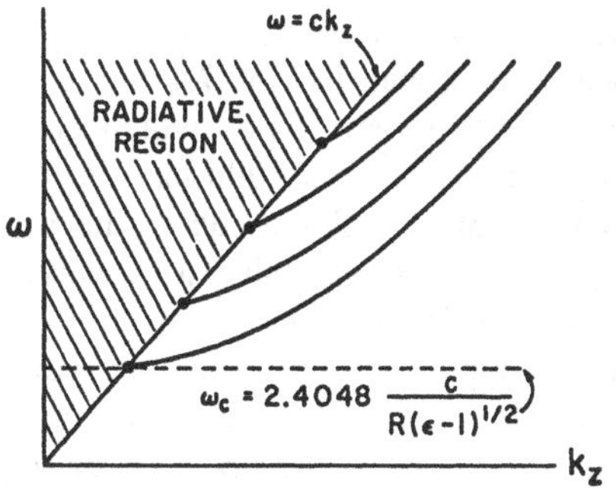

Fig. 7.3. The dispersion relation of the various normal modes of an optical fiber, with radius R and dielectric constant ε. There are no modes with frequency below the cutoff frequency ω_c

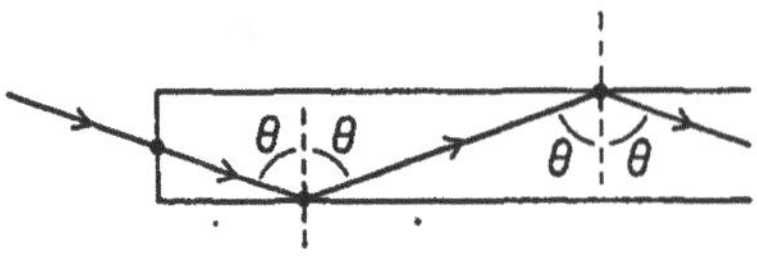

Fig. 7.4. The multiple reflections suffered by an optical beam injected into a dielectric cylinder

with the z axis parallel to that of the cylinder. The beam will strike the lower surface of the cylinder, with θ the angle between the beam, and the local normal to its surface. It reflects off the lower surface, to propagate to the upper surface where it again strikes the surface, making the angle θ with respect to the normal. By a sequence of repeated reflections, the beam travels down the cylinder, as illustrated. The wave vector of the light in the cylinder is $k = \omega\sqrt{\varepsilon}/c$, and the z component of the wave vector is $k_z = (\omega\sqrt{\varepsilon}/c)\sin\theta$.

In general, such a beam will be attenuated after a relatively small number of reflections. The reason is that at each reflection, a portion of the radiation is transmitted through the surface, into the air outside, assumed here to have a dielectric constant of unity. By Snell's law, the angle θ_T between the transmitted beam and the surface normal is given by $\sin\theta_T = \sin\theta/\sqrt{\varepsilon}$. Now if $\theta > \theta_c$, where $\sin\theta_c = 1/\sqrt{\varepsilon}$, there is no transmitted beam. This is the condition that allows total internal reflection of a beam incident on an interface. The sequence of multiple reflections allow the light beam to propagate down the cylinder unattenuated. The light is "trapped" within the cylinder, and is guided by the cylinder. Notice that when $\theta > \theta_c$, we have $k_z > \omega/c$, precisely the condition that emerged from our analysis of Maxwell's equations, in our search for waves bound to the cylinder. We saw also that when $k_z < \omega/c$, the waves outside had radiative character. Textbook discussions [7.4] of the phenomenon of total internal reflection from a planar interface show that just outside the dielectric surface, there is an electromagnetic disturbance that decays to zero exponentially as one moves into the air above. In our full discussion of waves guided by the dielectric cylinder, outside the cylinder the wave fields exhibit a variation with radial distance ρ given by $K_0(\kappa_0\rho)$ or $K_1(\kappa_0\rho)$. Both functions decay to zero exponentially with ρ. We have, for the cylindrical geometry, the same behavior encountered in the theory of total internal reflection from the flat surface. In essence, the analysis we have given of the normal modes of the fiber based on Maxwell's equations is the full treatment of the multiple reflection-induced trapping of an optical wave illustrated in Fig. 7.4.

Notice also the light rays contain electric or magnetic fields which have components parallel to the axis of the cylinder. If the incident light has electric field parallel to the xz plane, $E_z \neq 0$ and we realize a TM mode in the ray tracing picture. Similarly, if $\boldsymbol{B}$ is parallel to the xz plane, we realize a TE wave. Both require the condition $k_z > \omega/c$ to be trapped in the fiber.

From the guided wave solutions described above, we can construct wave packets that describe the propagation of pulses down the optical fiber. We may do this by superimposing solutions associated with one of the branches of the dispersion relation illustrated in Fig. 7.3. Let κ_i be the value of the parameter $\kappa = (\omega^2\varepsilon/c^2 - k_z^2)^{1/2}$ associated with the ith branch. The z component of the electric field associated with the solution of frequency ω can be written, from (7.12 a), $E_0 J_0(k_i\rho)\exp[\mathrm{i}(k_z - \omega t)]$ inside the cylinder, and the ρ component of the electric field is generated readily through use of (7.12 b).

We may then synthesize a wave packet by superimposing such waves, by writing for the electric fields

$$E_z(\rho, z; t) = \int \frac{d\Omega}{2\pi} E(\Omega) J_0[\kappa_i(\Omega)\rho] \mathrm{e}^{\mathrm{i}[(k_z+Q_z)z-(\omega+\Omega)t]} \tag{7.18 a}$$

and

$$E_\rho(\rho, z; t) = -\mathrm{i}\int \frac{d\Omega}{2\pi} \frac{(k_z + Q_z)}{\kappa_i(\omega)} E(\Omega) J_1[\kappa_i(\Omega)\rho] \mathrm{e}^{\mathrm{i}[(k_z+Q_z)-(\omega+\Omega)t]} . \tag{7.18 b}$$

Here, given Ω, Q_z is chosen from the dispersion relation in (7.17). That is, in the dispersion relation ω is replaced by $\omega + \Omega$, k_z by $k_z + Q_z$, and then Q_z is found as a function of Ω. Then $\kappa_i(\Omega) = [\varepsilon(\omega + \Omega)^2/c^2 - (k_z + Q_z)^2]^{1/2}$.

If we form an extended wave packet many wavelengths long, then $E(\Omega)$ is peaked around $\Omega = 0$, and we have $\Omega \ll \omega$, and $Q_z \ll k_z$. In (7.18), $\kappa_i(\Omega)$ may be replaced to excellent approximation by $\kappa_i(0) \equiv \kappa_i$, and the factor $(k_z + Q_z)/\kappa_i(\Omega)$ in (7.18 b) may be replaced by simply k_z/κ_i. The Bessel functions and the prefactor may then be extracted from the integral over Ω, and we have

$$E_z(\rho, z; t) = J_0(\kappa_i\rho) E(z, t) \mathrm{e}^{\mathrm{i}(k_z z-\omega t)} \tag{7.19 a}$$

and

$$E_\rho(\rho, z; t) = -\mathrm{i}\frac{k_z}{\kappa_i} J_1(\kappa_i\rho) E(z, t) \mathrm{e}^{\mathrm{i}(k_z z-\omega t)} , \tag{7.19 b}$$

where

$$E(z, t) = \int \frac{d\Omega}{2\pi} E(\Omega) \mathrm{e}^{\mathrm{i}(Q_z z-\Omega t)} . \tag{7.20}$$

The function $E(z, t)$ is an envelope function that varies slowly in space and time, to produce a wave packet or pulse which propagates down the waveguide. The transverse variation of the field components, however, to excellent approximation is identical to that in the perfectly monochromatic normal mode solution. The boundary conditions force the transverse profile of the wave form to be "rigid," and the wave has only one degree of freedom, so to speak, its deformability in the z direction.

The self-interaction effects discussed qualitatively just after (7.3) can thus lead to modifications of the profile of a pulse in the z direction, but will leave the radial or transverse variation unaffected. Thus, in the fiber, we may realize one-dimensional wave propagation in this sense, and self-focusing effects are suppressed by the "rigidity" of the transverse wave profile. This conclusion is correct as long as the nonlinearity provided by $\bar{\chi}^{(3)}$ is weak in the following sense: When $\bar{\chi}^{(3)}$ is zero, of course the normal modes are the guided waves as discussed above. When $\bar{\chi}^{(3)}$ is "turned on," the full solution will contain admixtures from branches of the guided wave spectrum other than the ith branch discussed above. These admixtures will be small, if the splitting between adjacent branches is large compared to the effective matrix element proportional to $\bar{\chi}^{(3)}$ that is responsible for the mixing. At sufficiently low powers, such interbranch mixing can be expected to be small, and we can achieve the one-dimensional limit.

We now turn to a discussion of the self-interaction effects in one-dimensional wave propagation, with application to optical fibers in mind.

7.2 Nonlinear Schrödinger Equation

We start with the basic wave equation, for a wave propagating parallel to the z direction. This reads, with $E(z, t)$ the electric field,

$$\frac{\partial^2 E}{\partial z^2} - \frac{1}{c^2}\frac{\partial^2 D}{\partial t^2} = \frac{4\pi}{c^2}\frac{\partial^2}{\partial t^2} P^{(\mathrm{NL})}(z, t) \,. \tag{7.21}$$

Following the description in (7.1), and then assuming that the contribution to $P^{(\mathrm{NL})}(z, t)$ of interest is the self-interaction term displayed in (7.3), (7.21) will be written

$$\frac{\partial^2 E}{\partial z^2} - \frac{1}{c^2}\frac{\partial^2 D}{\partial t^2} = -\frac{12\pi\omega^2\bar{\chi}^{(3)}}{c^2}|E_\omega(z, t)|^2 E_\omega(z, t)\mathrm{e}^{ikz}\mathrm{e}^{-i\omega t} \,, \tag{7.22}$$

where in calculating the time derivatives of $P^{(\mathrm{NL})}(z, t)$, we ignore the influence of the slow time variation of the envelope function $E_\omega(z, t)$.

We wish to treat the left-hand side of (7.21) in the slowly varying envelope approximation, but we wish to retain terms second order in the space and time variation of the envelope function. Also, a central role in what follows will be played by the frequency variation of the dielectric function. A simple calculation gives

$$\frac{\partial^2 E}{dz^2} = -\left[k^2 E_\omega - 2ik\frac{\partial E_\omega}{\partial z} - \frac{\partial^2 E_\omega}{\partial z^2}\right]\mathrm{e}^{ikz}\mathrm{e}^{-i\omega t} \,. \tag{7.23}$$

We now need $\partial^2 D/\partial t^2$. We begin with the relation between $D(z, t)$ and the electric field, taking the finite response time of the medium into account. We have, in the present notation,

$$D(z, t) = e^{ikz} \int dt' \varepsilon(t - t') E_\omega(z, t') e^{-i\omega t'} , \tag{7.24}$$

which may be written after a change of variable

$$D(z, t) = e^{ikz} e^{-i\omega t} \int_{-\infty}^{+\infty} d\tau \varepsilon(\tau) E_\omega(z, t - \tau) e^{+i\omega\tau} . \tag{7.25}$$

Since our envelope function varies in time slowly, we may expand $E_\omega(z, t - \tau)$ in powers of τ:

$$E_\omega(z, t - \tau) = E_\omega(z, t) - \tau\left(\frac{\partial E_\omega}{\partial t}\right) + \frac{1}{2}\tau^2\left(\frac{\partial^2 E_\omega}{\partial t^2}\right) + \cdots . \tag{7.26}$$

When this is inserted into (7.25), and we note the definition of the frequency dependent dielectric constant $\varepsilon(\omega)$ given in (2.17), we have the following relation between $D(z, t)$, and the electric field:

$$D(z, t) = \left[\varepsilon(\omega) E_\omega(z, t) + i\left(\frac{\partial \varepsilon}{\partial \omega}\right)\left(\frac{\partial E_\omega}{\partial t}\right) - \frac{1}{2}\left(\frac{\partial^2 \varepsilon}{\partial \omega^2}\right)\left(\frac{\partial^2 E_\omega}{\partial t^2}\right)\right] e^{ikz} e^{-i\omega t} . \tag{7.27}$$

Upon taking the second derivative of (7.27) with respect to time, and ignoring all terms higher order than second in the differentiation of the envelope function $E_\omega(z, t)$ with respect to time, we have for the moment a rather complex expression:

$$\frac{\partial^2 D}{\partial t^2} = -\left[\omega^2 \varepsilon E_\omega + 2i\omega\left(\varepsilon + \frac{1}{2}\omega\frac{\partial \varepsilon}{\partial \omega}\right)\frac{\partial E_\omega}{\partial t} - \left(\varepsilon + 2\omega\frac{\partial \varepsilon}{\partial \omega} + \frac{1}{2}\omega^2\frac{\partial^2 \varepsilon}{\partial \omega^2}\right)\frac{\partial^2 E_\omega}{\partial t^2}\right] e^{ikz} e^{-i\omega t} . \tag{7.28}$$

One proceeds by substitutng (7.23, 28) into the left-hand side of the wave equation (7.21), then choosing the k to be that of an electromagnetic wave of frequency ω in the linear dielectric. Thus, k, a function of frequency, is given by [recall (2.36)].

$$k^2(\omega) = \frac{\omega^2}{c^2}\varepsilon(\omega) . \tag{7.29}$$

The wave equation then becomes

$$2\mathrm{i}\left[k\frac{\partial E_\omega}{\partial z}+\frac{\omega}{c^2}\left(\varepsilon+\frac{1}{2}\omega\frac{\partial\varepsilon}{\partial\omega}\right)\frac{\partial E_\omega}{\partial t}\right]+\frac{\partial^2E_\omega}{\partial z^2}\cdot$$

$$-\frac{1}{c^2}\left[\varepsilon+2\omega\left(\frac{\partial\varepsilon}{\partial\omega}\right)+\frac{1}{2}\omega^2\frac{\partial^2\varepsilon}{\partial\omega^2}\right]\frac{\partial^2E_\omega}{\partial t^2}=-\frac{12\pi\omega^2\bar{\chi}^{(3)}}{c^2}|E_\omega|^2E_\omega. \tag{7.30}$$

We can rewrite the left-hand side of (7.30) in terms of physically meaningful quantities. In Chap. 2, when the properties of electromagnetic waves in a medium with frequency dependent dielectric constant were discussed, we made mention of a characteristic velocity, the group velocity $v_g(\omega) = \partial\omega/\partial k$. [See the discussion which follows (2.38).] Upon differentiating both sides of (7.29) with respect to ω, one finds the identity

$$\frac{\omega}{c^2}\left(\varepsilon+\frac{1}{2}\omega\frac{\partial\varepsilon}{\partial\omega}\right)=\frac{k}{v_g} \tag{7.31}$$

while differentiation a second time gives

$$\frac{1}{v_g^2}+k\frac{\partial}{\partial\omega}\left(\frac{1}{v_g}\right)=\frac{1}{c^2}\left(\varepsilon+2\omega\frac{\partial\varepsilon}{\partial\omega}+\frac{1}{2}\omega^2\frac{\partial^2\varepsilon}{\partial\omega^2}\right). \tag{7.32}$$

With these two results, the left-hand side of (7.30) may be expressed in terms of physically meaningful quantities. We recall the definition of the phase velocity [once again, see the discussion which follows (2.38)] $v_p = \omega/k$. Then one has

$$\left(\frac{\partial}{\partial z}+\frac{1}{v_g}\frac{\partial}{\partial t}\right)E_\omega-\frac{\mathrm{i}}{2k}\left(\frac{\partial^2}{\partial z^2}-\frac{1}{v_g^2}\frac{\partial^2}{\partial t^2}\right)E_\omega+\frac{\mathrm{i}}{2}\frac{\partial}{\partial\omega}\left(\frac{1}{v_g}\right)\frac{\partial^2E_\omega}{\partial t^2}$$

$$=\mathrm{i}\frac{6\pi\omega v_p\bar{\chi}^{(3)}}{c^2}|E_\omega|^2E_\omega\,. \tag{7.33}$$

Note the identity

$$\left(\frac{\partial}{\partial z}+\frac{1}{v_g}\frac{\partial}{\partial t}\right)E_\omega-\frac{\mathrm{i}}{2k}\left(\frac{\partial^2}{\partial z^2}-\frac{1}{v_g^2}\frac{\partial^2}{\partial t^2}\right)E_\omega$$

$$=\left[1-\frac{\mathrm{i}}{2k}\left(\frac{\partial}{\partial z}-\frac{1}{v_g}\frac{\partial}{\partial t}\right)\right]\left(\frac{\partial}{\partial z}+\frac{1}{v_g}\frac{\partial}{\partial t}\right)E_\omega\,. \tag{7.34}$$

Within the spirit of our slowly varying envelope approximation, the square bracket on the right-hand side of (7.34) may be replaced by unity, since the envelope function $E_\omega(z, t)$ varies little over a spatial region the size of a wavelength, and varies little over one cycle of the carrier wave at frequency ω (one expects v_g to be comparable to the speed of light in the medium). Hence, the term $(2k)^{-1}(\partial/\partial z - v_g^{-1}\partial/\partial t)$ inside the square bracket provides only a small correction to unity, and can be dropped.

We thus arrive at our primary equation,

$$\left(\frac{\partial}{\partial z} + \frac{1}{v_g}\frac{\partial}{\partial t}\right)E_\omega(z,t) + \frac{i}{2}\frac{\partial}{\partial \omega}\left(\frac{1}{v_g}\right)\frac{\partial^2 E_\omega}{\partial t^2} = i\,\frac{6\pi\omega v_p \bar{\chi}^{(3)}}{c^2}|E_\omega|^2 E_\omega \tag{7.35}$$

which is the principal equation we shall analyze in this section.

If we set $\bar{\chi}^{(3)}$ to zero, and if furthermore we overlook the term which arises from the frequency variation of the group velocity, then (7.35) reduces to the elementary statement

$$\left(\frac{\partial}{\partial z} + \frac{1}{v_g}\frac{\partial}{\partial t}\right)E_\omega(z,t) = 0\,. \tag{7.36}$$

Any function of the form

$$E_\omega(z,t) = E_\omega(z - v_g t) \tag{7.37}$$

will satisfy this equation. Thus, the envelope function propagates through the medium unchanged in form, with a speed equal to the group velocity v_g. This is the basis of the assertion in Chap. 2 that a pulse propagates with a velocity equal to the group velocity, not the phase velocity.

Our next goal will be to study the influence of the effect of frequency variation of the group velocity, in combination with that of the nonlinearity. Before we do this, we rearrange (7.35). We define

$$\lambda = \frac{6\pi\omega v_p \bar{\chi}^{(3)}}{c^2}\,, \tag{7.38}$$

a parameter that can be either positive or negative depending on the sign of $\bar{\chi}^{(3)}$, and also

$$\frac{\partial}{\partial \omega}\frac{1}{v_g} = -\frac{1}{v_g^2}\frac{\partial v_g}{\partial \omega} = -\sigma\mu\,, \tag{7.39}$$

where μ is a positive number always, and $\sigma = \pm 1$. Thus, if $\partial v_g/\partial \omega$ is positive, we choose $\sigma = +1$, while $\sigma = -1$ if this quantity is negative. Thus, (7.36) is written

$$-\frac{1}{2}\sigma\mu\frac{\partial^2 E_\omega}{\partial t^2} - \lambda|E_\omega|^2 E_\omega = i\left(\frac{\partial}{\partial z} + \frac{1}{v_g}\frac{\partial}{\partial t}\right)E_\omega\,. \tag{7.40}$$

We now change variables, from (z, t) to a new set (ξ, τ), where

$$\tau = t - \frac{1}{v_g}z \tag{7.41 a}$$

and

$$\xi = z \tag{7.41 b}$$

Then (7.40) is transformed into

$$-\frac{1}{2}\sigma\mu\frac{\partial^2 E_\omega}{\partial \tau^2} - \lambda|E_\omega|^2 E_\omega = i\frac{\partial E_\omega}{\partial \xi}\,, \tag{7.42}$$

which is a classical differential equation of nonlinear physics, the nonlinear Schrödinger equation.

If $\lambda = 0$, this has the form of the ordinary Schrödinger equation for a free particle, whose mass is inversely proportional to μ. The variable τ enters as an effective spatial coordinate, and ξ an effective time. When $\lambda \neq 0$, the combination $-\lambda|E_\omega|^2$ plays the role of an effective potential energy. The presence of the wave modifies the medium in a manner that may be represented as a potential energy that affects the wave form itself. If $\sigma = +1$ (group velocity increases with increasing frequency), and $\lambda > 0$, the effective potential energy is attractive. The wave "digs a hole" in the medium, and can be trapped or confined to this self-induced hole. If $\sigma = -1$, and $\lambda < 0$, self-trapping can also occur. The solutions that describe such states will be the solitons of the nonlinear Schrödinger equation.

Before we examine the influence of the nonlinearity, we set $\lambda = 0$, and explore the influence of the frequency variation of the group velocity on wave packet propagation, in the limit where linear theory applies.

7.3 Linear Theory of Pulse Propagation in a Dispersive Medium: Application to Optical Fibers

If $\bar{\chi}^{(3)} = 0$, then the space and time evolution of the envelope function E_ω is controlled by (7.42), with $\lambda = 0$. Thus, we have

$$-\frac{\sigma\mu}{2}\frac{\partial^2 E_\omega}{\partial\tau^2} = \mathrm{i}\frac{\partial E_\omega}{\partial\xi}\,. \tag{7.43}$$

Suppose we are supplied with the behavior of the function $E_\omega(\xi, \tau)$ as a function of τ at $\xi = 0$. We write $E_\omega(0, \tau)$ in the form

$$E_\omega(0, \tau) = \int_{-\infty}^{+\infty} \frac{d\Omega}{2\pi} F(0, \Omega)\mathrm{e}^{-\mathrm{i}\Omega\tau}\,. \tag{7.44}$$

The evolution of the envelope function with the variable ξ can then be written

$$E_\omega(\xi, t) = \int_{-\infty}^{+\infty} \frac{d\Omega}{2\pi} F(\xi, \Omega)\mathrm{e}^{-\mathrm{i}\Omega\tau}\,, \tag{7.45}$$

where (7.43) implies

$$\frac{\partial F(\xi, \Omega)}{\partial\xi} = -\frac{\mathrm{i}}{2}\sigma\mu\Omega^2 F(\xi, \Omega), \tag{7.46}$$

or

$$F(\xi, \Omega) = F(0, \Omega)\exp\left(-\frac{i}{2}\sigma\mu\Omega^2\xi\right). \tag{7.47}$$

Hence, we have

$$E_\omega(\xi, \tau) = \int_{-\infty}^{+\infty} \frac{d\Omega}{2\pi} F(0, \Omega)e^{-i\Omega\tau}\, e^{-i\sigma\mu\Omega^2\xi/2}\,. \tag{7.48}$$

We now recognize $\tau = t - z/v_g$, and $\xi = z$, so in terms of the original variables in space and time,

$$E_\omega(z, t) = \int_{-\infty}^{+\infty} \frac{d\Omega}{2\pi} F(0, \Omega)e^{-i\Omega[t-(z/v_g)]}e^{-i\sigma\mu\Omega^2 z/2}\,. \tag{7.49}$$

We see easily that if we are given the profile of the pulse in time at the point $z = 0$, then $F(0, \Omega)$ is the frequency Fourier tranform of this pulse:

$$F(0, \Omega) = \int_{-\infty}^{+\infty} dt E_\omega(0, t)e^{+i\Omega t}\,. \tag{7.50}$$

The relations in (7.49,50) allow us to folloow the subsequent evolution of any pulse whose time profile is known as $z = 0$. Clearly, if the dielectric constant is independent of frequency, $\mu = 0$, and the pulse propagates through the medium unchanged in shape, with the group velocity v_g (which, in this case equals the phase velocity v_p).

To illustrate the influence of dispersion, expressed here as the origin of the frequency dependence of the group velocity, one may consider the behavior of a pulse of Gaussian profile, for which

$$E_\omega(0, t) = E_0 e^{-t^2/t_0^2}\,, \tag{7.51}$$

where t_0 is a measure of its width. The various integrations may be carried out quite easily, to find

$$E_\omega(z, t) = \frac{t_0 E_0}{t_0^2 + 2i\sigma\mu z}\exp\left(-\frac{(t - z/v_g)^2}{t_0^2 + 2i\sigma\mu z}\right). \tag{7.52}$$

The energy density in the pulse is proportional to $|E_\omega(z, t)|^2$, which has the form

$$|E_\omega(z, t)|^2 = \frac{t_0^2 E_0^2}{t_0(z)^2}\exp\left[-\frac{2}{v_g^2 t_0(z)^2}(z - v_g t)^2\right], \tag{7.53}$$

where we have defined

$$t_0(z) = \left(t_0^2 + \frac{4\mu^2}{t_0^2}z^2\right)^{1/2}. \tag{7.54}$$

The expression in (7.53) describes a Gaussian pulse centered about the position $z = v_g t$. Thus, the energy density in the pulse still propagates with the group velocity, as before. However, as z increases, the pulse broadens, and its peak intensity decreases. The width of the pulse is $t_0(z)$ when the pulse has reached the position z. Notice that the pulse broadens, independently of the sign of the parameter μ. As discussed in Chap. 2, the various Fourier components from which the pulse was synthesized propagate at different phase velocities so as the pulse propagates down the material, necessarily it broadens in profile.

We can estimate the propagation distances that the pulse travels before the broadening becomes appreciable. We do this with optical fibers in mind. Before we turn to numerical estimates, a few comments on optical fibers will prove useful.

The optical fibers of interest are made from SiO_2 (quartz), and have diameters the order of a few microns. To realize very long propagation lengths in these transparent materials, one seeks a minimum in $\varepsilon_2(\omega)$. In Chap. 2 we discussed the general behavior of $\varepsilon_2(\omega)$ in insulators. One has absorption in the far infrared from the interaction of radiation with lattice vibrations. As the frequency of the radiation increases above those characteristics of the lattice vibrations, we saw that $\varepsilon_2(\omega)$ decreases very rapidly with increasing frequency, in an exponential manner. This is the multiphonon regime. Then, as frequency is increased further, we encounter the "Urbach tail," which is precursor of the electronic absorption edge. This behavior was illustrated in Fig. 2.6.

In the quartz fibers, the minimum in the absorption constant occurs near a wavelength of 1.5 μm [7.5], which corresponds to a photon energy of roughly 0.8 eV. This lies in the near infrared. Most experiments are carried out in this wavelength regime, and in fact practical optical fibers utilize wavelengths in the one micron realm, since very long optical paths (kilometers) are desired.

The frequency variation of the group velocity can be inferred from the principles of Chap. 2. Consider the dispersion curves of an electromagnetic wave in a medium with a sharp, well-defined resonance in $\varepsilon(\omega)$, at the frequency ω_0, see (2.38). The dispersion relation is illustrated in Fig. 2.4. For frequencies below ω_0, v_g decreases with increasing frequency, and frequencies above ω_0, v_g increases as frequency increases; the group velocity v_g is the slope of the dispersion curves displayed in Fig. 2.4. In an optical fiber, the behavior of $\varepsilon(\omega)$ is more complex, but the various lattice contributions to $\varepsilon(\omega)$ lie below the frequency ω, for ω in the near vicinity of the gap. The electronic transitions lie above. Somewhat below the minimum in $\varepsilon_2(\omega)$, the frequency of variation of $\varepsilon(\omega)$ is controlled importantly by the lattice contributions, and $\partial v_g/\partial\omega < 0$. As one moves through the minimum in $\varepsilon_2(\omega)$, one senses the higher frequency electronic contributions, and when these control the frequency variation of $\varepsilon(\omega)$, one has $\partial v_g/\partial\omega > 0$. It follows there is a point in the near vicinity of the absorption minimum where $\partial v_g/\partial\omega$ vanishes. In the quartz fibers, this occurs at the wavelength of 1.3 μm [7.5] Near this frequency, pulses can be propagated in the linear regime over appreciable distances, with minimal distortion.

Near, but not far from the zero in $\partial v_g/\partial\omega$, say at 1.5 μm, the value of $\partial v_g/\partial\omega$ is roughly 10^{-5} cm, which gives our parameter μ the value 2×10^{-26} s^3/cm, assuming $v_g \sim 2 \times 10^{10}$ cm/s in this spectral region. If the width of the pulse is t_0 at $z = 0$, the distance the pulse must travel for its width to increase to $2t_0$ is $z = \sqrt{3}t_0^2/4\mu$. If t_0 is 10 ps, a rather narrow pulse, then for the pulse to double in width, the propagation distance must be in the range of 2.5 km.

7.4 Solitons and the Nonlinear Schrödinger Equation

The optical fiber materials employed in current applications have $\bar{\chi}^{(3)} > 0$, so that our parameter $\lambda > 0$. The interesting regime is then that for which $\partial v_g/\partial\omega > 0$ which, from the remarks in the previous section, will be satisfied for wavelengths in the near infrared, longer than 1.3 μm. In what follows, we suppose the parameter $\sigma = +1$.

We consider (7.42), and let $\tau = \sqrt{\mu}y$, and also rescale the amplitude of the field so that $E_\omega = u/\sqrt{\lambda}$. We then have

$$-\frac{1}{2}\frac{\partial^2 u}{dy^2} - |u|^2 u = \mathrm{i}\frac{\partial u}{\partial \xi} . \tag{7.55}$$

We seek a solution of (7.55) with the form

$$u(y, \xi) = \Phi(y)\exp(\mathrm{i}\kappa\xi) , \tag{7.56}$$

where $\Phi(y)$ is real, and satisfies

$$-\frac{1}{2}\frac{d^2\Phi}{dy^2} + \kappa\Phi - \Phi^3 = 0 , \tag{7.57}$$

which is readily integrated subject to the boundary condition that $\Phi(y)$ and its derivative $d\Phi/dy$ vanish as $y \to \pm\infty$. To do this, upon multiplying (7.57) by $d\Phi/dy$, one has the statement

$$\frac{d}{dy}\left[+\frac{1}{4}\left(\frac{d\Phi}{dy}\right)^2 - \frac{1}{2}\kappa\Phi^2 + \frac{1}{4}\Phi^4\right] = 0 \tag{7.58}$$

which, with the boundary condition noted leads to

$$\left(\frac{d\Phi}{dy}\right)^2 = \Phi^2(2\kappa - \Phi^2) , \tag{7.59}$$

or

$$\int \frac{d\Phi}{\Phi(2\kappa - \Phi^2)^{1/2}} = \pm y , \tag{7.60}$$

which gives

$$\Phi(y) = \frac{\sqrt{2\kappa}}{\cosh(\sqrt{2\kappa}y)} . \tag{7.61}$$

Thus, we have a stable, localized pulse, a soliton which emerges as the solution of the nonlinear Schrödinger equation:

$$u(y, \xi) = \frac{\sqrt{2\kappa}}{\cosh(\sqrt{2\kappa}y)} \exp(\mathrm{i}\kappa\xi) . \tag{7.62}$$

If one traces one's way back through the various transformations we have made, to find $E_\omega(z, t)$, the result is

$$E_\omega(z, t) = c\left(\frac{\kappa}{3\pi\omega v_p \bar{\chi}^{(3)}}\right)^{1/2} \frac{\mathrm{e}^{\mathrm{i}\kappa z}}{\cosh\left[\frac{1}{v_g}\left(\frac{2\kappa}{\mu}\right)^{1/2}(z - v_g t)\right]} . \tag{7.63}$$

We thus have a pulse of stable shape which propagates through the system at the group velocity v_g. The quantity κ is a parameter, and in fact we have a solution for any choice of κ, so long as $\kappa > 0$. Thus, we can have stable pulses of any desired width. Which width is realized in practice depends on solving the full differential equation subject to the boundary condition that at $t = 0$, the solution match the input pulse. We comment on the means of doing this later.

One can generate a whole family of soliton solutions to the nonlinear Schrödinger equation as follows: Suppose that the function $u_0(y, \xi)$ is a solution of (7.55). Then after a short calculation, one may show that, for any choice of the parameter λ,

$$u(y, \xi) = \exp\left(\mathrm{i}\lambda y - \frac{\mathrm{i}}{2}\lambda^2\xi\right)u_0(y - \lambda\xi, \xi) \tag{7.64}$$

is also a solution of (7.55). Through use of this result, one can display a rich range of soliton solutions to (7.63). The particular choice

$$\lambda = \frac{1}{\sqrt{\mu}}\left(\frac{1}{v} - \frac{1}{v_g}\right) \tag{7.65}$$

provides a soliton which propagates at any desired velocity v:

$$E_\omega(z, t) = c\left(\frac{\kappa}{3\pi\omega v_p \bar{\chi}^{(3)}}\right)^{1/2} \frac{\exp\left\{\mathrm{i}\left(\frac{1}{v} - \frac{1}{v_g}\right)\frac{t}{\mu} + \mathrm{i}\left[\kappa + \frac{1}{2\mu}\left(\frac{1}{v_g^2} - \frac{1}{v^2}\right)\right]z\right\}}{\cosh\left[\frac{1}{v}\left(\frac{2\kappa}{\mu}\right)^{1/2}(z - vt)\right]} . \tag{7.66}$$

Clearly, the nonlinear Schrödinger equation is very rich in content. Indeed, this discussion has touched on only a small number of special solutions. The general mathematical problem one wishes to address, rather than simply extract special solutions from the equation, is to analyze the evolution of a pulse, given initial conditions on its form. For example, referring to (7.55), if we are provided $u(y, \xi = 0)$, we would like to calculate $u(y, \xi)$ for all values of z. To return to our original variables, noting $\xi = z$, this means given the time profile of the pulse at $z = 0$, we would like to generate the profile of the pulse "downstream," for all values of z, and the time. In the literature on nonlinear differential equations, there is a very powerful technique known as the inverse scattering method that allows one to achieve such an aim. The method has been applied to the nonlinear Schrödinger equation, with application to optical pulse propagation in mind, by *Zakharov* and *Mikailov* [7.6], and by *Satsuma* and *Yajima* [7.7].

The paper by Satsuma and Yajima contains a number of explicit results. Most particularly, these authors extract detailed information about the evolution with ξ (our coordinate z, recall) of pulses whose initial form is given by

$$u(y, 0) = \frac{A}{\cosh y}. \tag{7.67}$$

The example we considered above has $A = 1$, for $\sqrt{2\kappa} = 1$.

If A is an integer, say $A = n$, the pulse has the character of a localized clump of n solitons; the pulse changes shape and "breathes," as the solitons within it move relative to each other. One views these entities as bound states of solitons. It is also possible to construct explicit analytic expressions for $u(y, \xi)$ when A is an integer. For example, when $A = 2$, Satsuma and Yajima find that

$$u(y, \xi) = \frac{4\exp(-i\xi/2)[\cosh(3y) + 3\exp(-4i\xi)\cosh(y)]}{\cosh(4y) + 4\cosh(2y) + 3\cos(4\xi)}, \tag{7.68}$$

a function periodic in ξ with period $\pi/2$. The relative motion of the bound soliton pair is thus periodic in character. In Fig. 7.5, we reproduce from their paper the evolution with ξ of $|u(y, \xi)|$ for the cases $N = 1$, where the soliton shape is rigid, and also for $N = 2$ and $N = 3$.

If A is not an integer, the evolution of the pulse is more complex. If, however, the initial form can be described by (7.67), with $N - 1/2 < A < N + 1/2$, Satsuma and Yajima show that the final form of the pulse is an N soliton state, plus a piece whose amplitude decays in time. We thus have an analogue of the area theorem encountered in the discussion of self-induced transparency. Zakharov and Mikailov show quite generally that a pulse of arbitrary input shape evolves into an N soliton state, after sufficient time has passed.

Experimental studies have generated data in remarkable accord with the picture of high powered pulse propagation, described by the nonlinear Schrödinger equation. We illustrate this in Fig. 7.6, where we reproduce data reported by

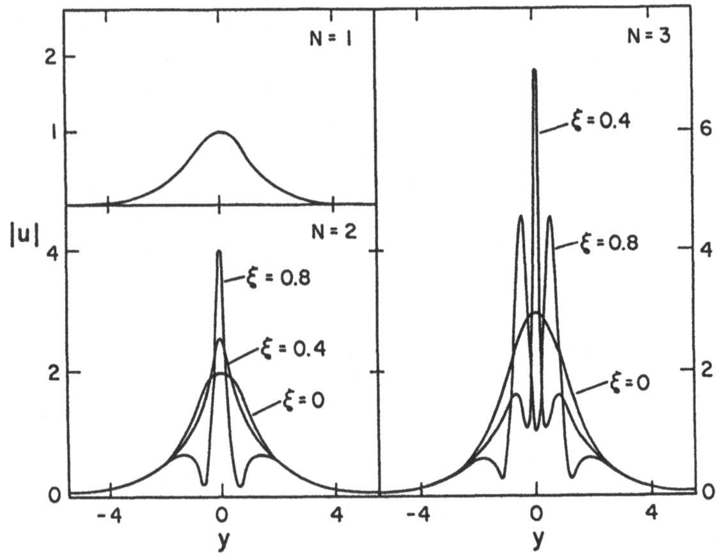

Fig. 7.5. From the work of *Satsuma* and *Yajima* [7.7], we reproduce the evolution with ξ of the solution of the nonlinear Schrödinger equation, for the case where $u(y, 0)$ has the form in (7.67), for $A = 1$, $A = 2$, and $A = 3$

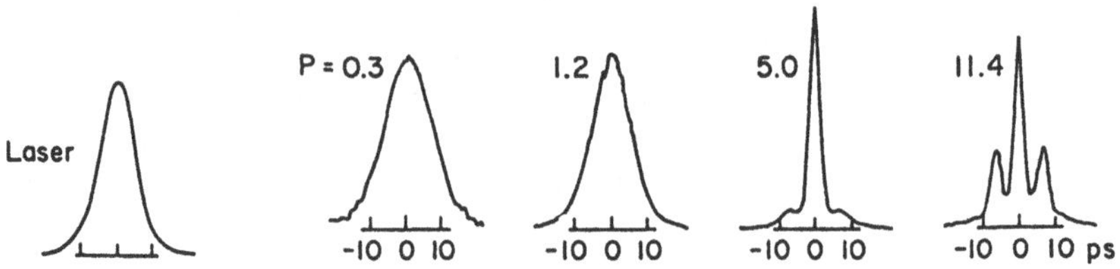

Fig. 7.6. The time profile of the incident laser pulse (labeled "laser"), and output pulses for propagation through a 700 meter long quartz optical fiber. The output pulse shape is shown for various laser powers, from 0.3 watts to 11.4 watts. The data is reproduced from [7.8]

Mollenauer et al. [7.8], which is also discussed a bit more extensivley by *Mollenauer* [7.5].

These authors propagated laser pulses with profile close to the sech y form displayed in (7.67), in a quartz fiber whose length (700 meters) was such that the multisoliton solutions could complete one half period of oscillation before the output end of the fiber was reached. In Fig. 7.6, the topmost figure shows the profile of the initial laser pulse, which has a halfwidth in the range of 10 ps. The wave labeled $W = 0.3$ is the output of the fiber at low power, 0.3 watts, where nonlinearity plays little role. Clearly, the output pulse is broader than the input pulse. The degree of broadening is that expected from the linear theory of wave packet propagation, as described in Sect. 7.2 of this chapter. The curve labeled 1.2 is a pulse with an input power of 1.2 watts. There is no broadening of this pulse at all. This is the $N = 1$ soliton pulse, which propagates with profile stabilized by the nonlinearity. Then when the power is raised to 5 watts, we see output characteristic of the two (bound) soliton pulse. At 11.4 watts, the three soliton state is found.

The multisoliton solutions exhibit rather narrow spikes, in time, at the output end of the fiber, as one can see from both Figs. 7.5, 6. Thus, one can achieve pulse compression by propagating suitable input pulses through optical fibers. In an experiment that uses a two stage arrangement, *Nikolaus* and *Grischowsky* [7.9] have produced an output pulse over 50 times narrow than the input pulse.

7.5 Gap Solitons in Nonlinear Periodic Structures

In this section, we explore another system which supports solitons, in the presence of the optical nonlinearity similar to that just described. This is a superlattice structure, which consists of a periodic array of dielectric films. We may imagine the structure $\cdots ABABAB \cdots$, where the symbol A refers to a thin film of thickness d_A with dieletric constant ε_A, and the symbol B refers to a thin film of thickness d_B with dielectric constant ε_B. In what follows, the frequency variation of ε_A and ε_B plays no essential role, and will be ignored. In our analysis, we will be led to explore a certain class of soliton which has been referred to as a gap soliton [7.10]. There is an extensive theoretical literature on this topic; at the time of this writing, the first experimental study of these objects has appeared recently [7.11].

Before we proceed with the description of the influence of the optical nonlinearities in the properties of periodic dielectric structures, we need to understand the nature of wave propagation in the linear limit. We explore this question first.

As described above, we consider a structure of the form $\cdots ABAB \cdots$, which is periodic with the spatial period $d = d_A + d_B$. Here d_A and d_B are the thickness of film A and film B, respectively. The z axis is oriented normal to the interfaces, and we confine attention to a simple plane wave which propagates parallel to the z direction. The electric field $E(z, t)$, assumed to have frequency ω, then obeys the simple differential equation

$$\frac{d^2E(z)}{dz^2} + \frac{\omega^2}{c^2}\varepsilon(z)E(z) = 0 , \tag{7.69}$$

where $\varepsilon(z) = \varepsilon_A$ when we are in a film of type A, and $\varepsilon(z) = \varepsilon_B$ when we are in a film of type B. Here we assume ε_A and ε_B are both real. The boundary conditions are that $E(z)$, and dE/dz be continuous at each interface. The latter condition insures the conservation of tangential components of the magnetic field B in the electromagnetic wave.

Solutions of (7.69), where $\varepsilon(z)$ is any periodic function with period d, have the character of Bloch funtions, whose properties are derived by a short argument. Any solution $E(z)$ for this periodic structure, within which all unit cells are identical, must have the property that the intensity of the wave, $|E(z)|^2$, is identical in each unit cell. For this to be true, $E(z)$ and $E(z + d)$ can differ by only a phase factor,

$$E(z + d) = e^{i\phi}E(z) . \tag{7.70}$$

Thus, one has $E(z + nd) = \exp(in\phi)E(z)$.

Now imagine we have a superlattice with a very large number of unit cells N. Suppose the structure is deformed into a very large circle, so the structure is in fact a closed continuous ring. Then the cell $N + 1$ is identical to cell 1. If the solution from $(n - 1)d < z < nd$ is written as

$$E(z) = e^{i(n-1)\phi}u[z - (n - 1)d] , \tag{7.71}$$

where $u(z)$ is defined on the interval $0 < z < d$, then we have a form consistent with (7.70), and the statement which follows: Since the cell $(N + 1)$ and 1 are the same, we require $E(z)$ to be identical for the choice $n = 1$, and $n = N$, recognizing that the interval $0 < z < d$ and $Nd < z < (N+1)d$ are identical. We then must have

$$e^{iN\phi} = 1 , \tag{7.72}$$

or $\phi = 2\pi m/N$, where $m = 0, 1, 2, \ldots (N - 1)$ is an integer. [Notice, the choice $m = rN + s$, with r a nonzero integer and s any integer, leads to a solution identical to that with $\phi = 2\pi s/N$, hence, we need confine our attention only to $0 \le m \le (N - 1)$.]

Define $\kappa_m = 2\pi m/L$, where $L = Nd$ is the length of our long superlattice. Then we may write the solution of our problem in the form, when the above constraints are assembled

$$E_{\kappa_m}(z) = \exp(i\kappa_m x_n)\phi_{\kappa_m}(z) ; \quad nd < z < (n + 1)d , \tag{7.73}$$

where $\phi_{\kappa_m}(z)$ is perfectly periodic with period d, and $x_n = nd$.

In the limit $L \to \infty$, κ may be viewed as a continuous variable, which ranges in value from 0 to $2\pi/d$. It is conventional, in fact, to choose κ to lie in the parameter regime $-\pi/d < \kappa < +\pi/d$, which in the theory of wave propagation in periodic structures is referred to as the first Brillouin zone. This range of κ provides exactly the same set of functions, labeled differently, as the choice $0 < \kappa < 2\pi/d$. One can see this from the parenthetical remark which follows (7.72).

The solution, then, as $L \to \infty$ is written

$$E_\kappa(z) = \exp(i\kappa x_n)\phi_\kappa(z) ; \quad nd < z < (n + 1)d , \tag{7.74}$$

where $\phi_\kappa(z) = \phi_\kappa(z + md)$, i.e. $\phi_\kappa(z)$ is periodic in z. This object is a Bloch function.

The basic equation, (7.69), is an eigenvalue equation for the frequency ω. For each Bloch function, described by a particular choice of κ in the first Brillouin zone, one realizes solutions for particular eigenfrequencies. Consider a specific unit cell, with $\varepsilon(z) = \varepsilon_A$ for $0 < z < d_A$, and $\varepsilon(z) = \varepsilon_B$ with $d_A < z < d_A + d_B$. The most general form for $\phi_\kappa(z)$ is

$$\phi_\kappa(z) = \begin{cases} A_+ e^{ik_A z} + A_- e^{-ik_A z}; & 0 < z < d_A , \\ B_+ e^{ik_B z} + B_- e^{-ik_B z}; & d_A < z < d_A + d_B , \end{cases} \tag{7.75}$$

where $k_A = \omega\sqrt{\varepsilon_A}/c$ and $k_B = \omega\sqrt{\varepsilon_B}/c$. Four homogeneous equations relating A_+, A_-, B_+, B_- are generated by requiring $\phi_\kappa(z)$ and $d\phi_\kappa/dz$ be continuous at $z = d_A$, and that both these quantities have the property that $\phi_\kappa(z = 0) = \exp(i\kappa d)\phi_\kappa(z = d)$, and similarly for $d\phi_\kappa/dz$. Recall that $d = d_A + d_B$, the length of the unit cell. Setting the determinant to zero leads to a relation between κ, k_A and k_B, which is an implicit dispersion relation for ω as a function of κ. One finds

$$\cos(\kappa d) = \cos(k_A d_A)\cos(k_B d_B) - \frac{1}{2}\left[\left(\frac{\varepsilon_A}{\varepsilon_B}\right)^{1/2} + \left(\frac{\varepsilon_B}{\varepsilon_A}\right)^{1/2}\right]\sin(k_A d_A)\sin(k_B d_B) . \tag{7.76}$$

To explore the behavior of the solutions to (7.76), consider first very low frequencies where both $k_A d_A \ll 1$, and $k_B d_B \ll 1$. The right-hand side of (7.76) then approaches unity, so the solution also has $\kappa d \ll 1$. We write $\cos(\kappa d) \cong 1 - (\kappa d)^2/2$ on the left-hand side, and expand the right-hand side in powers of $k_A d_A$ and $k_B d_B$ to find, with $f_A = d_A/d$ and $f_B = d_B/d$,

$$\kappa^2 = \frac{\omega^2}{c^2}(f_A\sqrt{\varepsilon_A} + f_B\sqrt{\varepsilon_B})^2 , \tag{7.77}$$

which is the dispersion relation of an electromagnetic wave in a medium whose index of refraction is $f_A\sqrt{\varepsilon_A} + f_B\sqrt{\varepsilon_B}$, which is an average of the index of refraction of two constituents, the contribution of each material weighted by the volume it occupies. In the limit of long wavelengths, the electromagnetic wave just averages over the structure.

At short wavelengths, where the wavelength in the medium approaches that of the unit cell length d, the wave senses details of the structure. To proceed, we write

$$\cos(\kappa d) = \Phi(\omega) , \tag{7.78 a}$$

where $\Phi(\omega)$ is the function on the right hand-side of (7.76). In the limit $k_A d_A \ll 1$, $k_B d_B \ll 1$ just considered, $\Phi(\omega) < 1$, and we have (7.77) as the solution. The wave vector κ is thus real. Whenever $\Phi(\omega)$ lies between -1 and $+1$, we have two solutions of (7.78 a), $+\kappa(\omega)$ and $-\kappa(\omega)$. We thus obtain a dispersion relation $\omega(\kappa)$ upon inverting this relation, with the property that as $\kappa \to 0$, $\omega \to 0$, in accordance with (7.77).

As ω increases, we encounter frequency bands where $\Phi(\omega) > +1$, or $\Phi(\omega) < -1$. This is illustrated in Fig. 7.7, for the choice $\varepsilon_A = 2.25$, $\varepsilon_B = 4.5$, $d_A = d_B = d/2$. When $|\Phi(\omega)| > 1$, there are no solutions for the dispersion relation for real values of κ. There are then forbidden gaps in the dispersion relation, within which there are no solutions to the wave equation. Right at the gap, we have $\kappa d = \pm\pi/d$, or $\kappa d = 0$, since $\Phi(\omega) = \pm 1$ at these points. As the frequency approaches a gap edge, the group velocity, $d\omega/d\kappa$ of the Bloch wave approaches zero. One see this easily upon differentiating (7.78 a), and noting that

$$\frac{d\omega}{d\kappa} = -\left(\frac{d\Phi}{d\omega}\right)\sin(\kappa d)\,, \tag{7.78 b}$$

and the fact that $d\Phi/d\omega$ is nonzero as $\Phi(\omega)$ passes through ± 1. For the structure illustrated in Fig. 7.7, we show the dispersion relation in Fig. 7.8.

When $\Phi(\omega)$ lies outside the region $-1 \le \Phi \le +1$, there are solutions for $\kappa(\omega)$. For $\Phi(\omega) > +1$, we have $\kappa(\omega) = \pm \mathrm{i}\alpha(\omega)$, noting that $\cos(\kappa d) \equiv \cosh(\alpha d)$. For $\Phi(\omega) < -1$, one has $\kappa(\omega) = \mathrm{i}\pi/d \pm \alpha(\omega)$, so then $\cos(\kappa d) = -\cosh(\alpha d)$. From (7.74), one sees that these solutions do not have the Bloch character assumed in (7.70), but instead have an envelope that varies exponentially as one moves into the superlattice. Thus, if a semi-infinite superlattice is illuminated from the outside with radiation whose frequency lies within a forbidden gap, the electromagnetic field inside the structure must decay to zero ex-

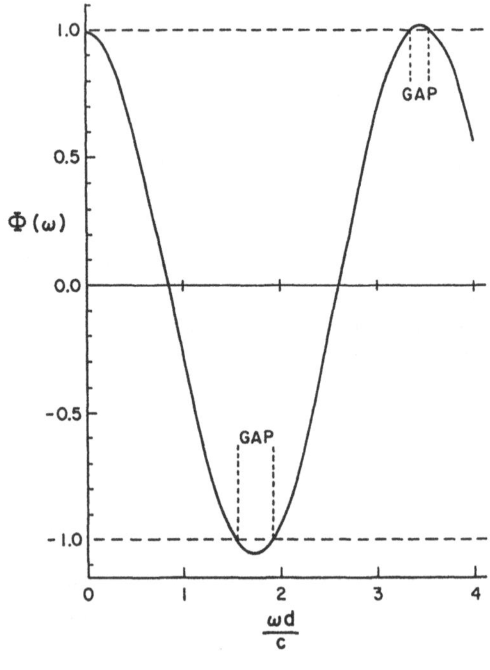

Fig. 7.7. The function $\Phi(\omega)$, on the right-hand side of (7.78 a), as a function of the variable $\omega d/c$. The calculations assume $\varepsilon_a = 2.25$, $\varepsilon_B = 4$, and $d_A = d_B$. The regions where $|\Phi(\omega)| > 1$ which lead to gaps in the dispersion relation are indicated

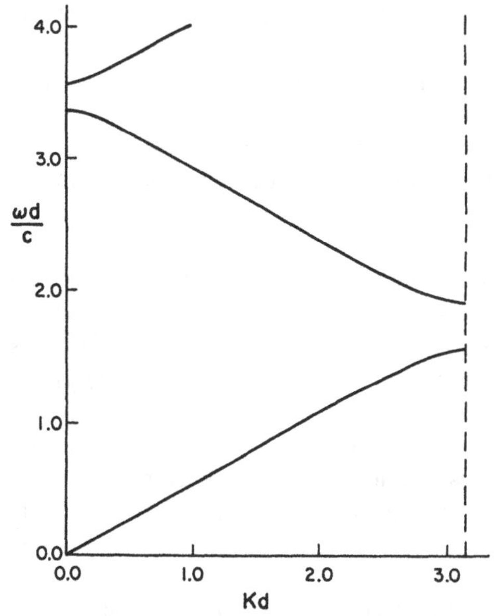

Fig. 7.8. The dispersion curves for propagation of electromagnetic waves, in the superlattice considered in Fig. 7.7

ponentially, since this is the only solution to the wave equation well behaved far from the surface. It follows that the reflectivity of the superlattice must be unity, since energy cannot be propagated through the structure.

In Fig. 7.9, we show the transmission coefficient $|T|^2$ of a finite piece of the model superlattice, with 20 unit cells. It is assumed an incident wave $\exp(\mathrm{i}k_0z)$ strikes one side of the structure, where $k_0 = \omega/c$ is the wave vector of an electromagnetic wave in vacuum. There is then a reflected wave $R\exp(-\mathrm{i}k_0z)$, and a transmitted wave $T\exp(+\mathrm{i}k_0z)$ emerging out the opposite side. The two wide prominent dips in $|T|^2$ mark the location of the forbidden gaps displayed

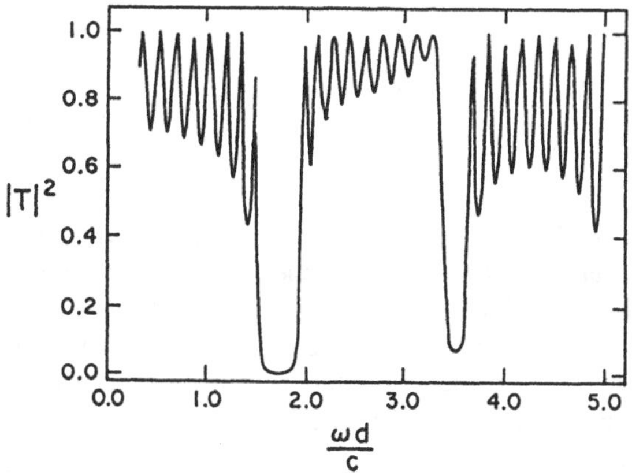

Fig. 7.9. The transmissivity, $|T|^2$, of the model superlattice considered in Fig. 7.7 as a function of frequency. The calculations explore a model superlattice twenty unit cells in length. The prominent dips in $|T|^2$ are produced by the gaps in the dispersion relation displayed in Fig. 7.8. These results have been provided by Prof. Leonard M. Kahn

in Fig. 7.8. One finds $|T|^2$ finite in these gaps, because the length L of the structure is finite; in the gaps one has $|T|^2 \sim \exp[-2\alpha(\omega)L]$, where $\alpha(\omega)$ was defined earlier. One sees fine structure in $|T|^2$ within the allowed frequency bands. These are produced by the various standing wave resonances of the superlattice structure as a whole. These are characterized by values of $\kappa \cong m\pi/L$. Often these structures are referred to as Fabry-Perot resonances of the sample.

We have one point left in our discussion of the linear electromagnetic response of the superlattice. This is contact with the theory of wave propagation in the elementary uniform dielectric. There we have a wave vector κ, and the frequency is connected with the wave vector by the relation $\omega = c\kappa/\sqrt{\varepsilon}$, as we have seen many times. It is easy to see that as $\varepsilon_A \rightarrow \varepsilon_B = \varepsilon$, (7.76) yields exactly such a solution.

However, in the discussion of the periodic lattice, we confined κ to the first Brillouin zone, $-\pi/d < \kappa < +\pi/d$, where in the simple dielectric, $-\infty < \kappa < +\infty$. We can rearrange the plot of the dispersion relation for the simple dielectric where κ covers the range $-\infty \leq \kappa \leq +\infty$ so that equivalent information is contained in a plot confined to the first Brillouin zone. This is done by taking the plot for $-\infty \leq \kappa \leq +\infty$, and dividing it up into segments whose length is $\Delta\kappa = 2\pi/d$, in the manner illustrated in Fig. 7.10 a. Then the piece of the dispersion curve which lies between $\kappa = \pi(2n + 1)d$ and $\kappa = \pi(2n + 3)/d$, is translated through the wave vector $-2\pi(n + 1)/d$. This is done for all integers n, positive or negative. The result is then the "shoelace" diagram in Fig. 7.10 b. Now, given any wave vector κ in the range $\pi(2n + 1)/d$ to $\pi(2n + 3)/d$, we can write

$$\kappa = \frac{2\pi}{d}(n + 1) + \kappa' , \tag{7.79}$$

where $-\pi/d < \kappa' < +\pi/d$. The solution of the wave equation when $\varepsilon_A = \varepsilon_B = \varepsilon$ is, of course, $\exp(i\kappa z)$. We may write this, formally, as a Bloch function

$$\exp(i\kappa z) = \phi(z)\exp(i\kappa' z) , \tag{7.80 a}$$

where

$$\phi(z) = \exp\left(i\frac{2\pi n}{d}z\right) \tag{7.80 b}$$

is periodic with period d.

This construction rearranges all information on wave propagation in the uniform dielectric into a plot that lies entirely within the first Brillouin zone. For $|\kappa| > \pi/d$, the information contained in Fig. 7.10 a has just been rearranged, and reinterpreted. From (7.10 b), it is easy to visualize the influence of imposing a periodic potential on the dispersion relation. The presence of the periodic modulation in the dielectric constant opens up gaps wherever two curves cross or touch in Fig. 7.10 b.

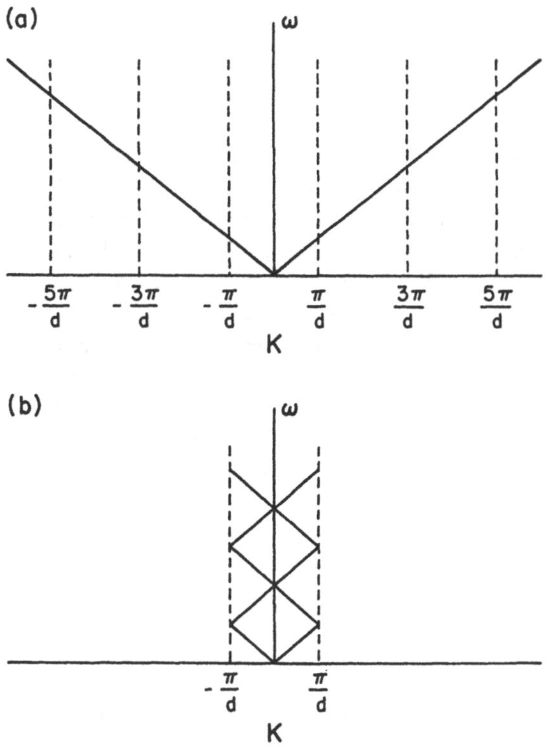

Fig. 7.10. The dispersion relation $\omega = c\kappa/\sqrt{\varepsilon}$ of an electromagnetic wave in a uniform dielectric, **a** plotted for $-\infty < \kappa < -\infty$, segmented as described in Chap. 7, and **b** the information in **a** has been rearranged by translating segments as described in the text into the first Brillouin zone

Constructions similar to that just described play an important role in the theory of electron energy bands in solids. Here, one examines a potential periodic in three spatial dimensions, for an extended crystal, so the description must be generalized [7.12].

We are now ready to turn to the influence of the presence of optical nonlinearities on electromagnetic propagation in periodic dielectric structures.

Our attention will be directed toward the behavior of solutions in which the electromagnetic disturbance is characterized by a well-defined frequency ω, as opposed to the pulses with finite time duration examined earlier. Thus, we examine the equations for solutions of the form

$$E(z,t) = E(z)e^{-i\omega t} + E^*(z)e^{+i\omega t}, \tag{7.81}$$

where the self-interaction effects that played a central role in our discussion of solitons in optical fibers will occupy our attention.

We are thus considering a periodic structure, responding to *cw* laser radiation. Ultimately, following [7.10] we shall consider a finite length sample of superlattice, similar to that used to calculate the transmissivity in Fig. 7.9. For the moment, we examine the basic solutions with the electric field as given by (7.81), for an infinitely extended superlattice.

The basic equation obeyed by the envelope function $E(z)$ is given by

$$\frac{d^2E(z)}{dz^2} + \frac{\omega^2}{c^2}\,\varepsilon(z)E(z) + \frac{12\pi\omega^2}{c^2}\,\bar{\chi}^{(3)}(z)|E(z)|^2E(z) = 0\,. \tag{7.82}$$

Here we examine the full equation, rather than a special limiting form generated by a slowly varying envelope approximation.

In the original work of *Chen* and *Mills* [7.10], through a numerical study of the solutions of (7.82), it was established that when $\bar{\chi}_3(z) \neq 0$, solitons are found within the forbidden gaps in the dispersion relation of linear wave propagation theory. These objects, referred to by these authors as gap solitons, were then shown to have a remarkable influence on the electromagnetic response of the finite superlattice excited by an external electromagnetic wave whose frequency ω lies within the gap. It is possible, in certain limits, to extract information from (7.82) by an analytic method discussed by *Mills* and *Trullinger* [7.13]. We first examine the properties of gap solitons in this limit, then summarize the results of the full study.

We consider the case where the dielectric constant $\varepsilon(z)$ has a spatial modulation of sinusoidal form,

$$\varepsilon(z) = \varepsilon + \Delta\varepsilon\cos\left(\frac{2\pi}{d}z\right), \tag{7.83}$$

where $\Delta\varepsilon \ll \varepsilon$. Also, for simplicity, we suppose $\bar{\chi}_3$ is independent of z.

As remarked above, we are interested in solutions for frequencies ω within a forbidden gap. The width of this gap will be small when $\Delta\varepsilon \ll \varepsilon$. As noted above, the gaps are opened at those frequencies ω where, in the scheme of Fig. 7.10 b, two solutions associated with the limit $\Delta\varepsilon = 0$ are exactly degenerate. We consider such a frequency, and suppose the point of degeneracy occurs at $k = \pi/d$. For such a frequency, we seek solutions with

$$E(z) = E_+(z)e^{ikz} + E_-(z)e^{i(k-2\pi/d)z}, \tag{7.84}$$

where also

$$k = \frac{\pi}{d} - \Delta k \tag{7.85}$$

and we will let $\Delta k \ll \pi/d$. Thus, we are "mixing" two plane waves, which are very nearly degenerate solutions to the wave equation in the limit $\Delta\varepsilon = \bar{\chi}_3 = 0$. Notice that we also have

$$E(z) = e^{-i\Delta kz}[E_+(z)e^{i(\pi/d)z} + E_-(z)e^{-i(\pi/d)z}]\,. \tag{7.86}$$

We shall imagine $E_+(z)$ and $E_-(z)$ vary slowly with z, so that when we insert (7.86) into (7.82), we ignore second derivatives of both E_+ and E_- with respect to z. Also, in $|E|^2$, we retain only terms with the spatial variation $\exp[i(\pm\pi/d - \Delta k)z]$, since only these will affect the solution strongly. One then finds two coupled differential equations for $E_+(z)$ and $E_-(z)$:

$$\left[\frac{\omega^2}{c^2}\varepsilon - \left(\frac{\pi}{d}\right)^2 + \frac{2\pi}{d}\Delta k\right]E_+ + \frac{1}{2}\frac{\omega^2}{c^2}\Delta\varepsilon E_- + \frac{2\pi \mathrm{i}}{d}\frac{dE_+}{dz}$$
$$+ \frac{12\pi\omega^2}{c^2}\bar{\chi}_3(|E_+|^2 + 2|E_-|^2)E_+ = 0 \tag{7.87 a}$$

$$\left[\frac{\omega^2}{c^2}\varepsilon - \left(\frac{\pi}{d}\right)^2 - \frac{2\pi}{d}\Delta k\right]E_- + \frac{1}{2}\frac{\omega^2}{c^2}\Delta\varepsilon E_+ - \frac{2\pi \mathrm{i}}{d}\frac{dE_-}{dz}$$
$$+ \frac{12\pi\omega^2}{c^2}\bar{\chi}_3(|E_-|^2 + 2|E_+|^2)E_- = 0\,. \tag{7.87 b}$$

There is a conservation law that may be derived from (7.87). One may show that the quantity

$$W = |E_+(z)|^2 - |E_-(z)|^2 \tag{7.88}$$

is in fact independent of position.

If $\bar{\chi}_3 \equiv 0$, (7.87) admit solutions with E_+ and E_- independent of z. A dispersion relation which links the frequency ω to the wave vector Δk measured form the Brillouin zone boundary at π/d follows upon setting the appropriate 2×2 determinant to zero. If we define

$$\omega_0 = \frac{c}{\sqrt{\varepsilon}}\frac{\pi}{d}, \tag{7.89}$$

the frequency at wave vector π/d when $\Delta\varepsilon = 0$, then we have two branches of the dispersion curve:

$$\omega_\pm^2(\Delta k) = \omega_0^2 \pm \frac{1}{2}\omega_0^2\left[\left(\frac{\Delta\varepsilon}{\varepsilon}\right)^2 + \left(\frac{2}{\pi}d\Delta k\right)^2\right]^{1/2} \tag{7.90}$$

There is thus a gap at $\Delta k = 0$, the zone boundary, bounded from below by

$$\omega_- = \omega_0\left(1 - \frac{1}{2}\frac{\Delta\varepsilon}{\varepsilon}\right)^{1/2} \tag{7.91 a}$$

and

$$\omega_+ = \omega_0\left(1 + \frac{1}{2}\frac{\Delta\varepsilon}{\varepsilon}\right)^{1/2} \tag{7.91 b}$$

Within the gap, $\omega_- < \omega < \omega_+$, we have solutions for which $E_+(z)$ and $E_-(z)$ exhibit the spatial variation $\exp(\pm\alpha(\omega)z)$. One finds

$$\alpha(\omega) = \frac{2\pi}{\omega_0^2 d}(\omega^2 - \omega_-^2)^{1/2}(\omega_+^2 - \omega^2)^{1/2}\,. \tag{7.92}$$

The results above are consistent with our more general discussion of the nature of linear wave propagation in periodic structures given earlier.

We next turn to solutions of soliton character. For simplicity we take $\Delta k = 0$. For any solution where $E_+(z)$ and $E_-(z)$ both vanish as $z \to \pm\infty$, we must have $W = 0$, in (7.88). Hence, $|E_+| = |E_-|$. Then (7.87) admit solutions with

$$E_+(z) = f(z)e^{+i\phi(z)} \tag{7.93 a}$$

and

$$E_-(z) = f(z)e^{-i\phi(z)}\,, \tag{7.93 b}$$

where both f and ϕ are real. Upon substituting these forms into (7.87), and separating real and imaginary parts, we obtain

$$\frac{2d}{\pi}\frac{d\phi}{dz} = \left(\frac{\omega^2 - \omega_0^2}{\omega_0^2}\right) + \frac{1}{2}\frac{\Delta\varepsilon}{\varepsilon}\cos(2\phi) + \frac{36\pi\bar{\chi}_3}{\varepsilon}f^2 \tag{7.94 a}$$

and

$$\frac{2d}{\pi}\frac{df}{dz} = \frac{1}{2}f\frac{\Delta\varepsilon}{\varepsilon}\sin(2\phi)\,. \tag{7.94 b}$$

One may differentiate (7.94 a) with respect to z one more time, then make use of (7.94 b) to generate an equation for the phase angle ϕ alone:

$$\left(\frac{2d}{\pi}\right)^2\frac{d^2\phi}{dz^2} + \left(\frac{\Delta\varepsilon}{\varepsilon}\right)\left(\frac{\omega^2 - \omega_0^2}{\omega_0^2}\right)\sin(2\phi) + \left(\frac{\Delta\varepsilon}{2\varepsilon}\right)^2\sin(4\phi) = 0\,. \tag{7.95}$$

Given a solution for $\phi(z)$ from (7.95), one may solve for $f^2(z)$ through use of (7.94 a).

If the term in $\sin(4\phi)$ were absent from (7.95), then the differential equation would be equivalent to the time independent version of the sine-Gordon equation encountered in our discussion of self-induced transparency, and also in Appendix B. With the $\sin(4\phi)$ term present, the equation is referred to as the double sine-Gordon equation. Like our other examples, this equation is encountered in various physical contexts, in the literature on ninlinear physics. A number of its properties are explored in a paper by *Campbell* et al. [7.14].

There are two basic soliton solutions to (7.95), which we refer to as type I and type II solitons. Each of these may be expressed in terms of a length $d(\omega)$, which diverges as ω approaches either ω_+, or ω_-. One has

$$d(\omega) = \frac{2d}{\pi}\frac{\omega_0^2}{(\omega^2 - \omega_-^2)^{1/2}(\omega_+^2 - \omega^2)^{1/2}}\,. \tag{7.96}$$

The type I gap soliton exists only when $\bar{\chi}_3 < 0$. We have for this case

$$\phi_I(z) = n\pi - \tan^{-1}\left\{\frac{(\omega^2 - \omega_-^2)^{1/2}}{(\omega_+^2 - \omega^2)^{1/2}} \tanh\left[\frac{z}{d(\omega)}\right]\right\} \tag{7.97 a}$$

$$f_I^2(z) = \frac{\varepsilon}{|8\pi\bar{\chi}_3|}\left(\frac{\omega^2 - \omega_-^2}{\omega_0^2}\right)\frac{\operatorname{sech}^2(z/d(\omega))}{1 + \left(\frac{\omega^2 - \omega_-^2}{\omega_+^2 - \omega^2}\right)\tanh^2\left(\frac{z}{d(\omega)}\right)}. \tag{7.97 b}$$

If $\bar{\chi}_3 > 0$, we realize the type II gap soliton. For this solution, we have

$$\phi_{II}(z) = \left(n + \frac{1}{2}\right)\pi + \tan^{-1}\left\{\left(\frac{\omega_+^2 - \omega^2}{\omega^2 - \omega_-^2}\right)^{1/2} \tanh\left[\frac{z}{d(\omega)}\right]\right\} \tag{7.98 a}$$

and

$$f_{II}^2(z) = \frac{\varepsilon}{|8\pi\bar{\chi}_3|}\left(\frac{\omega_+^2 - \omega^2}{\omega_0^2}\right)\frac{\operatorname{sech}^2(z/d(\omega))}{1 + \left(\frac{\omega_+^2 - \omega^2}{\omega^2 - \omega_-^2}\right)\tanh^2\left(\frac{z}{d(\omega)}\right)}. \tag{7.98 b}$$

Since these formulae are rather complex, we examine the behavior of the type II soliton as a function of ω, as one sweeps through the gap from ω_+ to ω_-.

(i) The case ω near ω_+:

We assume $(\Delta\varepsilon/\varepsilon) \ll 1$. Then after some algebra one has

$$d(\omega) \cong \frac{\sqrt{2}d}{\pi}\left(\frac{\varepsilon}{\Delta\varepsilon}\right)^{1/2}\left(\frac{\omega_0}{\omega_+ - \omega}\right)^{1/2}, \tag{7.99 a}$$

$$\phi_{II}(z) \cong \left(n + \frac{1}{2}\right)\pi + \tan^{-1}\left\{\left(\frac{2\varepsilon}{\Delta\varepsilon}\right)^{1/2}\left(\frac{\omega_+ - \omega}{\omega_0}\right)^{1/2} \tanh\left[\frac{z}{d(\omega)}\right]\right\}, \tag{7.99 b}$$

and

$$f_{II}^2(z) \cong \frac{\varepsilon}{4\pi\bar{\chi}_3}\left(\frac{\omega_+ - \omega}{\omega_0}\right)\operatorname{sech}^2\left[\frac{z}{d(\omega)}\right].$$

As $\omega \to \omega_+$, $d(\omega)$ becomes very large compared to d, so the "size" of the solitons is much larger than the length of the unit cell. In this limit, $\phi_{II}(z)$ varies little with z, while the envelope function $f_{II}(z)$ has the classical $\operatorname{sech}(z/d(\omega))$ form we have encountered earlier, for both the sine-Gordon equation, and also for the nonlinear Schrödinger equation.

(ii) The case ω near ω_-:

Here we have

$$d(\omega) \cong \frac{\sqrt{2}d}{\pi}\left(\frac{\varepsilon}{\Delta\varepsilon}\right)^{1/2}\left(\frac{\omega_0}{\omega - \omega_-}\right)^{1/2} \tag{7.100 a}$$

The behaviors of $\phi_{II}(z)$ and $f_{II}(z)$ are tricky to extract. The principal spatial variation present in both $\phi_{II}(z)$ and $f_{II}(z)$ takes place on the length scale $d(\varepsilon/\Delta\varepsilon)$, which is small compared to $d(\omega)$ in (7.100 a). Thus, the various formulae may be evaluated in the limit $[z/d(\omega)] \ll 1$. For $\phi_{II}(z)$, we find

$$\phi_{II}(z) = \left(n + \frac{1}{2}\right)\pi + \tan^{-1}\left[\frac{\pi}{2}\left(\frac{\Delta\varepsilon}{\varepsilon}\right)\frac{z}{d}\right] \tag{7.100 b}$$

and

$$f_{II}^2(z) = \frac{2\varepsilon^2 d^2}{9\pi^3\Delta\varepsilon\bar{\chi}_3}\frac{1}{z^2 + \left(\dfrac{2d}{\pi}\dfrac{\varepsilon}{\Delta\varepsilon}\right)^2} . \tag{7.100 c}$$

From these expressions we see, as remarked earlier, as $\omega \rightarrow \omega_+$ from below, $d(\omega)$ diverges, and the type *II* gap soliton, which exists only when $\bar{\chi}_3 > 0$ becomes spatially very extended. As ω is decreased from the near vicinity of ω_+, to the bottom of the gap near ω_-, the soliton shrinks until the measure of its size becomes $(2d/\pi)\varepsilon/\Delta\varepsilon$.

The type *I* soliton, which exists for $\bar{\chi}_3 < 0$, has a closely related behavior. It expands in size as ω approaches the bottom of the gap at ω_-, and shrinks as ω moves upward toward ω_+, to approach the limiting size $(2d/\pi)\varepsilon/\Delta\varepsilon$.

The above analytic treatment applies only in the limits stated, but in fact it reproduces nicely the principal features found in studies of gap solitons based on full numerical solutions of the equations which describe a superlattice fabricated from dielectric films, as discussed earlier in this section. In Fig. 7.11 a, we reproduce a gap soliton calculated by *Chen* and *Mills* [7.10], for a frequency near ω_-, in a model superlattice characterized by $\bar{\chi}_3 < 0$. Clearly, it extends over many unit cells of the structure. These authors found that as ω is lowered toward ω_-, the soliton broadens further, as expected from the above remarks.

Now that we understand that periodic structures which exhibit nonlinear response can support gap solitons, the next question is their influence on the optical response of such systems. The effects are most striking, as outlined by *Chen* and *Mills* [7.10, 15]. These authors considered a superlattice of finite length, rather like that whose transmissivity in the linear response regime is displayed in Fig. 7.9, and explored the dependence of the transmissivity on incident power, when the material is illuminated with radiation with fre-

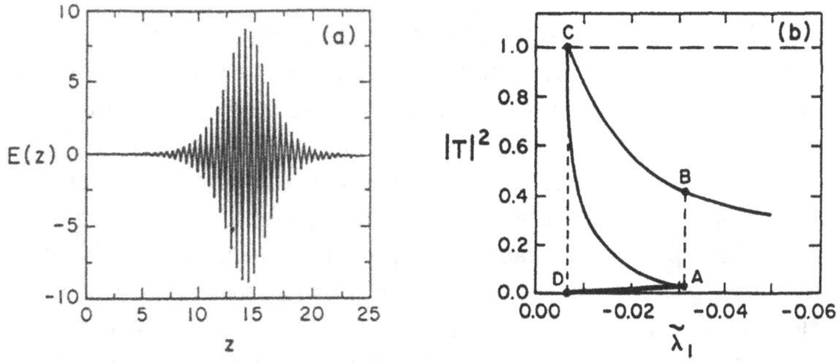

Fig. 7.11. **a** The electric field associated with a gap soliton in a very long superlattice. The frequency ω is just above the frequency ω_- of the lower edge of a forbidden gap, and the parameters are such that $\tilde{\chi}_3 < 0$. The figure is reproduced from [7.10]. **b** The transmissivity $|T|^2$ as a function of power, for a model superlattice with $\tilde{\chi}^{(3)} < 0$. The frequency ω lies just above ω_-. The parameter $\tilde{\lambda}$, in present notation, is $12\pi\tilde{\chi}_3(A)E_0^2/\varepsilon_A$, where E_0 is the magnitude of the incident field, $\tilde{\chi}_3(A)$ is the nonlinear susceptibility of film A in the unit cell of the superlattice, and ε_A its dielectric constant. For the second film in the unit cell, film B, it is assumed $\tilde{\chi}_3(B) = 0$. The figure is reproduced from [7.14]

quency in one of the forbidden gaps of linear response theory. A typical result is illustrated in Fig. 7.11 b.

At low powers, the transmissivity is very small, as expected from linear theory. As power is increased, one encounters a dramatic resonance in the transmissivity; at the peak, the transmissivity is unity. The nonlinearity has led to the possibility that within the forbidden gap, the material can become perfectly transparent! The curve which describes the transmissivity as a function of power is re-entrant, as illustrated in Fig. 7.11 b. A consequence is that the response of the material will be hysteretic. With increasing power, the transmissivity will rise gradually until point A in Fig. 7.11 b is reached. The system will jump discontinuously to a high transmissivity state at B. With decreasing power, the system will pass through the point where the transmissivity is unity until it reaches C. Then with further decrease in power, it will jump to the state of low transmissivity, state D.

The transmission resonance is associated with the fact that the incident radiation excites a gap soliton. A clue that this is the case is provided by the fact that the calculations in Fig. 7.11 b have been performed at a frequency quite close to ω_-, at the bottom of the gap. The transmission resonance occurs in the calculations only when $\tilde{\chi}_3 < 0$, which is the condition that insures a spatially extended gap soliton exists near the bottom of the forbidden gap.

The physical picture associated with the resonance is the following: In the infinitely extended superlattice, the gap soliton is a stable solution of the equations that describe eletromagnetic disturbances in the system, with infinite lifetime. In a sample of finite length, the soliton envelope function is finite at both surfaces. Thus, if a soliton is set up at time zero, it will radiate energy into the vacuum on each side of the finite superlattice, and thus acquire a finite lifetime. It may then be described as a nonlinear resonance of the structure. It then follows that since it couples to radiation fields outside the superlattice, it can be excited by an incident electromagnetic wave.

In Fig. 7.12, we show the field intensity inside a model superlattice, in response to an incident electromagnetic wave of unit amplitude [7.14]. In Fig. 7.12 a we show the field profile at a rather low power. We see the exponential decay of the envelope function, as expected from linear theory. Figure 7.12 b illustrates the field profile at the point where the nonlinearity has driven the transmissivity to unity. Rather than decay, the envelope initially increases as one moves into the structure. We clearly see that the gap soliton has been excited. The resonance character of the response can be appreciated by comparing the absolute magnitude of the intensities in Figs. 7.12 a and 7.12 b. A further increase in power reveals additional transmissivity resonances associated with multisoliton resonances.

As remarked earlier, at the time of this writing, experimental studies of gap solitons are in their early stages [7.11], though the theoretical literature is extensive. Particularly noteworthy is the analysis of *de Sterke* and *Sipe* [7.16], who analyze these objects within the framework of the slowly varying envelope approximation. They carry through an elegant analysis of the power dependent transmissivity by

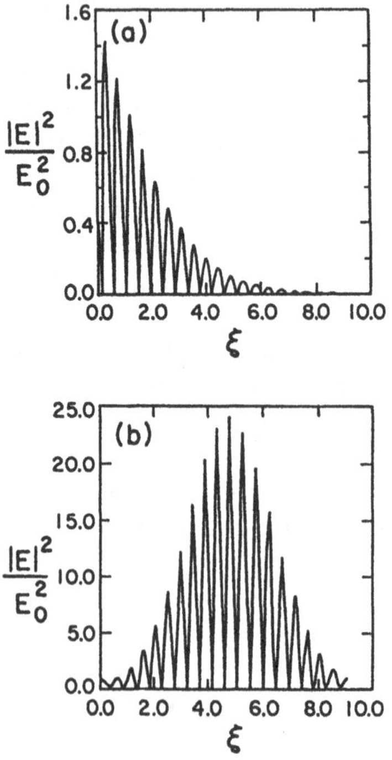

Fig. 7.12. The intensity of $|E|^2$ of the electric field inside a mode superlattice illuminated with an incident field whose strength is E_0. We show the spatial variation of the field when **a** E_0 is small, and power dependent effects on the optical response are very modest, and **b** the power is such that $|T|^2 = 1$ for a frequency in a forbidden gap, by virtue of the gap soliton mediated resonance. Further increases in power produce two soliton state resonances, as discussed in [7.14].

this means, to recover rather simply many of the results obtained by *Chen* and *Mills* in their numerical work.

Problems

7.1 Consider a dielectric slab of thickness D, whose two surfaces are parallel to the xy plane, and infinite in extent. Show that there are guided wave solutions to Maxwell's equation of TE and TM character, and find the implicit dispersion relation for each class of wave. (It is sufficient to consider waves which propagate parallel to the xy plane.)

7.2 Verify explicitly that the function on the right-hand side of (7.64) indeed satisfies the nonlinear Schrödinger equation, under conditions described in the text.

7.3 In this problem we examine the properties of the solution to (7.82) for a perfectly homogeneous medium where the dielectric constant $\varepsilon(z)$ is independent of z. We write the equation in the form

$$\frac{d^2E}{dz^2} + \frac{\omega^2}{c^2}(\varepsilon + \lambda|E(z)|^2)E(z) = 0 .$$

(a) Seek solutions of the form $E(z) = f(z)\mathrm{e}^{\mathrm{i}\psi(z)}$ where both f and ψ are real, and prove that

$$f^2\frac{d\psi}{dz} = W ,$$

where W is a constant independent of z. Write the Poynting vector in terms of ψ and f, and show this statement is equivalent to the requirement that the energy per unit are a per unit time which flows in the wave is independent of z.

(b) Through use of the result in part (a), derive a relation of the form

$$\left(\frac{df}{dz}\right)^2 + V(f) = A ,$$

where A is a constant of integration. Think of f as the displacement of a fictitious particle, and of z as a fictitious time. Show that the only solutions for $f(z)$ are those in which $f(z)$ is an oscillatory periodic function of z. That is, when ε is independent of z, (7.82) cannot admit soliton solutions.

7.4 Consider a semi-infinite nonlinear dielectric, described by the equation

$$\frac{d^2E}{dz^2} + \frac{\omega^2}{c^2}(\varepsilon + \lambda|E(z)|^2)E(z) = 0 ,$$

in response to a purely monochromatic plane wave. (This is (7.82) rearranged as in Problem 7.3.)

Find an equation from which one may determine the reflectivity of a semi-infinite piece of such material, as a function of laser power. (Note that within the dielectric, one has solutions with $|E(z)|$ independent of z.) For the choice $\varepsilon = 4$, plot the reflectivity as a function of the variable λE_0^2, where E_0 is the amplitude of the incident wave. Consider $\lambda > 0$, and $\lambda < 0$.

8. Nonlinear Optical Interactions at Surfaces and Interfaces

Our discussions so far have assumed that the nonlinear couplings responsible for the phenomena explored take place in bulk matter. Of course, the presence of bounding surfaces played a central role in the discussion of solitons in optical fibers presented in Chap. 7. But the role of the surface there is to guide the electromagnetic disturbance, and to confine its fields to a finite region of space. The nonlinearity responsible for soliton formation, as described in (7.2,3), is that characteristic of the bulk material from which the optical waveguide is fabricated.

On the surface of a material, or at the interface between two media, one encounters coupling constants for nonlinear optical interactions which differ distinctly and qualitatively from those in the bulk material. The low symmetry of the atomic or molecular sites at a surface are responsible for these effects. Consider, for example, the phenomenon of second-harmonic generation. The nonlinear electric dipole moment responsible for the signal is described in (4.4). As we have seen from our earlier discussions, in any material with an inversion center, the third-rank tensor $\chi^{(2)}_{\alpha\beta\gamma}$ must vanish. For such materials, there is then no second-harmonic generated by a laser beam. This is the case for a very wide class of materials. One thus sees no second-harmonic signal from liquids, gases, and solids such as silicon or germanium, whose crystal structure exhibits inversion symmetry.

Consider now an atom or molecule which resides on the surface of any material, regarded here as perfectly flat and smooth. If the surface is parallel to the xy plane, very clearly the xy plane is not a plane of reflection symmetry, and also an inversion center is absent. If this atom or molecule is exposed to a laser field, it will make a nonzero contribution to $\chi^{(2)}_{\alpha\beta\gamma}$. In essence, we may regard $\chi^{(2)}_{\alpha\beta\gamma}$ as a function of position. The tensor vanishes in the bulk, but is nonzero in the very near vicinity of the surface. One will realize a second-harmonic signal, with origin in the nonlinear electric dipole moments of those atoms and molecules right at the surface, and embedded in an interface between dissimilar materials as well.

It follows that such second-harmonic signals are a powerful probe of the very near vicinity of a surface and interface, and the atoms or molecules adsorbed there. For this reason, the study of second-harmonic generation from surfaces and interfaces, along with other nonlinear interactions unique to this environment, has been pursued intensely in recent years. We explore the central issues in this chapter. We shall begin with the example just discussed, the generation of the second-harmonic radiation from surfaces and interfaces.

8.1 Second-Harmonic Generation from Surfaces; General Discussion

In order to proceed with a discussion, it is useful to explore a model of the near surface region, with a clear physical picture in mind. We illustrate the example we examine in Fig. 8.1. We consider a semi-infinite substrate, whose surface coincides with the xy plane. The substrate is characterized by a complex dielectric constant $\varepsilon(\omega)$, so light which reflects from the material penetrates only a finite distance $\delta(\omega)$ into the substrate. When the imaginary part of the dielectric constant $\varepsilon_2(\omega)$ is nonzero, the penetration depth of the radiation is finite, by virtue of the energy dissipation discussed in Chap. 2. The substrate is supposed a material with an inversion center, so in the bulk of the material, $\chi_{\alpha\beta\gamma}^{(2,\infty)}$ vanishes. We append the symbol ∞ to the superscript to emphasize we are describing a bulk property.

We add to the surface a layer of adsorbed molecules, each cylindrically symmetric, and suppose their symmetry axis is aligned with the normal to the surface. The molecules have length d, which for typical adsorbates ranges from 2 Å to 10 Å. The length d is very small compared to the wavelength of light which, as we know, for visible radiation is the order of 5,000 Å.

Each molecule resides at a site for which inversion symmetry is absent, by virtue of the presence of the substrate. Thus, when exposed to radiation of frequency ω, symmetry allows the presence of a molecular electric dipole moment with the second-harmonic frequency 2ω. Clearly, if the adsorbate itself has inversion sym-

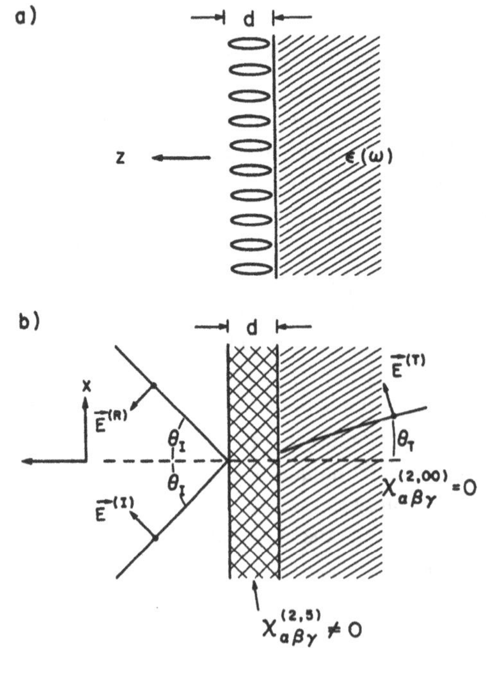

Fig. 8.1. The model considered for our discussion of second-harmonic generation from a surface layer. **a** We have a substrate with complex dielectric constant $\varepsilon(\omega)$, on which a layer of adsorbed molecules is present. Each molecule is cylindrically symmetric, and stands with symmetry axis normal to the surface. This system may be modeled as illustrated in **b**, by supposing the substrate is overlaid with a "second-harmonically active" layer of thickness d. Radiation of frequency ω strikes the surface at the angle of incidence θ_I.

metry, both the magnitude and frequency dependence of this second-harmonic contribution to the electric dipole moment is controlled by the nature of the bonding of the molecule to the substrate; if we remove the molecule from the substrate, symmetry considerations require the second-harmonic dipole moment to vanish.

We may model this system as illustrated in Fig. 8.1b. The wavelength of light is very long compared to both the size of the adsorbed molecules, and their separation. Thus, we may view the structure as a semi-infinite dielectric continuum, upon which a film of thickness d is superimposed, where d is the length of an adsorbed molecule. Within the overlayer only, the nonlinear susceptibility $\chi^{(2s)}_{\alpha\beta\gamma}$ is nonvanishing. The s in the superscript emphasizes this contribution is localized near the surface.

The discussion just given has supposed the second-harmonic signal has its origin in a layer of molecules adsorbed on the surface of a substrate. Suppose we consider instead simply a semi-infinite crystal with a flat, smooth surface, and no adsorbates present. The atoms in the outermost atomic layer of the crystal may be discussed with language very similar to that just employed for the molecular overlayer. They reside in sites which lack inversion symmetry, and when these atoms are exposed to an electromagnetic field of frequency ω, they will also have a second-harmonic electric dipole moment at frequency 2ω. The model illustrated in Fig. 8.1b thus applies equally well to a terminated crystal or liquid. The overlayer of thickness d describes the second-harmonic moment associated with near surface atoms or molecules.

We may describe the picture introduced above by introducing the nonlinear electric dipole moment per unit volume $P^{(NLs)}_{\alpha}(\boldsymbol{r}, t)$ which we write as

$$P^{(NLs)}_{\alpha}(\boldsymbol{r}, t) = \sum_{\beta\gamma} \chi^{(2s)}_{\alpha\beta\gamma}(z) E_{\beta}(\boldsymbol{r}, \omega) E_{\gamma}(\boldsymbol{r}, \omega) \mathrm{e}^{-\mathrm{i}2\omega t} + c.c., \tag{8.1}$$

where the second order susceptibility is nonvanishing just in the overlayer of thickness d, and $E_{\beta}(\boldsymbol{r}, \omega)$ is the appropriate cartesian component of the electric field of frequency ω felt by the molecules.

The nonzero elements of $\chi^{(2s)}_{\alpha\beta\gamma}$ are controlled by the symmetry of the overlayer environment. Thus, to proceed with an analysis of any given case, we require a statement about its structure. For the example illustrated in Fig. 8.1a, each molecule is assumed to be left unchanged by rotation about its symmetry axis, which is normal to the surface. We assume further the adsorbates are arranged on a two-dimensional lattice which renders the xz and yz planes a reflection plane. Let the lattice have fourfold symmetry, so the x and y directions are equivalent.

In this simple case, the nonzero elements of $\chi^{(2s)}_{\alpha\beta\gamma}$ are easily seen to be

$$\chi^{(2s)}_{zzz}(z) \equiv \chi^{(2s)}_{\perp\perp\perp}(z), \tag{8.2a}$$

$$\chi^{(2s)}_{zxx}(z) = \chi^{(2s)}_{\perp\parallel\parallel}(z), \tag{8.2b}$$

$$\chi^{(2s)}_{xxz}(z) = \chi^{(2s)}_{\parallel\parallel\perp}(z), \tag{8.2c}$$

and their equivalents by symmetry. Thus, if we suppose the incident radiation is p polarized with electric field parallel to the plane of incidence as illustrated in

Fig. 8.1b, we have

$$P_z^{(NLs)}(\mathbf{r}, t) = \chi_{\perp\perp\perp}^{(2s)}(z)[E_z(\mathbf{r}, \omega)]^2 e^{-i2\omega t} + 2\chi_{\perp\|\|}^{(2s)}(z)[E_x(\mathbf{r}, \omega)]^2 e^{-i2\omega t} + c.c. \quad (8.3a)$$

and

$$P_x^{(NLs)}(\mathbf{r}, t) = 2\chi_{\|\|\perp}^{(2s)}(z) E_x(\mathbf{r}, \omega) E_z(\mathbf{r}, \omega) e^{-i2\omega t} + c.c., \quad (8.3b)$$

while for s polarized incident radiation, where in the coordinate system of Fig. 8.1b) the laser field is perpendicular to the xz plane and thus parallel to y, the only nonzero component of the nonlinear polarization is

$$P_z^{(NLs)}(\mathbf{r}, t) = \chi_{\perp\|\|}^{(2S)}(z)(E_y(\mathbf{r}, \omega))^2 e^{-i2\omega t} + c.c.. \quad (8.4)$$

We see that for the high symmetry adsorbed overlayer examined here, use of either p polarized or s polarized incident radiation must lead to second-harmonic radiation of p polarized character, since $\mathbf{P}^{(NLs)}(\mathbf{r}, t)$ always is parallel to the plane of incidence. Violation of this selection rule will lead one to conclude that the surface environment has lower symmetry than supposed in this discussion. When the selection rule is violated, studies of the dependence of the second-harmonic on incident angle and polarization enable one to deduce which additional elements of $\chi_{\alpha\beta\gamma}^{(2s)}$ are nonzero, and by this means one may use second-harmonic generation as a probe of the symmetry of the near surface environment. We shall provide illustrations of such studies below.

A question which will emerge as an issue of major importance is the value of the electric field $E_\alpha(\mathbf{r}, \omega)$ to be employed in (8.1,3,4). In our earlier studies of nonlinear interactions, the incident waves were simply plane waves, whose field amplitudes are simply and directly related to those of the relevant incident beams. In the present circumstance, one must recognize that the presence of the substrate surface modifies the field amplitudes in its near vicinity, in a substantial fashion. We may appreciate this from Fig. 8.1b, where we see that just outside the substrate surface, the total electric field is in fact the vector sum of that of the incident and the reflected beam. The magnitude of the perpendicular and parallel components may thus differ very appreciably from those of the incident beam. It is this total field that should be employed to calculate the nonlinear dipole moment in a thin overlayer.

For a perfectly flat dielectric surface, it is a straightforward matter to find the amplitude of the reflected electric field just outside [8.1], and one may then form expressions for the total field amplitudes parallel and perpendicular to the surface. For p polarized radiation, one may write

$$E_\perp^{(TOT)} = \Lambda_\perp(p) E^{(I)} \quad (8.5a)$$

and

$$E_\|^{(TOT)} = \Lambda_\|(p) E^{(I)}, \quad (8.5b)$$

where

$$\Lambda_\perp(p) = \frac{\varepsilon(\omega)\sin(2\theta_I)}{\varepsilon(\omega)\cos(\theta_I) + [\varepsilon(\omega) - \sin^2\theta_I]^{\frac{1}{2}}} \quad (8.6a)$$

and

$$\Lambda_{\|}(p) = \frac{2\cos(\theta_I)[\varepsilon(\omega) - \sin^2\theta_I]^{\frac{1}{2}}}{\varepsilon(\omega)\cos(\theta_I) + [\varepsilon(\omega) - \sin^2\theta_I]^{\frac{1}{2}}}. \tag{8.6b}$$

For the s polarized case, we have

$$E_{\|}^{(TOT)} = \Lambda_{\|}(s)E^{(I)}, \tag{8.7}$$

where

$$\Lambda_{\|}(s) = \frac{2\cos(\theta_I)}{\cos\theta_I + [\varepsilon(\omega) - \sin^2\theta_I]^{\frac{1}{2}}}. \tag{8.8}$$

In the expressions above, $E^{(I)}$ is the amplitude of the electric field in the incident beam, far from the surface.

In Fig. 8.2, for a substrate with a dielectric constant characteristic of a semiconducting substrate such as Si or Ge, we illustrate the variation of the surface field intensities with angle of incidence. Quite clearly, field components parallel to the surface are suppressed substantially. For p polarized incident radiation, the curves suggest that the normal component of the incident electric field will be dominant at appreciable angles of incidence. This is indeed the case in practice, in many instances.

Later in the present chapter, we shall discuss methods of realizing very large enhancements of the near surface electric fields, over those associated with the incident beam itself. By exploiting these very large surface enhancements, one can achieve very substantial second-harmonic intensities from even a monolayer of material. Suppose, for example, we find a means of achieving a surface field enhancement of a factor of 10. We know from discussions in Chap. 4 that the intensity of the second-harmonic signal is proportional to the square of the intensity of the field which illuminates the material responsible for the signal. A factor of ten in field amplitude enhancement thus translates into factor of 10^4 in the output signal. Similar remarks apply to other nonlinear optical processes at surfaces and

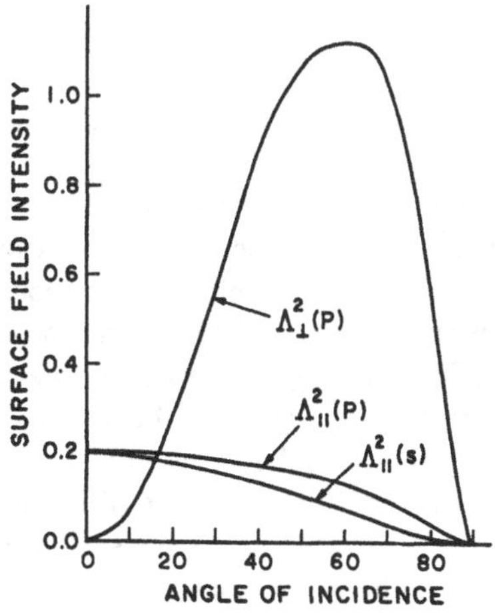

Fig. 8.2. The surface field intensities, calculated for a substrate with dielectric constant $\varepsilon = 12$. For p polarized radiation, $\Lambda_{\perp}(P)$ and $\Lambda_{\|}(P)$ are the amplitudes of the perpendicular and parallel components of the electric fields for p polarized radiation, and $\Lambda_{\|}(s)$ is the amplitude of the surface field for s polarized incident radiation. The fields are normalized to the amplitude of the incident field.

interfaces, such as sum or difference frequency generation in response to two incident beams, or Raman scattering. We shall discuss means of achieving large surface field enhancements later in the present chapter.

So far we have supposed that in our model structure, the second-harmonic nonlinear polarization arises from only the molecules (or atoms) in the outer atomic layer of the material, since these "sense" the absence of an inversion center. Interior atoms have, for the model explored here, a local environment in which inversion symmetry is present, and thus fail to contribute to $\chi^{(2s)}_{\alpha\beta\gamma}$ as a consequence.

Even in this circumstance, however, there will be second-harmonic radiation generated by the interior atoms, through the wave vector dependence of the nonlinear response tensor. For a material which possesses an inversion center, terms such as those described in (3.23) are nonzero. As one sees from (3.25), these generate a bulk second-harmonic polarization which we may write, for present purposes,

$$P_{\alpha}^{(NL\infty)}(\boldsymbol{r},t) = \sum_{\beta\gamma\delta} \Lambda_{\alpha\beta\gamma\delta} E_{\beta}(\boldsymbol{r},\omega) \frac{\partial E_{\gamma}}{\partial r_{\delta}}(\boldsymbol{r},\omega) \mathrm{e}^{-\mathrm{i}2\omega t}, \tag{8.9}$$

where $\Lambda_{\alpha\beta\gamma\delta} = (\partial\chi^{(2)}_{\alpha\beta\gamma}/\partial k_{\delta})_0$ is the derivative of $\chi^{(2)}_{\alpha\beta\gamma}$ with respect to the wave vector component k_{δ} of the incident radiation.

The fourth-rank tensor $\Lambda_{\alpha\beta\gamma\delta}$ is nonzero in any medium, including those with an inversion center. One refers to contributions to the nonlinear electric dipole moment in (8.9) as electric quadrupole contributions, since in quantum mechanical perturbation theory, they are generated by going beyond the electric dipole approximation for the various matrix elements which enter the microscopic theory.

In a cubic material, there are four nonzero elements of the tensor. These are Λ_{xxxx}, Λ_{xxyy}, Λ_{xyxy}, Λ_{xyyx}, and their equivalents formed by permutation of the subscripts. It is a short exercise to show that for such a case, one may group terms to read

$$\begin{aligned} P_{\alpha}^{(NL\infty)} &= \Lambda_a E_{\alpha}(\boldsymbol{r},\omega)\frac{\partial E_{\alpha}}{\partial r_{\alpha}}\mathrm{e}^{-\mathrm{i}2\omega t} + \Lambda_b \frac{\partial}{\partial r_{\alpha}}[E^2(\boldsymbol{r},\omega)]\mathrm{e}^{-\mathrm{i}2\omega t} \\ &\quad + \Lambda_c[\boldsymbol{E}(\boldsymbol{r},\omega)\cdot\boldsymbol{\nabla}]E_{\alpha}(\boldsymbol{r},\omega)\mathrm{e}^{-\mathrm{i}2\omega t} \\ &\quad + \Lambda_d[\boldsymbol{\nabla}\cdot\boldsymbol{E}(\boldsymbol{r},\omega)]E_{\alpha}(\boldsymbol{r},\omega)\mathrm{e}^{-\mathrm{i}2\omega t}, \end{aligned} \tag{8.10}$$

where $\Lambda_a = \Lambda_{xxxx} - \Lambda_{xxyy} - \Lambda_{xyxy} - \Lambda_{xyyx}$, $\Lambda_b = \Lambda_{xyyx}/2$, $\Lambda_c = \Lambda_{xyxy}$ and $\Lambda_d = \Lambda_{xxyy}$.

One's first reaction is that since the wavelength of light is so long compared to the microscopic lengths which characterize the substrate, the contributions to the nonlinear dipole moment in (8.10) are quite small. From simple considerations of dimensional analysis, we may expect a term such as $\Lambda_{\alpha\beta\gamma\delta}E_{\beta}(\partial E_{\gamma}/\partial r_{\delta})$ to be smaller than those which enter (8.1) by a factor of roughly ka_0, where k is the wave vector of the laser radiation which illuminates the sample, and a_0 a bond length or lattice constant characteristic of the substrate. However, the terms in (8.1) are nonzero only over a microscopic distance the order of a_0, while those in (8.10) act throughout the full penetration depth of the substrate. In the end, as we shall see, the influence of the electric quadrupole terms cannot be overlooked.

While the expression in (8.10) is rather complex in structure, in fact when we apply it to the model system under consideration, it simplifies very considerably. In a cubic or isotropic medium, the electric field in the incident beam is purely transverse, as we have seen in Chap. 2. Thus, the term $[\boldsymbol{E}(\boldsymbol{r},\omega)\cdot\boldsymbol{\nabla}]E_\alpha(\boldsymbol{r},\omega)$ vanishes everywhere, since the electric field amplitude is independent of position everywhere in a plane parallel to $\boldsymbol{E}(\boldsymbol{r},\omega)$. Furthermore, in the interior of a charge free dielectric, $\nabla\cdot\boldsymbol{E}(\boldsymbol{r},\omega)$ must also vanish everywhere.

We are then left with only the terms proportional to Λ_a and Λ_b. If the substrate is not cubic, but in fact is isotropic in character, then the expression for $P_\alpha^{(NL\infty)}$ must be invariant in form under arbitrary rotations of the coordinate system. The term proportional to Λ_a is incompatible with this requirement, so symmetry considerations require $\Lambda_a \equiv 0$.

In the interest of simplicity, we assume the substrate to be isotropic, and take for $P_\alpha^{(NL\infty)}(\boldsymbol{r},\omega)$ the simple form

$$P_\alpha^{(NL\infty)}(\boldsymbol{r},t) = \Lambda_b\frac{\partial}{\partial r_\alpha}[E^2(\boldsymbol{r},\omega)]\mathrm{e}^{-\mathrm{i}2\omega t}\ . \tag{8.11}$$

We are now left with the task of analyzing the second-harmonic signal generated by the two sources of nonlinear polarization outlined above. We may regard the piece described by (8.3) or (8.4) as residing within a thin layer of thickness d placed just outside the substrate with dielectric constant $\varepsilon(\omega)$, and that in (8.11) is nonzero throughout the semi-infinite substrate. We shall carry through the calculation of the second-harmonic signal; to do this we will need to analyze the signal generated by a nonlinear polarization in a spatially non-uniform medium. We use this analysis as a means of introducing a method the present author has found most useful for a variety of applications.

We shall consider the configuration illustrated in Fig. 8.1b, where p polarized radiation is incident on the sample, which lies in the half space $z<0$. Then far from the surface the incident beam has an electric field we write as

$$\boldsymbol{E}^{(I)}(\boldsymbol{r},t) = E^{(I)}(\hat{x}\cos\theta_I+\hat{z}\sin\theta_I)\mathrm{e}^{\mathrm{i}k_\parallel^{(I)}x}\mathrm{e}^{-\mathrm{i}k_\perp^{(I)}z}\mathrm{e}^{-\mathrm{i}\omega t}\ , \tag{8.12}$$

where $k_\parallel^{(I)}=(\omega/c)\sin\theta_I$, and $k_\perp^{(I)}=(\omega/c)\cos\theta_I$. The fields within the layer of adsorbates may be calculated be setting $z=0$ in (8.12), then using (8.5a,b) to account for the fact that the surface molecules are exposed to a coherent superposition of the incident and reflected light. The nonlinear polarization at the surface then has an x and z component that may be written, when all factors are combined,

$$P_x^{(NLs)}(\boldsymbol{r},t) = \tilde{\chi}_{\parallel TOT}^{(2s)}(z)(E^{(I)})^2\mathrm{e}^{\mathrm{i}2k_\parallel^{(I)}x}\mathrm{e}^{-\mathrm{i}2\omega t} \tag{8.13a}$$

and

$$P_z^{(NLs)}(\boldsymbol{r},t) = \tilde{\chi}_\perp^{(2s)}(z)(E^{(I)})^2\mathrm{e}^{\mathrm{i}2k_\parallel^{(I)}x}\mathrm{e}^{-\mathrm{i}2\omega t}\ , \tag{8.13b}$$

where we have defined

$$\tilde{\chi}_{\parallel TOT}^{(2s)}(z) = 2\chi_{\parallel\ \parallel\perp}^{(2s)}(z)\Lambda_\parallel(p)\Lambda_\perp(p) \tag{8.14a}$$

and

$$\tilde{\chi}^{(2s)}_{\perp TOT}(z) = \chi^{(2s)}_{\perp\perp\perp}(z)\Lambda^2_\perp(p) + 2\chi^{(2s)}_{\perp\parallel\parallel}(z)\Lambda^2_\parallel(p) . \tag{8.14b}$$

The quadrupole contributions to the nonlinear polarization require knowledge of the transmitted fields, which are calculated easily from the Fresnel coefficients [8.1]. We write the transmitted field in the form

$$E^{(T)}(r,t) = E^{(I)}T(p)\left[\frac{ck^{(T)}_\perp}{\omega}\hat{x} + \sin\theta_I\hat{z}\right] e^{ik^{(I)}_\parallel x} e^{-ik^{(T)}_\perp z} e^{-i\omega t} , \tag{8.15}$$

where we define

$$T(p) = \frac{2\cos\theta_I}{\varepsilon(\omega)\cos\theta_I + [\varepsilon(\omega) - sin^2\theta_I]^{\frac{1}{2}}} \tag{8.16a}$$

and also

$$k^{(T)}_\perp = \frac{\omega}{c}[\varepsilon(\omega) - \sin^2\theta_I]^{\frac{1}{2}} . \tag{8.16b}$$

Later it will be useful to note we can write $T(p)$ in the equivalent form

$$T(p) = \frac{2k^{(I)}_\perp}{\varepsilon(\omega)k^{(I)}_\perp + k^{(T)}_\perp} . \tag{8.16c}$$

The quantity $k^{(T)}_\perp$ is the perpendicular component of the wave vector of the transmitted beam, in the substrate, and $k^{(T1)}_\perp$ is the real part of its wave vector. In general, as we know, the dielectric constant $\varepsilon(\omega)$ of the substrate is complex. Thus, this is true of $k^{(T)}_\perp$ as well. At later points in our discussion, it will be convenient to write

$$k^{(T)}_\perp = k^{(T1)}_\perp + \frac{i}{2\delta^{(T)}} , \tag{8.17}$$

where $\delta^{(T)}$ is the penetration depth of the transmitted wave in the substrate. The factor of two has its origin in the convention that the penetration depth is defined as the distance required for the intensity to decrease to (1/e) of its initial value. The intensity is proportional to the square of the electric field.

The electric quadrupole contribution to the nonlinear electric dipole moment then has the form, for the model developed here,

$$\begin{aligned} P^{(NL\infty)}(r,t) &= 2i\theta(-z)[T(p)E^{(I)}]^2\varepsilon(\omega)\Lambda_b \\ &\times \left[k^{(I)}_\parallel\hat{x} - k^{(T)}_\perp\hat{z}\right] e^{i2k^{(I)}_\parallel x} e^{-i2k^{(T)}_\perp z} e^{-i2\omega t} . \end{aligned} \tag{8.18}$$

In (8.18), $\theta(-z)$ is unity when its argument is positive ($z < 0$), and zero otherwise. This simply reminds us that $P^{(NL\infty)}(r,t)$ is nonzero only within the substrate, which occupies the half space $z < 0$.

To generate expressions for the second-harmonic electric fields which are produced by the electric dipole moments described in (8.13,18), we need to solve the analogue of (4.2b), recognizing now that the dielectric constant which appears on the left-hand side depends on z. The dielectric constant is unity in the vacuum $z > 0$, and for the second-harmonic field assumes the value $\varepsilon(2\omega)$ for $z \leq 0$.

We write the spatially dependent dielectric constant as $\varepsilon(2\omega, z)$, and the second-harmonic field is

$$E_x(\boldsymbol{r}, t) = \mathcal{E}_x^{(2)}(z) e^{i2k_\parallel^{(I)}x} e^{-i2\omega t} , \tag{8.19}$$

and similarly for $E_z(\boldsymbol{r}, t)$. We then need to solve

$$\left[\frac{d^2}{dz^2} + \frac{(2\omega)^2}{c^2}\varepsilon(2\omega, z)\right]\mathcal{E}_x^{(2)}(z) - 2ik_\perp^{(I)}\frac{d\mathcal{E}_z^{(2)}}{dz}(z)$$
$$= -\frac{16\pi\omega^2}{c^2}\left[\tilde{\chi}_{\parallel TOT}^{(2s)}(z) + 2i\theta(-z)\Lambda_b\varepsilon(\omega)k_\parallel^{(I)}T^2(p)e^{-2ik^{(T)}{}_\perp z}\right](E^{(I)})^2 \tag{8.20a}$$

and

$$\left[\frac{(2\omega)^2}{c^2}\varepsilon(2\omega, z) - (2k_\parallel^{(I)})^2\right]\mathcal{E}_z^{(2)}(z) - 2ik_\parallel^{(I)}\frac{d\mathcal{E}_x^{(2)}(z)}{dz}$$
$$= -\frac{16\pi\omega^2}{c^2}\left[\tilde{\chi}_{\perp TOT}^{(2s)}(z) - 2i\theta(-z)\Lambda_b\varepsilon(\omega)k_\perp^{(T)}T^2(p)e^{-2ik_\perp^{(T)}z}\right](E^{(I)})^2 . \tag{8.20b}$$

It will be kept in mind that the surface nonlinear susceptibilities in (8.20a,b) are nonzero only in a thin layer of thickness d just outside the substrate.

The solution of these equations is rendered cumbersome by the presence of the substrate surface. From the mathematical perspective, this feature enters the left-hand side through the spatially varying dielectric constant $\varepsilon(2\omega, z)$.

In the mathematics of inhomogeneous differential equations, the method of Green's functions proves a powerful means of obtaining general solutions [8.2]. An extension of this technique to the case where we are describing multicomponent vector fields allows one to solve the problem under consideration. This method, in fact, is readily applied to the study of a diverse array of optical interactions in films, or in films and near interfaces. As remarked earlier, the present author and his colleagues have employed the method on numerous occasions [8.3]. We present a general discussion of the form of the matrix of Green's functions needed for this class of problems in Appendix C. We sketch their use here.

We begin by rewriting (8.20a,b) in a matrix notation. Introduce the two component vector object

$$\boldsymbol{\mathcal{E}}^{(2)}(z) = \begin{pmatrix} \mathcal{E}_x^{(2)}(z) \\ \mathcal{E}_z^{(2)}(z) \end{pmatrix} . \tag{8.21}$$

and a 2×2 array of differential operators

$$\boldsymbol{\mathcal{L}} = \begin{pmatrix} \mathcal{L}_{xx} & \mathcal{L}_{xz} \\ \mathcal{L}_{zx} & \mathcal{L}_{zz} \end{pmatrix} \tag{8.22}$$

defined by

$$\mathcal{L}_{xx} = \frac{d^2}{dz^2} + \frac{(2\omega)^2}{c^2}\varepsilon(2\omega, z) , \tag{8.23a}$$

$$\mathcal{L}_{xz} = \mathcal{L}_{zx} = -2ik_\parallel^{(I)}\frac{d}{dz} , \tag{8.23b}$$

and

$$\mathcal{L}_{zz} = -(2k_{\parallel}^{(I)})^2 + \frac{(2\omega)^2}{c^2}\varepsilon(2\omega, z) \,. \tag{8.23c}$$

We may then write (8.20a,b) in the matrix form

$$\mathcal{L}\mathcal{E}^{(2)}(z) = -\frac{16\pi\omega^2}{c^2}\boldsymbol{P}^{(NL)}(z, 2\omega) \,. \tag{8.24}$$

through suitable definition of $\boldsymbol{P}^{(NL)}(z, 2\omega)$.

The solution of the system of equations in (8.24) is found by introducing a matrix of Green's functions, constructed explicitly in Appendix C.

These have the form

$$\mathcal{G}(2k_{\parallel}^{(I)}, 2\omega|z, z') = \begin{pmatrix} \mathcal{G}_{xx} & \mathcal{G}_{xz} \\ \mathcal{G}_{zx} & \mathcal{G}_{zz} \end{pmatrix} , \tag{8.25}$$

and they satisfy, with I the 2×2 identity matrix,

$$\mathcal{L}\mathcal{G} = +4\pi I\delta(z - z') \,. \tag{8.26}$$

The differential equations (8.26) are supplemental by the appropriate boundary conditions, as discussed in Appendix C.

The solutions to (8.20a,c) then have the form

$$\mathcal{E}_x^{(2)}(z) = -\frac{(2\omega)^2}{c^2}\left[\int_{-\infty}^{+\infty} \mathcal{G}_{xx}(2k_{\parallel}^{(I)}, 2\omega|zz')P_x^{(NL)}(z', 2\omega)dz' + \int_{-\infty}^{+\infty} \mathcal{G}_{xz}\,(2k_{\parallel}^{(I)}, 2\omega|zz')P_z^{(NL)}(z', 2\omega)dz'\right] . \tag{8.27}$$

A similar expression describes $\mathcal{E}_z^{(2)}(z)$.

Equation (8.27) describes both the second-harmonic fields in the vacuum above the substrate ($z > 0$), and also those in the substrate ($z < 0$). To calculate either, we insert the relevant forms for the Green's functions, and carry out the required integration, which is elementary. Before we do this, however, we comment on the physical content of the Green's function method.

Consider the contribution to the nonlinear electric dipole moment displayed in (8.13a). This describes a very thin sheet of radiating dipoles, placed just above the substrate surface. Within this sheet, the dipoles are modulated in strength by the phase factor $\exp(i2k_{\parallel}^{(I)}x)$, and the dipoles oscillate with frequency 2ω, of course. Upon writing out (8.26), one sees that $\mathcal{G}_{xx}(2k_{\parallel}^{(I)}, 2\omega|z, z')$ and $\mathcal{G}_{zx}(2k_{\parallel}^{(I)}, 2\omega|zz')$ describe the x and z components of the electric field radiated by a thin sheet of dipoles of suitable strength (with dipole moments parallel to the x direction), placed not just above the surface, but rather on the plane $z = z'$. Similarly, $\mathcal{G}_{xz}$ and $\mathcal{G}_{zz}$ describe the electric fields generated by such a sheet of dipoles, but with dipole moments parallel to z.

The actual nonlinear electric dipole moment distribution has an extended spatial extent, given by the various expressions displayed above. The integrals in (8.27) calculate the total electric field generated by the distributed dipole array by summing over the entire dipole array. This is an expression of the superposition principle, which applies to any linear theory such as electrodynamics. The electric field generated by a given sheet of dipoles is affected by its proximity to the surface. Such effects are contained in the Green's functions.

In (8.27), we have both the surface and the volume contributions to the nonlinear dipole moment. The surface terms are calculated by letting $z' = 0+$ in the Green's functions, and replacing both $\tilde{\chi}^{(2s)}_{\parallel TOT}(z)$ and $\tilde{\chi}^{(2s)}_{\perp TOT}(z)$ by their values within a thin layer of thickness d above the surface. When this is done, and we make explicit use of the expression for the quadrupole contribution to the nonlinear moment, we have

$$\begin{aligned}\mathcal{E}^{(2)}_x(z) &= -\frac{(2\omega)^2}{c^2}\left(E^{(I)}\right)^2 \exp\left(\mathrm{i}2k^{(I)}_\parallel x - \mathrm{i}2\omega t\right)\\ &\times \Big\{d\left[\tilde{\chi}^{(2s)}_{\parallel TOT}\mathcal{G}_{xx}(2k^{(I)}_\parallel, 2\omega|z, 0+) + \tilde{\chi}^{(2s)}_{\perp TOT}\mathcal{G}_{xz}(2k^{(I)}_\parallel, 2\omega|z, 0+)\right]\\ &+2\mathrm{i}\Lambda_b T^2(p)\varepsilon(\omega)\Big[k^{(I)}_\parallel \int_{-\infty}^{0} \mathcal{G}_{xx}(2k^{(I)}_\parallel, 2\omega|zz')\mathrm{e}^{-\mathrm{i}2k^{(T)}_\perp z'}dz'\\ &-k^{(T)}_\perp \int_{-\infty}^{0} \mathcal{G}_{xz}(2k^{(I)}_\parallel, 2\omega|zz')\mathrm{e}^{-\mathrm{i}2k^{(T)}_\perp z'}dz'\Big]\Big\} \ . \end{aligned} \tag{8.28}$$

The expression in (8.28) describes the second-harmonic field in the vacuum above the crystal, and also that within the substrate. We shall explore here only the field above the substrate in the vacuum. Formally, we may do this by supposing z is always larger than z', which never exceeds $z' = 0+$. Then we have, from Appendix C, for $z > z'$

$$\mathcal{G}_{xx}(2k^{(I)}_\parallel, 2\omega|zz') = \frac{4\pi}{W_p(2k^{(I)}_\parallel, 2\omega)}\mathcal{E}^{>}_x(2k^{(I)}_\parallel, 2\omega|z)\mathcal{E}^{<}_x(2k^{(I)}_\parallel, 2\omega|z') \tag{8.29a}$$

and

$$\mathcal{G}_{xz}(2k^{(I)}_\parallel, 2\omega|zz') = -\frac{4\pi}{W_p(2k^{(I)}_\parallel, 2\omega)}\mathcal{E}^{>}_x(2k^{(I)}_\parallel, 2\omega|z)\mathcal{E}^{<}_z(2k^{(I)}_\parallel, 2\omega|z'). \tag{8.29b}$$

In the vacuum, for $z > 0$, we have

$$\mathcal{E}^{>}_x(2k^{(I)}_\parallel, 2\omega|z) = -\frac{k^{(2>)}_\perp}{2k^{(I)}_\parallel}\mathrm{e}^{+\mathrm{i}k^{(2>)}_\perp z}, \tag{8.30}$$

where

$$k^{(2>)}_\perp = \left[\frac{(2\omega)^2}{c^2} - (2k^{(I)}_\parallel)^2\right]^{\frac{1}{2}} \equiv \frac{2\omega}{c}\cos\theta_I \tag{8.31}$$

is the perpendicular component of the wave vector of the second-harmonic signal, in the vacuum above the crystal. We shall also find, in our various formulae, the perpendicular component of the wave vector of the second-harmonic wave in the substrate. We call this $k^{(2<)}_\perp$, which is defined by

$$k^{(2<)}_\perp = \left[\frac{(2\omega)^2}{c^2}\varepsilon(2\omega) - (2k^{(I)}_\parallel)^2\right]^{\frac{1}{2}} \ . \tag{8.32}$$

After resorting to expressions in Appendix C, and a bit of algebra, one finds for $z > 0$

$$\mathcal{E}_x^{(2)}(z) = \frac{8\pi \mathrm{i}(E^{(I)})^2 k_\perp^{(2>)}}{[\varepsilon(2\omega)k_\perp^{(2>)} + k_\perp^{(2<)}]} \exp\left[\mathrm{i}(2k_\parallel^{(I)}x + k_\perp^{(2>)}z - 2\omega t)\right]$$

$$\times \left\{ d\left[k_\parallel^{(I)}\tilde{\chi}_{\perp TOT}^{(2s)} - \frac{1}{2}k_\perp^{(2<)}\tilde{\chi}_{\parallel TOT}^{(2S)}\right]\right.$$

$$\left. -\mathrm{i}k_\parallel^{(I)}\Lambda_b\varepsilon\,(\omega)T^2(p)\left[k_\perp^{(2<)} + 2k_\perp^{(T)}\right]\int_{-\infty}^{0} \mathrm{e}^{-\mathrm{i}[k_\perp^{(2<)}+2k_\perp^{(T)}]z'}dz'\right\} . \quad (8.33)$$

The integral is elementary, and yields our final form:

$$\mathcal{E}_x^{(2)}(z) = \frac{8\pi \mathrm{i}(E^{(I)})^2 k_\perp^{(2>)}}{[\varepsilon(2\omega)k_\perp^{(2>)} + k_\perp^{(2<)}]} \exp\left[\mathrm{i}(2k_\parallel^{(I)}x + k_\perp^{(2>)}z - 2\omega t)\right]$$

$$\times \left\{ d\left[k_\parallel^{(I)}\tilde{\chi}_{\perp TOT}^{(2s)} - \frac{1}{2}k_\perp^{(2<)}\,\tilde{\chi}_{\parallel TOT}^{(2s)}\right]\right.$$

$$\left. +k_\parallel^{(I)}\,\Lambda_b\varepsilon(\omega)T^2(p)\right\} . \quad (8.34)$$

While we have obtained (8.34) for a rather special model, nonetheless its form illustrates several general points which enter discussions of nonlinear interactions at surfaces. We discuss these general points, then turn to a description of some key experiments.

First of all, consider the direction of the second-harmonic output beam in the region above the crystal. In the plane of the surface, the wave vector component of the second-harmonic is $2k_\parallel^{(I)} = 2(\omega/c)\sin\theta_I$, while the perpendicular component is $2(\omega/c)\cos\theta_I$, as we see from (8.31). The second-harmonic beam thus makes the angle θ_I with respect to the normal, exactly as does the specular beam at the primary frequency ω. One must "look" in the direction of the specular beam to see this signal. If one considers the second-harmonic beam transmitted through a thin film, a similar condition applies. A discussion of schemes for detecting the second-harmonic signal may be found in the excellent review article by *Heinz* [8.4].

If we consider other nonlinear interactions at the surface mediated by $\chi_{\alpha\beta\gamma}^{(2)}$, in general the output beam will emerge at an angle distinctly different from the specularly reflected input radiation. The fact that the output beam coincides with the specular component of the input beam is unique to second- (or third-)harmonic generation. Consider, for example, the generation of sum-frequency radiation produced by the mixing of two beams, one of frequency ω_1 and one of frequency ω_2. Let each incident beam lie in the same plane, making angles θ_1 and θ_2 with respect to the surface normal, respectively. The output beam at the sum frequency $\omega_1 + \omega_2$ will have a wave vector component parallel to the surface given by $k_\parallel^{(+)} = (\omega_1/c)\sin\theta_1 + (\omega_2/c)\sin\theta_2$, while the perpendicular component of its wave vector is easily found to be

$$k_\perp^{(+)} = \frac{1}{c}\left[\omega_1^2\cos^2\theta_1 + \omega_2^2\cos^2\theta_2 + 2\omega_1\omega_2(1 - \sin\theta_1\sin\theta_2)\right]^{\frac{1}{2}} . \quad (8.35)$$

So long as $\theta_1 \neq \theta_2$, the beam will emerge from the surface at an angle distinctly different than the specular beam associated with either input beam.

In the case of second-harmonic generation, our discussion shows that the polarization of the second-harmonic signal contains information about the symmetry of the near surface environment. For the simple example considered, where cylindrically symmetric molecules stand perpendicular to the underlying isotropic substrate, either s or p polarized incident radiation will produce a p polarized second-harmonic signal, independent of the orientation of the plane of incidence. Suppose, for example, the molecules are tilted at an angle with respect to the surface normal. Then one will realize pure p polarized second-harmonic radiation only when the plane of incidence is aligned so the symmetry axis of the molecules lies in this plane. Also, if the substrate is a single crystal, as the plane of polarization is rotated about the normal to the surface, the symmetry of the underlying crystalline matter will leave a signature on the polarization of the second-harmonic signal. We will illustrate these issues in the next subsection.

In the case of second-harmonic generation, one may explore the polarization properties of the output beam, as a function of that of the single input beam. If one considers sum or difference-frequency generation, then one has the polarizations of both input beams as parameters. At least in principle, stronger constraints on surface or interface symmetry can be placed as a consequence of such studies.

Next, we consider the relative importance of the electric quadrupole terms, which arise from that portion of the substrate illuminated by the incident beam, to the near surface contributions described by $\tilde{\chi}^{(2s)}_{\perp TOT}$ and $\tilde{\chi}^{(2s)}_{\parallel TOT}$. As we have seen, the quadrupole terms have their origin in the wave vector dependence of the third-rank tensor $\chi^{(2)}_{\alpha\beta\gamma}$; in a substrate with inversion symmetry, $\chi^{(2)}_{\alpha\beta\gamma}$ vanishes, and the coupling constant Λ_b in (8.34) has its origin in $(\partial\chi^{(2)}_{\alpha\beta\gamma}/\partial k_\delta)$, where k_δ is a cartesian component of the wave vector of the incident radiation. We recall this from (8.9). It follows, as remarked earlier, that on physical grounds, we expect Λ_b to be the order of magnitude of the nonzero elements of the surface nonlinear susceptibility tensor introduced in (8.1), multiplied by a microscopic length a_0 the order of the lattice constant, or a typical bond length in the substrate. We see that in our final expression displayed in (8.34), the surface terms are also multiplied by the microscopic length d, which in general is comparable in magnitude to a_0. Thus, one concludes that unless a special circumstance prevails, the electric quadrupole terms with bulk origin contribute to the final signal at roughly the same level as the surface terms. One must take due account of both effects in the analysis of data.

One may appreciate the reason why this is so by going back one step, and examining the structure of (8.33). We see the prefactor of the bulk term (the piece proportional to Λ_b) involves one more factor of the wave vector of the radiation than do the surface terms. This is because the electric quadrupole term involves the gradient of the incident field, as we see from (8.10,11). The electric quadrupole signal from a selected very small volume dV is thus smaller than that generated by the surface terms in a volume of the same size by a factor of roughly ka_0, where k is the wave vector of the radiation. The effective volume over which the bulk signal is generated, however, is controlled by the coherence length of the non linear interaction, as we recall from Sect. 4.1. The second-harmonic radiation which emerges from the surface propagates in the $+z$ direction, while the incident beam propagates along $-z$. The coherence length ℓ_c of the nonlinear interaction is thus, in the language of Sect. 4.1, $\ell_c = 1/|\Delta k| = 1/|2k_\perp^{(T)} + k_\perp^{(2<)}|$. This coherence length is much larger than d, in general, and in the mathematical expression for the second-harmonic signal, this larger coherence length precisely compensates for the reduction factor introduced by the fact that the electric quadrupole term is proportional to the gradient of the electric field.

The formalism developed and applied here is readily extended to more complex models of the surface region, or to the description of other surface mediated nonlinear interactions such as sum and difference frequency generation. We direct the reader to the review article by *Heinz* [8.4] cited earlier, and also to the review by *van Driel* [8.5]. Next, we turn to the description of specific experimental studies which illustrate the principles outlined in this section.

8.2 Nonlinear Optical Interactions at Surfaces and Interfaces; Examples

8.2.1 Second-Harmonic Generation from Clean Crystal Surfaces

The discussion in Sect. 8.1 was based on a model in which we envisioned an overlayer of adsorbed molecules was placed on the crystal surface. If in the bulk of the crystal an inversion center is present, a second-harmonic signal results only from the rather weak electric quadrupole terms in (8.9), while the adsorbed molecules acquire a second-harmonic contribution to the electric dipole moment to leading order, by virtue of the breakdown of inversion symmetry at the surface.

We may apply precisely the same physical picture to the surface of a terminated perfect crystal, with no adsorbates present, as mentioned earlier. Consider the very outermost layer of atoms in such a crystal. These sit in sites that lack inversion symmetry, and thus respond to an incident electromagnetic wave in a manner identical to the adsorbate molecules that were the focus of our attention in Sect. 8.1. To excellent approximation, the second layer of atoms in the crystal sit in sites whose local environment is very similar to atoms in the bulk. We may thus view the terminated crystal as possessing a surface layer of microscopic thickness which senses the absence of inversion symmetry, placed on a substrate which may be viewed as bulk matter. The phenomenology of Sect. 8.1 may be applied directly.

An early description of the origin of second-harmonic signals from metal surfaces was put forward in a fundamental paper by *Rudnick* and *Stern* [8.6]. They focussed on the role of the conduction electrons in generating near surface second-harmonic currents. As we have seen in Chap. 2, for frequencies well below the plasma frequency of the conduction electrons, a condition satisfied in the visible frequency range for many simple metals, the conduction electrons contribute importantly to the optical response, as we see from the last term in (2.46c). Right at the surface, where the conduction electron density falls to zero, the absence of an inversion center asserts itself, and one finds second-harmonic currents in the conduction electrons. *Rudnick* and *Stern* presented elegant arguments which suggest that the nonlinear surface dipole moment may be written, for p polarized radiation incident in the xz plane

$$P_z^{(NL)}(r,t) = a\frac{e\omega_p^2}{4m\omega^2}\,[E_z(0-)]^2\,\delta(z+)e^{i2k_\parallel x}e^{-i2\omega t} \tag{8.36a}$$

and

$$P_x^{(NL)}(r,t) = b\frac{e\omega_p^2}{4m\omega^2}\,[E_z(0-)]\,[E_x(0-)]\,\delta(z+)e^{i2k_\parallel x}e^{-i2\omega t}\,. \tag{8.36b}$$

In these expressions, a and b are dimensionless constants, assumed of order unity, ω_p is the conduction electron plasma frequency encountered in Chap. 2, e and m are the electron charge and mass, and ω is the frequency of the incident radiation. Here $E_z(0-)$ and $E_x(0-)$ are the relevant components of the incident field evaluated just inside the surface, and $\delta(z+)$ is a Dirac delta function whose argument vanishes just above the surface $z = 0$.

The expressions in (8.36) have precisely the form which entered the discussion of Sect. 8.1. Here we do not have a layer of atoms or molecules, but instead the conduction-electron continuum whose density falls from the bulk value to zero, within the last Ångstrom or two of the surface. Again we may think of second-harmonic currents confined to a very thin layer just above the surface.

The detection of second-harmonic radiation from the surface of Ag and Al films was reported by *Quail* and *Simon* [8.7], who found excellent agreement with the theory of *Rudnick* and *Stern* [8.6]. For the case of Al, they found agreement with the theory by choosing $a = 1.5$.

Recently the theory of the surface response of conduction electrons in metals has reached a very high level of sophistication. A consequence is that it is now possible to calculate the parameter a from microscopic theory [8.8]. For the surface of Al, one finds $a \approx 35$, a value enormously larger than the value found in [8.7]. Theory predicts the surface second-harmonic signal to be larger than reported in [8.7] by a factor of 400, since the signal scales as a^2! A new experimental study of second-harmonic generation from very clean Al surfaces prepared in ultra-high vacuum provides results in remarkable quantitative accord with theory [8.9]. Evidently the surfaces studied by *Quail* and *Simon* [8.7] were oxidized. The difference between the two sets of experiments provides a dramatic illustration of the sensitivity of surface second-harmonic signals to the microscopic details of the surface environment.

This method can also be used to probe the outer surface layer of semiconducting or insulating materials, where again experiments display a strong sensitivity to the outermost atomic layer. This is demonstrated elegantly in the study of second-harmonic generation reported by *Heinz* et al. [8.10], who examined a most important surface, the (111) surface of crystalline silicon. This is the cleavage surface of the material.

Numerous past studies of this surface demonstrate the arrangement of the atoms in the outermost few atomic layers is dramatically different than the simple hexagonal pattern one expects from the bulk silicon structure. The stable surface phase is referred to as the 7×7 structure. The atomic arrangement is very complex; the unit cell in the plane is seven times larger than the bulk unit cell in the two principal directions in the plane, and thus its area is forty nine times larger. There is also a metastable surface 2×1 phase, where the unit cell is two times longer in one direction, and equal in length in the other, to the bulk unit cell. *Heinz* et al. [8.10] studied the polarization of second-harmonic radiation emitted by this surface, in response to plane polarized radiation which strikes the surface at normal incidence. The plane of polarization of the incident radiation was rotated about the normal to the surface, and the pattern produced by the second-harmonic radiation was a reflection of the low symmetry of the atomic arrangement in the outermost atomic layer. The azimuthal variation of the second-harmonic intensity allowed an accurate determination of the orientation of the reflection planes associated with the two low-symmetry surface structures.

Experimental studies such as those just described constitute a most impressive achievement. Second-harmonic generation is a feeble optical phenomenon, with its origin in a very weak nonlinear effect. In Chap. 4, we saw how the exploitation of phase matching over optical paths of suitable length allowed a large fraction of the incident beam to be converted to the second-harmonic. The experiments outlined here use second-harmonic generation to probe a single atomic monolayer of matter, under conditions where phase matching is absent, as we have seen.

8.2.2 Second-Harmonic Generation from Adsorbate Layers on Surfaces

The discussion of the previous subsection already informs us that second-harmonic generation is most sensitive to the outer layer of matter, if a sample is studied where (to leading order) second-harmonic generation in the bulk is forbidden. We saw that the signals reported in early work by *Quail* and *Simon* [8.7] were vastly weaker than that found for the same surfaces, held very clean and free from oxidation in ultra high vacuum [8.9]. A careful and quantitative study of the influence of oxidation, and adsorption of CO and sodium on the Rh (111) surface has been reported [8.11]. *Tom* et al. found that they can easily detect the influence of 5% of a monolayer! Clearly, we have in hand a phenomenon remarkably sensitive to small influences on the surface environment.

The simple model explored in Sect. 8.1 examined a particularly simple geometry, wherein we have cylindrically symmetric molecules with symmetry axis normal to the surface. Quite frequently, molecular overlayers are canted, with symmetry axis inclined with respect to the surface normal. This tendency may be appreciated from a simple physical argument. Consider an isolated molecule in free space, cigar shaped as in Fig. 8.1. If the two ends are dissimilar, then there is no reflection plane normal to the long axis. It follows the molecule then must have a static electric dipole moment. Even if the free molecule has such a reflection plane, when it bonds to the surface, reflection symmetry is absent, and it acquires an electric

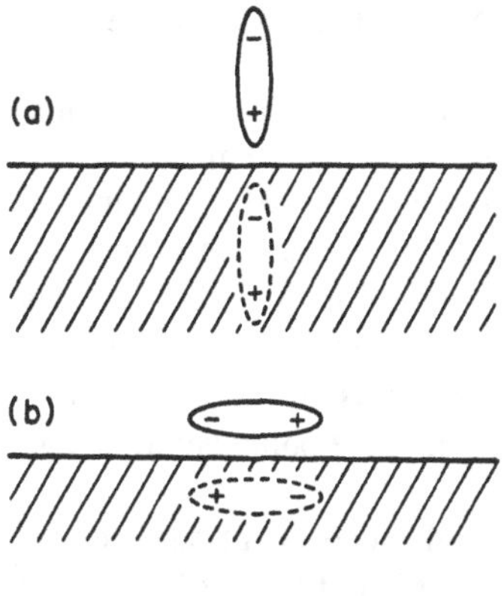

Fig. 8.3. A schematic illustration of the interaction of the electric dipole moment of an adsorbed molecule, and its image charges in a dielectric or metal substrate, for **a** perpendicular orientation, and **b** parallel orientation.

dipole moment by virtue of bonding to the surface. There will be image charges in the substrate, in response to the dipole moment. In Fig. 8.3 we illustrate an electric dipole and its image array, for a dipole perpendicular to, and parallel to the surface. Clearly, the electrostatic energies favor the parallel orientation. There are other energies important in a real system, and for a dense overlayer one encounters steric inhibitions on the possibility that all molecules lie parallel. A consequence is that not uncommonly the symmetry axis is canted.

As we remarked earlier, the polarization and angular variations of the second-harmonic generated from a molecular overlayer can be used as a probe of the degree of canting. For the purposes of interpreting the data, it is necessary to develop a microscopic model. We refer the reader to a recent study where this has been done [8.12].

Not only canting of the symmetry axis, but the low symmetry of the molecule itself (in the adsorbed state) may be probed by the method of second-harmonic generation. For example, in the simple example explored in Sect. 8.1, we saw that both s and p polarized incident radiation generated a second-harmonic of p polarized character. In a recent study, the adsorption of chiral molecules has been shown to lead to a violation of this rule [8.13].

Quite clearly, second-harmonic generation serves as a powerful probe of various aspects of the environment of a surface or an interface. The literature by now is very extensive, so in the present discussion we can provide only a very few illustrative examples. In modern surface science, other powerful surface probes have been developed to a high level of sophistication. Examples include photoemission, and various probes based on the scattering of electrons and atoms from surfaces. Nonlinear-optical probes, of which second-harmonic generation is one example, can be applied to diverse situations where these classic probes of surface science either fail, or are of limited utility. Examples are provided by the complex molecular overlayers just discussed, or by interfaces buried under a macroscopic layer of material transparent to electromagnetic radiation suitable for use in nonlinear optics. Electrons or atoms cannot be transmitted through such macroscopic layers.

8.2.3 The Generation of Sum Frequencies from Adsorbates on Surfaces

While the theoretical development in Sect. 8.1 concentrated on the description of second-harmonic radiation from the surface region of materials, we remarked that the entire range of nonlinear optical phenomena may be explored, including the generation of sum- and difference-frequency radiation. Of course, such experimental studies require the use of two input laser beams, one with frequency ω_1 and one with frequency ω_2. While this is a considerable cost in experimental complexity, the fact is there is a great deal to gain. As noted earlier, for example, the symmetry of the near surface environment may be probed by altering the polarization of each of the two input beams.

An interesting experimental study of molecules adsorbed on Pt surfaces has been reported by *Daum* et al. [8.14]. In this study, an infrared beam of frequency ω_1 is employed in combination with visible light of frequency ω_2, and the sum frequency $\omega_1 + \omega_2$ is monitored as the output.

One may use this technique as a form of surface vibrational spectroscopy as follows: The infrared frequency ω_1 is scanned through a vibrational resonance of the molecule of interest. As one passes through resonance, one sees an enhancement of the sum frequency output.

This process is illustrated schematically by our earlier discussion of Sect. 3.1, where we considered an anharmonic oscillator driven by two superimposed fields, one at frequency ω_1 and one of frequency ω_2. In this picture, the magnitude of the electric dipole moment at the sum frequency is given by $ex^{(2)}(t)$, where the relevant contribution to $x^{(2)}(t)$ is that provided by (3.21d). If ω_2 (the visible frequency beam in the present example) is very large compared to the resonance frequency ω_0 of the oscillator, and if in addition $\omega_2 \gg \omega_1$, we have

$$ex^{(2)}(t) = -\frac{ae^3}{2M^2\omega_2^4} \frac{E_1 E_2 e^{-i(\omega_1+\omega_2)t}}{(\omega_0^2 - \omega_1^2 - i\omega_1 \Gamma)} \tag{8.37}$$

which displays a resonance as the infrared beam frequency ω_1 is swept through the resonant vibrational frequency ω_0 of the molecule.

In the study reported in [8.14], the method just described was applied to the study of CO vibrations, for molecules adsorbed on a single crystal of Pt, held in ultrahigh vacuum. In addition, CN molecules adsorbed on a Pt surface in an electrochemical cell environment were studied in considerable detail. Here the surface is covered by an electrolytic aqueous solution, and sum-frequency spectroscopy was thus utilized to probe a "buried interface" inaccessible to the many methods used in surface science. For example, electron mean free paths in dense matter are so short that all electron based spectroscopies fail, as remarked earlier.

8.3 Resonant Enhancement of Electromagnetic Fields Near Surfaces and Interfaces and Their Role in Surface Nonlinear Optics

In the discussion presented in Sect. 8.1, we saw that the electromagnetic field "sensed" by an atom or molecule at or near a surface can be substantially different than that associated with the incident beam or beams. For the perfectly smooth, ideal surface considered there, a molecule just outside the surface is exposed to a total electric field which is a superposition of that of the incident beam, and the reflected beam. As we see from Fig. 8.2, the resulting field can be very different both in magnitude and direction from that in the incident beam, far from the surface.

Under circumstances we shall now discuss, it is possible to realize very large near-surface enhancements in the electric fields associated with the various optical waves involved in interactions there. In the particular case of second-harmonic generation, we now know well that the signal is proportional to the fourth power of the strength of the incident electric field in the interaction region. An enhancement of the incident electric field by an order of magnitude thus leads to a 10^4 enhancement in the signal level, as we noted above.

In general, for large enhancements to occur, it is necessary for the surface to be rough, or have protrusions. It is the case also that one may realize large field enhancements in the near vicinity of small particles, such as those found in colloidal suspensions or aerosols.

We may appreciate this by a simple example. Imagine a perfectly flat surface with a single imperfection which we idealize as a cone with an apex angle α, as indicated in Fig. 8.4a. If the linear dimensions of the structure are small compared to the wavelength of light, then if this feature is illuminated by a beam incident on the surface, the electric field in the beam may be regarded as spatially uniform, in the near vicinity of such a feature. We may then use electrostatics to calculate the electric field near the protrusion. If the surface is a metal, and if we assume it to be a perfect conductor, one may show that near the tip of the cone, the components of the electric field diverge as $(1/r^{1-\varepsilon})$, where ε is a number small compared to unity [8.15]. We thus realize very strong electric fields near the tip of such a protrusion. Similarly, one may show that the electric field also diverges near the upper edge of a step, modeled as a sequence of two perfectly sharp, ninety degree bends on a surface of infinite conductivity, as shown in Fig. 8.4b.

Of course, the tip of any imperfection on the surface which assumes the form of a bump or protrusion is never perfectly sharp, but is rounded with a finite radius of curvature. The field will thus never diverge near a real feature, but it can be enhanced very strongly over the value appropriate to the incident beam, far from the surface. Similarly a step is never perfectly sharp. Its minimum radius of curvature is set by the atomic diameters of the atoms from which the substrate is formed. The argument suggests, however, that strong field enhancements are realized in the near vicinity of steps. It is the case that even the most carefully prepared single crystal surfaces will have a high density of steps. It is difficult to achieve surfaces with terraces wider than a few hundred Ångstroms in the very best circumstances.

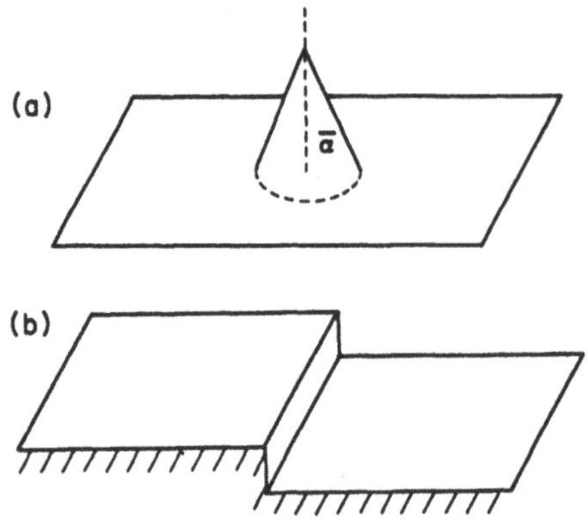

Fig. 8.4. **a** A perfectly flat surface, with an imperfection idealized as a conical structure, with angle α at the apex. **b** A step on an otherwise smooth surface.

Thus, we see that on real surfaces that are not perfectly smooth and flat, as in our idealized models, defects and steps produce regions within which we have very substantial field enhancements in the incident beam. These "hot spots" will increase the strength of the couplings responsible for the generation of second-harmonic radiation, the Raman scattering of light by adsorbed species, and the radiation produced by other nonlinear interactions of interest. It is the case that the fields of the signal wave generated by the nonlinear interaction (the second-harmonic output, the Raman scattered photon, ...) are enhanced near the defect as well. This, we shall see below, also enhances the coupling and hence the output.

With this effect in mind, and upon recalling that modest field enhancements may produce large enhancements in signal, it is possible to design structures which result in strongly enhanced signals from nonlinear optical processes at surfaces or interfaces. One may roughen surfaces, produce surfaces with high step densities, etch gratings onto surfaces, and so on. This is a very active field of current research.

The remarks above assume that the origin of surface field enhancements is purely geometrical, with large field enhancements on conducting surfaces realized near the tips of protrusions or bumps, and steps. The physics that underlies field enhancements on conducting surfaces is actually much more interesting than we have suggested. Near bumps and protrusions, the conducting electrons in metals exhibit a resonant response to the incident field, in certain frequency domains described below. Even a perfectly flat surface exhibits a form of resonant mode, though if the surface is truly perfectly flat, an incident electromagnetic wave cannot excite this resonance. Roughening the surface allows such coupling, and results in very substantial field enhancements.

We turn next to the discussion of the resonances of conducting structures, and their role in enhancing nonlinear signals generated from interactions at or near surfaces. To keep the discussion simple, we consider two extreme limiting cases: an isolated conducting sphere, and the perfectly flat surface, roughened very slightly. These two examples will enable us to appreciate the physics which underlies experiments on real materials.

8.3.1 Resonant Enhancement of Electric Fields Near Small Conducting Spheres

Consider a small isolated conducting sphere, whose radius R is very small compared to the wavelength of radiation we suppose illuminates the sphere. We shall suppose the sphere is fabricated from material whose frequency dependent dielectric constant is described by (2.46c). We have in mind perhaps the constituents of a colloidal suspension, where the colloidal particles are a simple metal such as silver, gold or copper. From a physical point of view, our discussion of the nature of the response of this isolated sphere applies to a spherical particle resting on a substrate, or a bump on a surface, though in these cases we must reckon with a complex geometry.

If the sphere's radius is indeed very small compared to the wavelength of the radiation to which it is exposed, the electric field may then be regarded as spatially uniform in its near vicinity. As in our discussions above, we may use electrostatics to describe the electric fields in the near vicinity of the sphere. This is a standard problem, discussed in all texts on electromagnetic theory [8.15].

Outside the sphere, the electric field is a superposition of the incident electric field $\boldsymbol{E}_0$, and that of a pure electric dipole $\boldsymbol{P}$ whose dipole moment is parallel to $\boldsymbol{E}_0$. Here $\boldsymbol{P}$ is the dipole moment induced in the sphere by the external field. If the external field E_0 is aligned along the z direction, and $\hat{r}$ is a unit vector in the radial direction, the field outside the sphere is

$$\boldsymbol{E}^{>}(\boldsymbol{r}) = \hat{z}E_0 + P(\omega)\left(\frac{3(\hat{z}\cdot\hat{r})\hat{r}-\hat{z}}{r^3}\right), \tag{8.38a}$$

where

$$P(\omega) = \left(\frac{\varepsilon(\omega)-1}{\varepsilon(\omega)+2}\right)E_0R^3 \tag{8.38b}$$

is the magnitude of the electric dipole moment of the sphere. Inside the sphere, the electric field is spatially uniform, with value [8.16]

$$\boldsymbol{E}^{<}(\boldsymbol{r}) = \frac{3}{\varepsilon(\omega)+2}E_0\hat{z}\ . \tag{8.39}$$

The frequency dependent dielectric constant displayed in (2.46c) contains two contributions, as discussed in Sect. 2.1. The first $\varepsilon_{\mathrm{INTER}}(\omega)$ has its origin in interband transitions, and the second in the contribution from the conducting electrons. Let us simplify our discussion for the moment by ignoring the interband transitions. We do this by setting $\varepsilon_{\mathrm{INTER}}(\omega)$ to unity. When this is done, after a line or two of algebra, we find

$$P(\omega) = \left(1+\frac{\omega(\omega+\mathrm{i}/\tau)}{\frac{\omega_p^2}{3}-\omega^2-\mathrm{i}\frac{\omega}{\tau}}\right)R^3E_0\ . \tag{8.40}$$

We see from (8.40) that the sphere exhibits a *resonant* response to the incident field. If $\omega\tau \ll 1$, a condition realized in high quality metals in the visible frequency range, there is a resonance at the frequency $\omega_p/\sqrt{3}$. The induced dipole moment $P(\omega)$ becomes very large for frequencies near the resonance, called the Mie resonance of the conducting sphere. From (8.38a) we see that the incident electric field is enhanced also.

In a simple metal, conduction electron density is in the range of 5×10^{22} electrons/cm^3. If we use the free-electron mass in the formula, this density gives $\hbar\omega_p/\sqrt{3} = 5$ eV, a frequency above the visible range. To discuss real metals, we should take due account of the presence of interband transitions, since typically $\varepsilon_{\mathrm{INTER}}(\omega)$ is substantially larger than unity. When this is done, in many cases, the resonance is shifted into the visible. For example, the Mie resonance of small gold particles lies in the red region of the visible spectrum.

The Mie resonance is responsible for dramatic and interesting effects. For example, the dipole moment $P(\omega)$ radiates electromagnetic waves; this is the scattering of the incident light by the metal sphere. There is thus resonant scattering of light, as one sweeps through the Mie resonance [8.17]. One may make beautifully colored glass by imbedding a colloidal suspension of fine metal particles in the glass matrix. The color arises from the strongly enhanced scattering produced in a narrow spectral region by the Mie resonance. The famous rich red color of ancient Venetian glass has its origin in gold particles imbedded in the glass. The tourist may rest assured that the inexpensive red glass objects sold currently in Venice use a different means of generating the red color.

If we have an ellipsoidal particle, the frequency of the Mie resonance depends on its shape. By varying the shape of the particle, one may thus "tune" the resonance over a wide spectral range. For an ellipsoid of revolution around the z axis, filled with free electrons whose plasma frequency is ω_p, we show the variation of the resonance frequency with particle shape in Fig. 8.5. It is assumed here that the particle is a figure of revolution around the z axis, and that the electric field which excites the resonance is parallel to z, as shown in the figure. A long needle shaped particle, with field parallel to the long axis, has a resonance at a frequency far below that $\omega_p/\sqrt{3}$, while an oblate, disc shaped spheroid has a resonance close to ω_p itself.

One may obtain results such as these by resorting to the description of the electrostatic response of a dielectric ellipsoid [8.18]. Application of a spatially uniform external field such as that illustrated in Fig. 8.5 produces a spatially uniform internal field which may be written $E_0 - 4\pi n_\parallel P_\parallel(\omega)$, where $P_\parallel(\omega)$ is the dipole moment induced by the electric field applied parallel to the axis of rotation. Here $n_\parallel$, which depends only on the shape of the particle, is referred to as the depolarization coefficient. A resonance in the particle occurs when the condition

$$1 + [\varepsilon(\omega) - 1]n_\parallel = 0 , \tag{8.41}$$

as one sees from [Ref. 8.18, Eq. (8.8)]. For a dielectric sphere filled with free electrons, where $\varepsilon(\omega) = 1 - \omega_p^2/\omega^2$, the resonance frequency is given by

$$\frac{\omega_r}{\omega_p} = \sqrt{n_\parallel} . \tag{8.42}$$

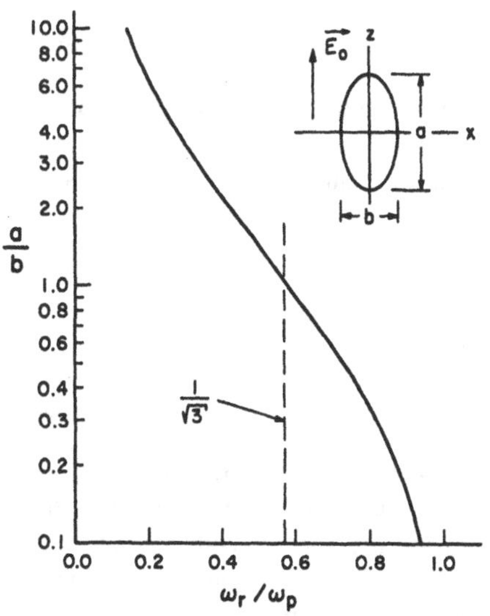

Fig. 8.5. The resonance frequency ω_r of an ellipsoid of revolution, with electric field applied along the symmetry axis. We assume the ellipsoid is filled with free electrons of plasma frequency ω_p. The particle is a figure of revolution about the z axis, and the electric field that excites the resonance is applied parallel to the z direction.

For an ellipsoid of revolution, elementary expressions are provided for $n_{\parallel}$ in [8.18]. It is a straightforward matter to obtain an expression for the resonance frequency of an ellipsoid of revolution in response to an exciting field that is not parallel to the symmetry axis. This will depend on the angle between the field and the symmetry axis, of course.

A simple physical picture of the origin of the resonance is easily obtained. To see this, we return to our spherical particle, filled with free electrons. We assume their density is n, and we treat the lattice of ions as a rigid sphere of positive charge, with charge density ne.

Imagine we hold the ion lattice fixed, and make a rigid body displacement $\boldsymbol{u}$ of the conduction electrons. As illustrated in Fig. 8.6, the upper half of the sphere will acquire a negative charge density, and the lower half of the sphere will acquire a positive charge density. If u is the magnitude of the displacement (parallel to the z direction) and θ the polar angle measured relative to the z direction, the surface charge density is $d\sigma = -neu\cos\theta$, where e, a positive number, is the magnitude of the electron charge. The electrostatic potential $\phi(\boldsymbol{r})$ at any point in space produced by this charge density is

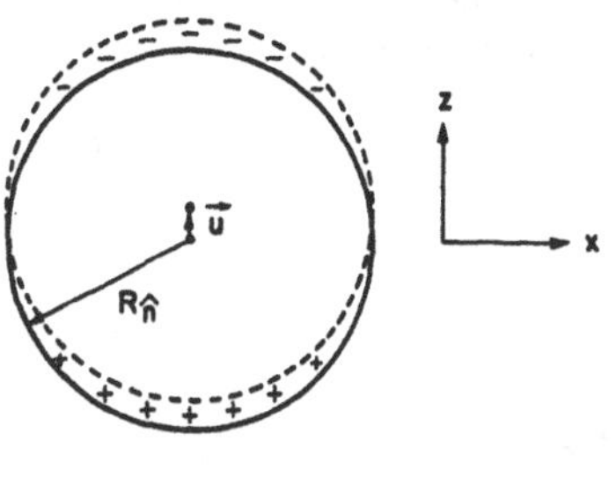

Fig. 8.6. A spherical particle which contains free electrons. The electrons have been subjected to a rigid body displacement $\boldsymbol{u}$ relative to the rigid, positive ion background.

$$\phi(\boldsymbol{r}) = -R^2 neu \int \frac{\cos\theta'}{|\boldsymbol{r} - R\hat{\boldsymbol{n}}'|} d\Omega(\hat{\boldsymbol{n}}') , \tag{8.43}$$

where $R\hat{\boldsymbol{n}}'$ is a vector from the sphere's center to a point on the sphere's surface, and $d\Omega(\hat{\boldsymbol{n}}') = \sin\theta' d\theta' d\phi'$ is the element of solid angle. If we choose $\boldsymbol{r}$ to lie inside the sphere, standard methods allow one to evaluate the integral in (8.43). One finds

$$\phi(\boldsymbol{r}) = -\frac{4\pi neu}{3} r\cos\theta , \tag{8.44}$$

which generates an electric field inside the sphere

$$\boldsymbol{E}^{<}(\boldsymbol{r}) = +\frac{4\pi neu}{3}\hat{z} = \frac{4\pi ne}{3}\boldsymbol{u} . \tag{8.45}$$

Thus, a rigid body displacement of the electrons, as illustrated in Fig. 8.6, produces a spatially uniform electric field everywhere inside the sphere. This field exerts a force on each and every electron inside the sphere, with magnitude independent of the electron's position. Thus, if we pick one electron, and e is the magnitude of its charge ($e > 0$), it obeys the equation of motion

$$m\ddot{\boldsymbol{u}} = -e\boldsymbol{E}^{<} = -\frac{4\pi ne^2}{3}\boldsymbol{u} . \tag{8.46}$$

Thus, with $\omega_p^2 = 4\pi ne^2/m$, the displacement $\boldsymbol{u}(t)$ satisfies the simple harmonic oscillator equation

$$\ddot{\boldsymbol{u}} + \frac{\omega_p^2}{3}\boldsymbol{u} = 0 . \tag{8.47}$$

The electrons thus exhibit a spontaneous oscillation at the frequency $\omega_p/\sqrt{3}$, which is the resonance frequency which appears in (8.40).

This is an example of a collective excitation of the conduction electrons in the sphere. Once they are set in motion, they oscillate coherently at the frequency $\omega_p/\sqrt{3}$, according to (8.47). The external electric field excites this collective mode which, as we see from Fig. 8.6, has a net dipole moment. This dipole moment generates an additional field which can be very large, provided $\omega\tau \gg 1$, and the resonance is excited strongly.

We next turn to a simple argument which demonstrates there is an optimum particle size which yields the maximum field enhancement. First, when $\omega\tau \gg 1$, we may obtain a simple expression for the magnitude of the enhanced field on the particle's surface. Consider the field E_s at the "north pole" of the sphere, $\theta = 0$, just a bit outside the particle. From (8.38a,b), we find at this point

$$\frac{E_s}{E_0} = 1 + \frac{2P(\omega)}{R^3} \tag{8.48}$$

or

$$\frac{E_s}{E_0} = \frac{3\varepsilon(\omega)}{\varepsilon(\omega) + 2} . \tag{8.49}$$

This gives, for frequencies very close to the point where $\varepsilon(\omega) + 2 = 0$,

$$\frac{E_s}{E_0} \cong \frac{2\omega^2}{\frac{\omega_p^2}{3} - \omega^2 - i\frac{\omega}{\tau}} . \tag{8.50}$$

We suppose also $\omega\tau \gg 1$, so the resonance is sharp and narrow. Right on resonance, (8.50) then gives

$$\frac{E_s}{E_0} \cong \frac{2}{i\sqrt{3}}\omega_p\tau . \tag{8.51}$$

To obtain very large field enhancements for a given material, where ω_p is fixed, we want the relaxation time τ to be as long as possible. We thus turn to the various contributions to the relaxation rate for a very small metal particle. There are three distinct contributions. Two are size dependent. The optimum particle size follows upon minimizing the overall relaxation rate $(1/\tau)$.

The definition of the relaxation time has been given in (2.42). We imagine setting up a coherent current in which the electrons acquire a drift velocity $\langle v \rangle$, to decay to zero. From (2.42), we see the key quantity is the relaxation rate $(1/\tau)$. When there are several independent processes which result in the decay of $\langle v \rangle$, we add the rates of decay from each to the right-hand side of (2.42). Thus, to calculate the total relaxation rate for the small particle, we add together the contribution from each relevant process.

There are three independent processes we consider. As an individual electron propagates through the volume of the particle, it is scattered by defects and thermal motions of the atoms, very much as in bulk crystalline matter. This gives a contribution we write as $(1/\tau_\infty)$ to the relaxation rate. This parameter may be deduced from the electrical resistivity of the bulk material. At room temperature, typical values of τ_∞ are in the range of 5×10^{-15} seconds or so. In the small particle, the electron is scattered from the surface, which in general will be rough and irregular. Electrons in metals travel with a speed given by the Fermi velocity v_F, which lies in the range 3×10^8 cm/s. The time between successive surface scatterings is then R/v_F, so we have a contribution to the relaxation rate we write as v_F/R. We ignore a numerical prefactor here controlled by the geometrical details we overlook in this simple picture. At this point,

$$\frac{1}{\tau} = \frac{1}{\tau_\infty} + \frac{v_F}{R} . \tag{8.52}$$

There is a third contribution. When the resonance mode is excited, we have seen there is coherent collective motion of the electrons, and as a consequence the particle acquires an oscillatory electric dipole moment, as we see from Fig. 8.6. This dipole moment radiates energy into space. We thus have radiation damping. So long as the radius of the particle is small compared to the wavelength of the emitted radiation, an elementary argument provides the radiative damping rate.

We consider the spherical particle, filled with free electrons, again with n their number density. If the dipole moment of the particle is P, energy is radiated at the rate [8.19]

$$\frac{dE_R}{dt} = \frac{\omega^4}{3c^3}P^2 = \frac{\omega_p^4}{27c^3}P^2 , \tag{8.53}$$

where the frequency ω is $\omega_p/\sqrt{3}$ in our case. If u is the displacement associated with the coherent motion of the electrons, then using the picture provided by Fig. 8.6, one calculates $P = 4\pi neR^3u/3$, so that

$$\frac{dE_R}{dt} = \frac{16\pi^2}{243}\frac{\omega_p^4 n^2 e^2 u^2 R^6}{c^3} . \tag{8.54}$$

The radiative decay time is found by dividing the radiation rate in (8.52) by the total energy of the motion. This is the kinetic energy of the electrons, given by $T = (4\pi R^3/3)(nmv^2/2) = (4\pi R^3/3)(nm\omega^2u^2/2)$, or $T = (2\pi nm\omega_p^2 R^3u^2/9)$. Then

$$\frac{1}{\tau_R} = \frac{1}{T}\left(\frac{dE_R}{dt}\right) = \frac{2}{27}\frac{\omega_p^4 R^3}{c^3} . \tag{8.55}$$

The total relaxation rate which then should be used in the analysis of the resonant response of the small spherical particle is then

$$\frac{1}{\tau} = \frac{1}{\tau_\infty} + \frac{v_F}{R} + \frac{2}{27}\frac{\omega_p^4 R^3}{c^3} . \tag{8.56}$$

As the particle size is decreased, surface scattering dominates and shortens the relaxation rate, thus decreasing the enhanced fields one may realize. As the particle size increases, radiation damping controls, and again τ is shortened. The optimum particle size is found by minimizing the right-hand side of (8.54). This gives

$$R_c = \left(\frac{9}{2}\frac{v_F}{c}\right)^{\frac{1}{4}}\frac{c}{\omega_p} \tag{8.57}$$

as the optimum. For a metal with 5×10^{22} electrons/cm^3, and with m being the free electron mass, we have $\omega_p = 1.6 \times 10^{16}$ radians/s. If we take $v_F = 3 \times 10^8$ cm/s as a typical Fermi velocity, then we find

$$R_c \cong 95 \text{ Å} \tag{8.58}$$

as the particle size for optimum field enhancement. The minimum value of $1/\tau$ is easily seen to be

$$\left(\frac{1}{\tau}\right)_{\min} = \frac{1}{\tau_\infty} + \frac{4}{3}\frac{v_F}{R} . \tag{8.59}$$

We may estimate the maximum field enhancement that may be achieved by inserting (8.57) into (8.49). We let $\tau_\infty \to \infty$, to make our estimate as optimistic as possible. When this is done, we have the remarkably simple result

$$\left|\frac{E_s}{E_0}\right|_{\max} = \frac{\sqrt{3}}{2}\frac{\omega_p R_c}{v_F} = \frac{3}{2^{5/4}}\left(\frac{c}{v_F}\right)^{\frac{3}{4}} . \tag{8.60}$$

The numbers above give

$$\left|\frac{E_s}{E_0}\right|_{\max} \cong 40 . \tag{8.61}$$

This estimate is at the "north pole" of our sphere. At the equator, the field enhancement is smaller than that in (8.59) by a factor of two. A more reasonable measure of the field enhancement is to compute the square of the field everywhere on the sphere, average this over its surface, and take the square root to get a rms average. This gives

$$\left|\frac{E_s}{E_0}\right|_{\text{rms, max}} = \frac{3}{2^{7/4}}\left(\frac{c}{v_F}\right)^{\frac{3}{4}} , \tag{8.62}$$

and we find

$$\left|\frac{E_s}{E_0}\right|_{\text{rms, max}} = 28 \, . \tag{8.63}$$

Clearly, through use of ideal conditions, very large surface-field enhancements can be achieved through excitation of the Mie resonance of a small metal particle. A very similar picture applies to bumps or protrusions from metallic surfaces. These will support localized electromagnetic resonances quite similar to the Mie resonance of isolated particles. The considerations of this section apply to such situations, in a qualitative sense. We have seen that intensities of a number of non-linear processes (second harmonic generation, sum or difference frequency generation, Raman scattering) scale as the fourth power of the field amplitudes. Thus the resonant enhancement mechanism can boost signals by factors in the range of 10^5 to 10^6, under optimal conditions.

The aim of the discussion above is to provide a semiquantitative guide to the physics which controls the magnitude of field enhancements that may be achieved through excitation of localized electromagnetic resonances. In the literature one may find calculations that produce field enhancements very much larger than the result displayed in (8.61). Such estimates must be viewed with some caution. As we shall emphasize again in the next section, through an example, it is essential to take due account of the role of radiative damping in limiting achievable field enhancements. To incorporate this into a complete calculation, it is necessary to solve Maxwell's equations in full, and not use the electrostatic approximation to the fields near the object of interest. Also, realistic estimates of other processes which limit the electron relaxation must be employed. In the view of the present author, it is difficult to achieve field enhancements in real materials even as large as displayed in (8.61), because the estimate there still contains idealizations. For example, we have ignored the influence of $(1/\tau_\infty)$, and we have thus overlooked the role of scattering processes within the body of the particle in limiting the field enhancement. In practice, of course, in a small particle $(1/\tau_\infty)$ may possibly be considerably larger than in a single, high quality crystal.

8.3.2 Resonant Response of a Slightly Roughened Surface to Electromagnetic Fields; The Role of Surface Polaritons

The discussion of the previous section applies to an isolated particle, and by extension to a protrusion or well formed bump on the surface. A very rough surface may be envisioned to be a collection of such structures, each of which supports its own local electromagnetic resonance whose frequency is controlled by its shape. The collection of resonances may extend over a wide range of frequencies, if we imagine the collection of surface features ranges from flat disc-like bumps, to needle like protrusions. We may appreciate this by supposing an analogue of Fig. 8.5 applies to features on such a surface, and not just isolated particles. We may thus realize resonant enhancements over a wide spectral range, in this circumstance.

8.3.3 Resonant Enhancement of Electromagnetic Fields Near Rough Surfaces of Conducting Media

We now turn to the opposite extreme, a nearly smooth surface with only small amplitude roughness present. We shall see again that under suitable conditions, we may realize substantial field enhancements in its near vicinity. The physical picture, however, is rather different.

We begin with one comment on the spectral regime where the Mie resonances of the isolated particle occur. If the imaginary part of the dielectric constant is very small (the condition required for the resonance to be "sharp"), from (8.41), the resonance occurs at the frequency where

$$\varepsilon_1(\omega) = -\left(\frac{1}{n_\parallel} - 1\right) . \tag{8.64}$$

The depolarizing factor $n_\parallel$ lies in the range $0 \leq n_\parallel \leq 1$, with $n_\parallel = 1$ appropriate to a sharp needle like object. For a perfect sphere, we have remarked that $n_\parallel = 1/3$.

For metallic particles of various shapes, the electromagnetic resonances thus always occur in the spectral regime where $\varepsilon_1(\omega)$ is *negative*. For a simple sphere of free electrons, where we may set $\varepsilon_{\text{INTER}}(\omega)$ in (2.46c) to unity, $\varepsilon_1(\omega)$ is negative whenever $\omega < \omega_p$. When interband transitions are incorporated into the picture, quite clearly from the form of (2.46c), the condition $\varepsilon_1(\omega) < 0$ will apply at sufficiently low frequencies. Thus, in conducting media, where the plasma frequency ω_p is nonzero, we may expect the dielectric constant $\varepsilon_1(\omega)$ to always be negative at sufficiently low frequency.

In this same spectral regime, we shall see that a perfectly smooth conducting surface will also possess an intrinsic electromagnetic resonance. These have the character of surface electromagnetic waves, which may propagate along the surface, with electromagnetic fields bound to it, and thus localized near the surface.

In Sect. 2.1, we discussed the nature of electromagnetic waves which propagate in a dielectric material, whose dielectric constant exhibits a strong frequency dependence, such as that displayed in (2.38). These modes were referred to as polaritons, in the discussion which follows (2.38,39). The surface electromagnetic waves of present interest are referred to often as surface polaritons. Notice that in (2.38), if we let the resonance frequency $\omega_0 \to 0$, the right-hand side may be viewed as the frequency dependent dielectric constant of a metal, whose plasma frequency is Ω_p. Thus, by letting $\omega_0 \to 0$ in the discussion which follows (2.39), we have a description of the bulk polaritons of a conducting medium. Notice from

(2.36), that for an idealized material with $\varepsilon(\omega)$ real, in the frequency regime where $\varepsilon(\omega) < 0$, the wave vector k of any electromagnetic disturbance is purely imaginary, since $k^2 < 0$. In this regime, where we shall see shortly we find surface polariton solutions of Maxwell's equations, there are no propagating plane wave solutions in the bulk of the material.

It is quite simple to obtain a description of the surface polaritons. We imagine a semi-infinite material whose surface is perfectly flat, and which coincides with the xy plane. The material lies in the lower half plane $z < 0$, and has the dielectric constant $\varepsilon(\omega)$.

Imagine we consider an electromagnetic disturbance which propagates parallel to the x axis, with fields that decay to zero as $z \to +\infty$, and also as $z \to -\infty$. We consider a disturbance of p polarized character, whose electric field lies in the xz plane. Then in the vacuum $z > 0$, we consider a wave whose electric field is given by

$$\boldsymbol{E}^{>}(x,z;t) = \left(\hat{x}E_x^{>} + \hat{z}E_z^{>}\right)\mathrm{e}^{\mathrm{i}k_\parallel x}\mathrm{e}^{-\alpha^{>}z}\mathrm{e}^{-\mathrm{i}\omega t} . \tag{8.65}$$

We must have $\boldsymbol{\nabla}\cdot\boldsymbol{E}^{>} = 0$ everywhere. This requires $E_z^{>} = (\mathrm{i}k_\parallel/\alpha^{>})E_x^{>}$, so that

$$\boldsymbol{E}^{>}(x,z;t) = E_x^{>}\left(\hat{x} + \mathrm{i}\frac{k_\parallel}{\alpha^{>}}\hat{z}\right)\mathrm{e}^{\mathrm{i}k_\parallel x}\mathrm{e}^{-\alpha^{>}z}\mathrm{e}^{-\mathrm{i}\omega t} . \tag{8.66}$$

It is the case also that each of the components of electric field must obey the wave equation in vacuum. Thus, we require

$$\left(\frac{\partial^2}{\partial x^2} + \frac{\partial^2}{\partial z^2} - \frac{1}{c^2}\frac{\partial^2}{\partial t^2}\right)E_{x,z}^{>}(x,z;t) = 0 . \tag{8.67}$$

The wave equation leads to a relation between the decay constant $\alpha^{>}$, and both $k_\parallel$ and ω. We have

$$\alpha^{>} = \left[k_\parallel^2 - \left(\frac{\omega}{c}\right)^2\right]^{\frac{1}{2}} . \tag{8.68}$$

The expression in (8.66) contains a constraint. We must necessarily have $k_\parallel > (\omega/c)$ for $\alpha^{>}$ to be real. This constraint will prove most important in what follows.

In the lower half space $z < 0$, we seek a solution which also decays away from the interface. We wish a solution for which

$$\boldsymbol{E}^{<}(x,z;t) = (\hat{x}E_x^{<} + \hat{z}E_z^{<})\mathrm{e}^{\mathrm{i}k_\parallel x}\mathrm{e}^{+\alpha^{<}z}\mathrm{e}^{-\mathrm{i}\omega t} . \tag{8.69}$$

We have again $\nabla\cdot\boldsymbol{E}^{<} = 0$, and the fields must satisfy the appropriate wave equations. One finds

$$\boldsymbol{E}^{<}(x,z;t) = E_x^{<}(\hat{x} - \mathrm{i}\frac{k_\parallel}{\alpha^{<}}\hat{z})\mathrm{e}^{\mathrm{i}k_\parallel x}\mathrm{e}^{+\alpha^{<}z}\mathrm{e}^{-\mathrm{i}\omega t} , \tag{8.70}$$

and also

$$\alpha^{<} = \left[k_\parallel^2 - \varepsilon(\omega)\frac{\omega^2}{c^2}\right]^{\frac{1}{2}} . \tag{8.71}$$

If we confine our attention to the case where $\varepsilon(\omega)$ may be supposed real, and to a frequency regime where $\varepsilon(\omega) < 0$, then $\alpha^<$ is always real.

The appropriate boundary conditions must be obeyed at the surface $z = 0$. The conservation of tangential components of $\boldsymbol{E}$ require $E_x^< = E_x^>$, while normal components of $\boldsymbol{D}$ are conserved only if

$$\varepsilon(\omega) = -(\alpha^</\alpha^>) . \tag{8.72}$$

Since $\alpha^<$ and $\alpha^>$ are both positive numbers, the ratio $(\alpha^</\alpha^>)$ in (8.72) is necessarily positive. Thus, we see the surface electromagnetic wave may exist only in the frequency regime where $\varepsilon(\omega) < 0$. We realize surface polaritons in precisely the same frequency domain where we encounter the Mie resonance of elliptical particles.

For a fixed frequency ω, we may regard the ratio $(\alpha^</\alpha^>)$ to be a function of $k_\parallel$. Thus, (8.70) in general will be obeyed only for a selected value of $k_\parallel$. In fact, we may solve (8.70) for $k_\parallel$ to obtain the dispersion relation of the surface polariton, or surface electromagnetic wave. After a few lines of algebra, we find

$$\frac{c^2k_\parallel^2}{\omega^2} = \frac{\varepsilon(\omega)}{1+\varepsilon(\omega)} . \tag{8.73}$$

For a simple metal with $\varepsilon(\omega) = 1 - \omega_p^2/\omega^2$, we illustrate the dispersion relation in Fig. 8.7. We see that for all frequencies below $\omega_p/\sqrt{2}$, we have surface polaritons which may propagate along the surface. The frequency $\omega_s = \omega_p/\sqrt{2}$ is often called the surface plasmon frequency of the metal. For aluminum metal, $\hbar\omega_s = 10.6$ eV, so the modes exist over a very wide frequency range, from the near ultraviolet to the visible. Many metals support these modes in the visible frequency range. In the present author's view, it is remarkable that these ubiquitous modes known for many decades, are mentioned only rarely in texts on electromagnetic theory.

The surface polaritons may be viewed, for the flat surface, as the analogue for the surface of the Mie resonance of the metallic particle. These are collective excitations of the electrons in the surface region, just as the Mie resonance is a collective excitation of small conducting particles. The properties of these modes at diverse surfaces and interfaces have been summarized in a volume devoted to this topic [8.20].

Strong enhancements of electromagnetic fields near surfaces and interfaces may be realized through the excitation of surface polaritons associated with such struc-

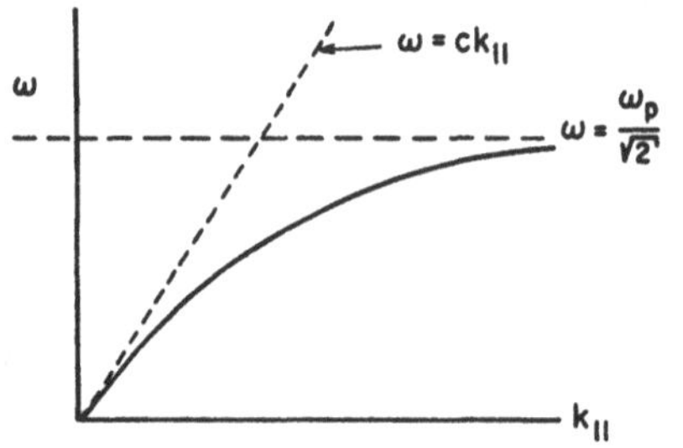

Fig. 8.7. The dispersion relation for surface polaritons on a simple metal surface, for which the frequency dependent dielectric constant $\varepsilon(\omega) = 1 - \omega_p^2/\omega^2$.

tures by an incident beam, just as large field enhancements near small particles may be achieved through excitation of the Mie resonance. There is one difficulty: an electromagnetic wave incident on the perfectly flat surface just described cannot excite surface polaritons. This possibility is excluded by kinematical considerations.

One may see this as follows: As we have seen, surface polaritons are wavelike excitations bound to the surface, characterized by the wave vector $k_\parallel$ parallel to the surface plane. Necessarily $k_\parallel$ must be greater than (ω/c), for the wave to be "bound" to the surface on the vacuum side, i.e., for $\alpha^>$ in (8.66) to be real. This is true not only for the simple semi-infinite dielectric discussed above, but for *any* perfectly planar structure, such as a surface covered by an adsorbed film or a multilayer. When $k_\parallel < (\omega|c), \alpha^>$ becomes purely imaginary, and the fields above the surface are radiative in character.

Now consider a plane wave incident on the surface. Its angle of incidence is θ_I, the projection of its wave vector onto the surface plane is $(\omega/c)\sin\theta_I$ which, of course is less than (ω/c). While the surface may support a surface polariton at this frequency, there is a clear mismatch between the wave vector of the surface polariton, and the projection of that of the incident wave onto the plane of the surface. There is thus no phase matching between the two waves, to use our earlier language. A consequence is that the incident plane wave is unable to excite the surface polariton. It reflects off the surface, without coupling to these modes at all.

Now if the surface is not perfectly smooth, but is roughened slightly, then the picture is very different. The incident wave may have the character of a perfect plane wave far from the surface, but as one approaches the surface, the presence of the roughness introduces spatial modulations into the electric field associated with the incoming wave. In effect, the standard electromagnetic boundary conditions apply at each point on the surface, but the boundary is now rough. Suppose the length scale parallel to the surface associated with the roughness is denoted by a. Then the roughness induced "ripples" in the electric field of the incident wave can be described by the appropriate Fourier transform representation, and one will be required to include wave vector components out to roughly π/a.

If $k_\parallel^{(s)}$ is the wave vector of the surface polariton whose frequency matches that of the incident wave, then if $k_\parallel^{(s)} \lesssim \pi/a$, the roughness will introduce spatial Fourier components into the incident wave which match $k_\parallel^{(s)}$. In this circumstance, the incident wave can excite surface polaritons. When this occurs, substantial enhancements of near surface electric fields are the consequence.

We shall not summarize the theory of roughness-induced coupling between incident radiation, and surface polaritons in this chapter. We note that the Green's function methods introduced above in our discussion of second-harmonic radiation from surfaces have proved most useful in the analysis of the influence of roughness on the optical response of surfaces [8.21]. So long as the amplitude of the roughness is modest, one may build a perturbation theoretic scheme to treat its influence by these methods.

We saw in our discussion of enhanced fields near small metallic particles that there is an optimum particle radius for achieving the maximum field enhancement. Similar considerations apply to roughness-induced enhancements of field amplitudes near metal surfaces. As we have seen, there is no coupling between an incident wave and surface polaritons on perfectly smooth surfaces. As the amplitude of the roughness increases, the field enhancement effect does as well. But a surface polariton propagating on a rough surface is damped by both radiative and nonradiative scattering processes [8.22]. Its lifetime thus shortens with increasing roughness amplitude, with the consequence the modes are excited less efficiently at larger roughness amplitudes.

These points are illustrated by calculations for metal surfaces upon which perfectly periodic gratings are emboss [8.23]. For such perfectly periodic structures, one may carry out calculations of near surface fields for structures of appreciable amplitude. For a grating on a Ag surface of sawtooth profile, the calculations yield enhanced field intensities of roughly 25 under optimal conditions, where the peaks and valleys of the sawtooth structure are in the range of 250 Å above or below the average surface. For gratings deeper than this, the field enhancement decreases with increasing groove depth, because the surface polariton mode responsible for the field enhancement is damped heavily by grating induced scattering processes.

8.4 Experimental Studies of Surface Enhanced Nonlinear Optical Interactions

There is by now a very large literature in which experiments are reported which exploit the very large field enhancements which result from coupling between the optical fields involved, and collective electronic excitations localized near surfaces or interfaces.

The original observation [8.24], and much of the subsequent experimental and theoretical literature has focused on the phenomenon of Raman scattering from molecules adsorbed on surfaces of metallic samples. Effective Raman cross sections enhanced by factors of 10^6 over values appropriate to the gas phase have been reported in many papers in the literature. Such enormous cross sections allow the Raman technique to explore very low coverages of adsorbed material. Under conditions where such large signals are realized, one refers to the phenomenon as SERS, an acronym for surface enhanced Raman scattering. These very large signals have proved very exciting to researchers who wish to explore adsorbed molecules on surfaces, at low coverages. The early literature on diverse experimental configuratures is discussed in the volume edited by *Chang* and *Furtak* [8.25], while *Otto* [8.26] has provided a more recent review. One realizes very large SERS signals both from rough metallic surfaces, and colloidal suspensions of metallic particles. (It is the view of the present author that care must be exercised in evaluating some theoretical contributions to the literature, wherein huge enhancements in field intensities are reported. For example, many of these papers do not incorporate the radiation damping of the collective modes which, as we have seen, play a crucial role in limiting the field enhancements that may be realized).

The role of surface roughness in generating very large Raman signals was apparent in the original experimental discovery of SERS [8.24]. The molecules responsible for the giant signals were pyradine molecules, adsorbed on a silver electrode in an electrochemical cell environment. At the start of the experiment, the silver surface is quite smooth, and the Raman signal very feeble. After a few electrochemical voltage cycles, the signals built up to the very huge levels now associated with SERS in diverse environments. As the voltage was cycled in the cell, silver atoms were removed and then redeposited on the surface to form a very rough, lumpy surface structure. One may then argue one has a diverse array of bumps and protrusions from the surface, which possess collective resonances qualitatively similar to the Mie resonances of the isolated spheroid.

In the theory of Raman scattering presented in Chap. 5, we considered an isolated molecule exposed to an incident photon viewed as a simple plane wave. Similarly, the scattered photon is a simple plane wave also. An expression for the scattering rate is given in (5.16). The number of Raman scattered photons is clearly proportional to n_I, the number of photons in the incident beam. The energy flow in the incident beam scales as $\hbar\omega_i n_I$; if we rewrite this in classical language, then we see that the number of Raman events scales, not surprisingly, as $|E_I|^2$, the intensity of the incident light that strikes the molecule. Thus, if through roughness induced coupling of the incident photon to surface polaritons, or to localized collective excitations near bumps or pits on the surface, the intensity of the incident beam is enhanced, the Raman intensity is enhanced by the same factor.

Notice that when the number of Raman photons n_s is large compared to unity, we see that the scattering rate scales as n_s, the number of scattered photons. In classical language, this means that the Raman intensity scales also as the intensity of the scattered light. Hence, if field enhancements are present in the scattered wave as well, then the Raman cross section is "boosted" by a factor which describes the enhancement of the intensity in the scattered radiation. Under typical conditions, the frequency shift of the scattered photon is small, so if the incident photon couples strongly to collective excitations on the surface, the scattered photon does as well. Then the Raman cross section is enhanced by the fourth power of that of the field amplitude, not just its square if only the intensity of the incident beam entered into the considerations.

It is possible to experimentally isolate, and study separately, the enhancement of the Raman cross section with origin in field enhancements of the incident radiation, and those of the scattered radiation. This has been done in an elegant study reported by *Tsang* et al. [8.27]. Rather than study Raman signals from molecules adsorbed on randomly rough surfaces, these authors prepared a diffraction grating, which provides coupling between a plane electromagnetic wave, and surface polaritons on the surface. This coupling, however, can occur only at very selected, well defined angles of incidence, as one may see from simple arguments [8.23]. These authors then placed their detector at an angle where coupling of the scattered photon to the surface polaritons vanishes, and swept the incident radiation through the critical angle where this radiation couples to the surface modes. By this means, they directly observe the field enhancement effect; it is possible to extract this from the data without knowledge of the Raman cross section of the adsorbed species. We shall shortly appreciate why this is significant.

Then by setting the incident radiation to an angle where coupling to the surface polaritons is absent, they may sweep the detector of the scattered radiation through the critical angle where coupling occurs. By this means, they can study directly the role of surface enhancements of the scattered radiation. The results of this experiment are in good accord with theoretical calculations of field enhancements that may be realized on grating structures [8.23]. In these calculations, it should be remarked, grating induced radiation damping play a central role in limiting the field enhancements one may realize.

As remarked above, in the literature it is reported in many studies that the effective Raman cross sections measured for adsorbed species can be larger than those of the same molecule in the gas phase by a factor of 10^6. Our estimates above suggest that enhancements of field amplitudes of roughly thirty can be achieved, but only under optimum conditions. Thus signal enhancements of 10^6 are difficult to achieve from only surface enhanced fields.

It is quite possible that the actual Raman tensor of the adsorbed species [the parameter $a^{(R)}_{\alpha\beta\gamma}$ in (5.2b)] may differ very substantially from that in the gas phase. This would be a consequence of the chemical bonds between the adsorbate and the substrate. In fact, the complex consisting of the molecule and the nearby substrate atoms should be regarded as a single quantum system, which may possess electronic excited states which lie much lower in energy than those in the isolated molecule. For example, there is clear experimental evidence for the presence of "charge transfer excitations" in such complexes. These are excitations wherein an electron is promoted from the molecule, into an empty level just above the Fermi energy of the metallic substrate. These states may have appreciable oscillator strength, and they may lead to new contributions to the Raman tensor not present in the gas phase which produce substantial enhancements of the Raman cross section. In this regard, it should be noted that a recent absolute Raman tensor determination for hydrogen on the (111) surface of silicon shows the cross section to be larger by a factor of 80 than found in the SiH_4 molecule [8.28]. In contrast to the experimental work reported in [8.27], in nearly all SERS studies it is impossible to separate the contribution from field enhancement effects, and those associated with the Raman tensor of the substrate. The experiments employ rough surfaces whose topology is known only crudely, very little is known about the precise nature of adsorption site, and one lacks the dramatic angular sensitivities exploited in [8.27] to isolate clearly and unambiguously the role of field enhancements in both the incident and the scattered radiation. The review by *Otto* [8.26] summarizes various mechanisms that may lead to enhanced Raman tensors for adsorbed species.

As remarked earlier, enhanced fields near roughened surfaces, or those associated with excitation of collective excitations of small particles, lead to very large signals for any nonlinear optical process at surfaces, not just the Raman scattering processes. For example, *Shen* and his colleagues have explored second-harmonic generation from roughened silver surfaces, to find signals 10^4 larger than those associated with a nominally smooth surface [8.29]. Once again, an enormous enhancement of signal is achieved. These authors attribute the entire effect to enhanced fields near the surface, as a consequence of roughness induced coupling to surface polaritons. In the experiments, the incident frequency was 1.17 eV, lower than the visible frequency used in most SERS studies. At these lower frequencies, roughness induced radiation damping will be less severe than in the visible, it should be remarked. *Farias* and *Maradudin* [8.30] have developed the theory of second-harmonic generation on periodic gratings, to find surface polariton mediated enhancements comparable to, and even larger than those observed in the experiments.

Quite clearly, enhanced fields near surfaces and interfaces play a major role in boosting the signals of feeble nonlinear optical processes there, with the consequence that one may probe these environments by a range of nonlinear optical methods.

Problems

8.1 In this chapter, we employed the Green's functions displayed in (8.29) and elsewhere to derive expressions for the second-harmonic radiation above the surface of a dielectric with adsorbed molecules present.

Consider the denominator $W_p(k_\parallel, \omega)$ which appears in these functions. Suppose the dielectric constant $\varepsilon(\omega)$ of the substrate is negative. Show that, when considered as a function of $k_\parallel$ for fixed ω, this denominator has a zero, and thus the relevant Green's functions have a pole, when $k_\parallel$ equals the wave vector of the surface polariton of frequency ω, see (8.73). From the physical interpretation of the Green's functions in the text, give a discussion of why this pole is present.

8.2 Consider a dielectric slab of thickness D. Its dielectric constant is $\varepsilon(\omega)$. (a) Construct the fields $\mathcal{E}_z^>(k_\parallel\omega|z)$, $\mathcal{E}_z^<(k_\parallel\omega|z)$, $\mathcal{E}_x^>(k_\parallel\omega|z)$ and $\mathcal{E}_x^<(k_\parallel\omega|z)$ required to construct the Green's functions $g_{\alpha\beta}(k_\parallel\omega|zz')$. (b) From the fields in (a), construct $W_p(k_\parallel, \Omega)$ and show it is independent of z by evaluating it inside and outside the slab.

8.3 Consider the slab described in (8.2), and suppose $\varepsilon(\omega)$ is real and negative. Find the dispersion of the polariton modes which propagate parallel to the slab surface, with fields which decay exponentially to zero as one moves away from either surface, into the vacuum outside. This latter condition means one explores the regime $k_\parallel > \omega/c$, notice. (Note: For each value of $k_\parallel$ there are two modes. For one, the z component of electric field normal to the slab surf aces has even parity, and for the other, it has odd parity). Sketch the dispersion relation of each mode. These two modes may be viewed as a surface polariton localized on each surface, which mix or interact, when $k_\parallel D$ is comparable to unity.

If you have solved Problem 8.2, show that $W_p(k_\parallel, \omega)$ has zeros when $k_\parallel$ satisfies the dispersion relation just derived.

8.4 Consider a sphere of radius R, made of conducting material whose dielectric constant $\varepsilon(\omega)$ is real and negative. Show that the sphere has a hierarchy of electromagnetic resonances, whose frequencies are found from the relation $\varepsilon(\omega) = -(\ell + 1)/\ell$, where $\ell = 1, 2, 3, \ldots$. Note that for $\ell = 1$, we have the condition $\varepsilon(\omega) + 2 = 0$, which entered centrally in the discussion of enhanced fields near small particles. The modes for larger values of ℓ are mulitpole resonances. (Hint: Consider an electrostatic description of the potential around the sphere, and show that the homogeneous equations $\nabla \cdot \boldsymbol{D} = 0$ and $\nabla \cdot \boldsymbol{E} = 0$ have solutions when the condition stated is obeyed. Seek solutions whose angular variation is controlled by the spherical harmonic $Y_{\ell m}$.)

9. Optical Interactions in Magnetic Materials

In this chapter, we turn our attention to a rich and elegant topic, the interaction of electromagnetic radiation with the magnetic degrees of freedom in materials. This is a field with a long history, and also a large number of phenomena are encountered in diverse classes of materials. Thus, we shall be content to explore a few key concepts in the influence of magnetism on the linear propagation characteristics of materials, and non linear interactions mediated by magnetic degrees of freedom.

A material may acquire a finite magnetization density $\boldsymbol{M}$ by two ways. One may apply an external dc magnetic field $\boldsymbol{H}_0$ to a nominally nonmagnetic material, and by this means induce a nonzero magnetization density. If the external field is applied along a principal axis, or if the material is cubic, one has $\boldsymbol{M} = \chi_0 \boldsymbol{H}_0$ if the external magnetic field is not too large, where χ_0 is the dc susceptibility. In ferromagnets, on the other hand, the magnetization $\boldsymbol{M}$ appears spontaneously when the material is cooled below the Curie temperature T_c, as noted briefly in Chap. 1. There are also magnetically ordered materials in which the net magnetization density $\boldsymbol{M}$ vanishes. An example is provided by antiferromagnets, in which the ordered array of magnetic moments may be partitioned into two sublattices A and B. Those on the A sublattice point upward, and those on the B sublattice point downward, so $\boldsymbol{M}$ vanishes. In this chapter, in the interest of brevity, we confine our attention to materials in which $\boldsymbol{M} \neq 0$, though the topic of magneto-optic phenomena in antiferromagnets and related materials is of great interest.

The magnetization density $\boldsymbol{M}$ consists of two components. The first, $\boldsymbol{M}^{(0)}$, is independent of time and here is assumed to be independent of position. We then have a piece $\boldsymbol{m}(\boldsymbol{r}, t)$ which depends on either position or time, by virtue of either the presence of an electromagnetic wave, or thermal fluctuations in the medium. Electromagnetic radiation may interact with both the static and time varying contributions to $\boldsymbol{M}$, so we shall consider each in turn. First, however, we begin with general remarks to provide the reader with an orientation on basic concepts.

9.1 Introductory Remarks

Consider a ferromagnet material, with spontaneous magnetization $\boldsymbol{M}^{(0)}$, which we assume to be independent of position. An external dc magnetic field $\boldsymbol{H}^{(0)}$ is present as well. If $\boldsymbol{M}^{(0)}$ is independent of position, then necessarily $\boldsymbol{M}^{(0)}$ is parallel to $\boldsymbol{H}^{(0)}$. If this were not so, there would be a torque $\boldsymbol{M}^{(0)} \times \boldsymbol{H}^{(0)}$ exerted on the magnetization which would reorient it in direction.

Before we proceed, we remark that in practice, the assumption that $M^{(0)}$ is uniform in space is nontrivial. A ferromagnetic material will in general break up into a pattern of domains. [9.1] Within each domain, the magnetization density will be uniform, but the overall domain pattern will be such that the magnetization density averaged over the entire sample is zero. The ferromagnet does this to reduce the energy stored in the macroscopic magnetic fields generated by a spatially uniform magnetization density which would extend over all space, in the absence of these structures. The particular domain pattern realized will depend on the shape of a given sample [9.1], and on the nature of "pinning centers" which can inhibit motion of the walls between adjacent domains. Application of an external magnetic field $H^{(0)}$ of sufficient strength eliminates the domains, to produce a sample with spatially uniform magnetization, as assumed in the first paragraph of the present section. Similar remarks apply to the ferroelectric materials mentioned very briefly in Chap. 1. Here, of course, one applies an electric field to eliminate domains.

Now suppose an electromagnetic disturbance propagates through the material. The time dependent magnetic field necessarily associated with it exerts a torque on the magnetization. We then have the magnetization given by

$$M(r,t) = M_0 + m(r,t) \ , \tag{9.1}$$

where for an electromagnetic disturbance that is very weak, the magnitude of m is small compared to M_0. We consider a disturbance of plane wave form, so we assume all quantities vary in space and time as $\exp(ik \cdot r - i\omega t)$. The complete Maxwell equations, with magnetization included, then become [from (1.1a,b) and (1.2a,b)]

$$k \cdot D = 0, \quad k \cdot B = 0 \ , \tag{9.2a}$$

$$k \times E = \frac{\omega}{c} B, \quad k \times H = -\frac{\omega}{c} D \ , \tag{9.2b}$$

and, in addition to the relation between D and E given in (1.2a), we now have

$$B(r,t) = H(r,t) + 4\pi m(r,t) \ , \tag{9.3}$$

where we consider the relations between the time dependent pieces of the various fields.

In Chap. 2, we presented a description of the relation between D and E in nonmagnetic dielectric media. Shortly we turn our attention to the form of this relation in magnetic materials. In addition, as we see from (9.3), we require also a link between $B(r,t)$ and $H(r,t)$, so only one of these quantities enters Maxwell's equation.

In the ferromagnet considered here, this relation follows from the equations of motion for the magnetization. Let $H_T(r,t) = \hat{z} H^{(0)} + H(r,t)$. There is then a torque on the magnetization given by $M \times H_T(r,t)$. If γ is the gyromagnetic ratio of the medium, then $(dM/dt)(1/\gamma)$ is the time rate of change of the angular momentum of the system. Thus, we have

$$\frac{d\boldsymbol{M}}{dt} = \gamma(\boldsymbol{M} \times \boldsymbol{H}_T) \tag{9.4}$$

as the equation of motion of the magnetization. This may be written as, noting $\boldsymbol{M}^{(0)}$ is directed parallel to $\hat{z}$,

$$-i\omega \boldsymbol{m}(\boldsymbol{r}, t) = \gamma H^{(0)}(\boldsymbol{m} \times \hat{z}) + \gamma M^{(0)}[\hat{z} \times \boldsymbol{H}(\boldsymbol{r}, t)] + \gamma[\boldsymbol{m} \times \boldsymbol{H}(\boldsymbol{r}, t)] \ . \tag{9.5}$$

Our first interest will be in the propagation of small amplitude electromagnetic waves and the description of the linear response of the medium. The first two terms on the right-hand side are linear in both $\boldsymbol{m}$ and $\boldsymbol{H}(\boldsymbol{r}, t)$, while the third term is quadratic in these small amplitudes. We thus set the third term aside. Then $\boldsymbol{m}$ lies in the xy plane necessarily. It is a simple matter to write (9.5) in component form, to find

$$m_\alpha = \sum_\beta \chi_{\alpha\beta}^{(M)}(\omega) H_\beta \ , \tag{9.6}$$

where $\chi_{xx}^{(M)}(\omega) = \chi_{yy}^{(M)}(\omega) = \chi_1^{(M)}(\omega)$, $\chi_{xy}^{(M)}(\omega) = -\chi_{yx}^{(M)}(\omega) = i\chi_2^{(M)}(\omega)$, and all other elements of $\chi_{\alpha\beta}$ vanish. If we define the characteristic frequencies $\omega_H = \gamma H^{(0)}$ and $\omega_M = \gamma M^{(0)}$, then

$$\chi_1^{(M)}(\omega) = \frac{\omega_M \omega_H}{\omega_H^2 - \omega^2} \tag{9.7a}$$

and

$$\chi_2^{(M)}(\omega) = \frac{\omega_M \omega}{\omega_H^2 - \omega^2} \ . \tag{9.7b}$$

It then follows that

$$B_\alpha = \sum_\beta \mu_{\alpha\beta}(\omega) H_\beta \ , \tag{9.8}$$

where

$$\mu_{xx}(\omega) = \mu_{yy}(\omega) = 1 + 4\pi \chi_1^{(M)}(\omega) \ , \tag{9.9a}$$

$$\mu_{xy}(\omega) = -\mu_{yx}(\omega) = 4\pi i \chi_2^{(M)}(\omega) \ , \tag{9.9b}$$

$$\mu_{xz}(\omega) = \mu_{yz}(\omega) = \mu_{zx}(\omega) = \mu_{zy}(\omega) = 0 \ , \tag{9.9c}$$

and

$$\mu_{zz}(\omega) = 1 \ . \tag{9.9d}$$

The tensor $\chi_{\alpha\beta}^{(M)}(\omega)$ is the magnetic susceptibility tensor, and $\mu_{\alpha\beta}(\omega)$ is the magnetic permeability tensor. With (9.8) in hand, combined with Maxwell's equations, we may address the nature of wave propagation in ferromagnetic media, once we explore the relation between $\boldsymbol{D}$ and $\boldsymbol{E}$ for these materials.

Before we proceed, some order of magnitude estimates of the quantities which enter (9.7) are in order. If we consider a free electron, with spin angular momentum

$\hbar/2$, the gyromagnetic ratio $\gamma = ge/2mc$, with g very close to two. One has $\gamma \cong 1.8 \times 10^7$ (Gauss-s)$^{-1}$. For nearly all magnetic moment bearing ions, the gyromagnetic ratio will be close to this in value. Commonly encountered laboratory magnetic fields are in the range of 10^4 Gauss, or one Tesla. Hence, $\omega_H \approx 2 \times 10^{11}$ radians/s in a typical situation, so $\nu_H = (\omega_H/2\pi) \approx 30$ GHz. In typical magnets, M_0 ranges from a few hundred to at most 4000 Gauss, so ω_M is less than the value of ω_H.

Thus, both ω_H and ω_M lie in the microwave frequency range. If we are interested in the properties of visible radiation, or even radiation in at infrared frequencies, then $\omega \gg \omega_H, \omega_M$. In this limit, $\chi_{\alpha\beta}^{(M)}(\omega) \cong 0$, and $\mu_{\alpha\beta}(\omega) = \delta_{\alpha\beta}$, so we have, to an excellent approximation,

$$\boldsymbol{B} = \boldsymbol{H} \qquad \text{(visible and infrared frequencies).} \tag{9.10}$$

The frequency of oscillation of the magnetic field is so very much higher than the natural frequencies of the array of magnetic moments which couple together to form the magnetization density $\boldsymbol{M}^{(0)}$ that the magnetization cannot follow these fields. We only need be concerned about the relation between $\boldsymbol{B}$ and $\boldsymbol{H}$ contained in (9.9) when we enter the microwave regime. We will have occasion to consider the microwave response of the ferromagnet later in the present chapter, but for now we are content with (9.10).

Very similar remarks apply to nonferromagnets, such as the antiferromagnets mentioned earlier. Here, exchange couplings between spins on the two sublattices "stiffen" the magnetic response of the system, to move the intrinsic resonance frequencies up into the near infra-red, in many cases. Above this range of frequencies, most certainly in the near infrared and visible range, (9.10) holds as well. We refer the reader to a discussion of the magnetic response of antiferromagnets given elsewhere [9.2].

We turn our attention next to the relation between $\boldsymbol{D}$ and $\boldsymbol{E}$. With the spatial dispersion effects discussed in Chap. 2 set aside, this has the form given by

$$D_\alpha = \sum_\beta \varepsilon_{\alpha\beta}(\omega; \boldsymbol{M}^{(0)}) E_\beta \,, \tag{9.11}$$

where we note that, in addition to its dependence on frequency, the dielectric tensor is influenced by the presence of the spontaneous magnetization $\boldsymbol{M}^{(0)}$.

A microscopic description of the influence of $\boldsymbol{M}^{(0)}$ on the dielectric tensor is a rather complex and technical undertaking. However, general principles allow us to draw conclusions about the structure of the dielectric tensor in this instance. We shall see that when $\boldsymbol{M}^{(0)} \neq 0$, even in a material whose underlying structure is cubic, the dielectric tensor has off-diagonal elements, and renders the medium optically anisotropic, in a sense discussed below. For the purpose of our discussion, we assume the medium is perfectly transparent.

A general constraint on the structure of the dielectric tensor of a transparent anisotropic medium follows by requiring the time averaged rate of energy dissipation to vanish. It is a straightforward matter to extend the expression in (2.22) to an anisotropic crystal. With a slight change of notation, to conform with (9.11), we have

$$\frac{\partial U_E}{\partial t} = -\frac{i\omega}{4\pi} \sum_{\alpha,\beta} \left[\varepsilon_{\alpha\beta}(\omega; \boldsymbol{M}^{(0)}) - \varepsilon^*_{\beta\alpha}(\omega; \boldsymbol{M}^{(0)})\right] E_\alpha E^*_\beta \ . \tag{9.12}$$

For dissipation to be absent for all electric field combinations, we must enforce the condition

$$\varepsilon_{\alpha\beta}(\omega; \boldsymbol{M}^{(0)}) = \varepsilon^*_{\beta\alpha}(\omega, \boldsymbol{M}^{(0)}) \ . \tag{9.13}$$

Thus, diagonal matrix elements of the dielectric tensor must be real, in a transparent medium. However, notice that off-diagonal elements may be complex.

There is a second relation, which follows from the use of time reversal symmetry to constrain the dielectric tensor. This is a special case of the well-known Onsager relations of nonequilibrium thermodynamics [9.3]. One has

$$\varepsilon_{\alpha\beta}(\omega; +\boldsymbol{M}^{(0)}) = \varepsilon_{\beta\alpha}(\omega, -\boldsymbol{M}^{(0)}) \ . \tag{9.14}$$

We can sketch a proof of this statement from the general formula for the dielectric tensor derived in Appendix A as follows: See, for instance, (A.27), and for our present purpose, we may let $\boldsymbol{k} \to 0$, noting then $J^{(P)}_\beta(0) = J^{(P)}_\beta(0)^+$. The ground-state with magnetization $-\boldsymbol{M}^{(0)}$ is obtained from that with $+\boldsymbol{M}^{(0)}$ by a time reversal operation. Thus, to calculate $\varepsilon_{\beta\alpha}(\omega; -\boldsymbol{M}^{(0)})$, we use (A.27) with α and β exchanged, and with the ground state wave function $|\psi_0(0)\rangle$ replaced by $T\ |\psi_0(0)\rangle$, with T the time reversal operator. One has [9.4] $T = \left\{\Pi_i \sigma_y^{(i)}\right\} K$, where K is the complex conjugation operator, and $\sigma_y^{(i)}$ the y component of the Pauli spin operator $\boldsymbol{\sigma}$ for the ith electron. For the intermediate states, we may use the complete set formed by applying the time reversal operator to all states $\psi_n(0)$, to generate the array $T\psi_n(0)$. The matrix elements in (A.27) may be written out, and one then notes the Pauli spin matrices commute with the current density $J^{(P)}_\beta(0)$, and also $J^{(P)}_\alpha(0)$. Thus, when we calculate $\varepsilon_{\beta\alpha}(\omega, -\boldsymbol{M}^{(0)})$, in place of $\langle\psi_0 \mid J^{(P)}_\beta(0) \mid \psi_n(0)\rangle$ we shall have $\langle K\psi_0(0) \mid J^P_\beta(0) \mid K\psi_n(0)\rangle$, and similarly for the remaining matrix elements in (A.27). But

$$\langle K\psi_0(0) \mid J^{(P)}_\beta \mid K\psi_n(0)\rangle = \langle\psi_n(0) \mid J^{(P)}_\beta(0) \mid \psi_0(0)\rangle \ . \tag{9.15}$$

The matrix element on the right-hand side of (9.15) is precisely that which enters the calculation of $\varepsilon_{\alpha\beta}(\omega, +\boldsymbol{M}^{(0)})$. The energy eigenvalues which enter the denominator of (A.27) are not affected by the time reversal operation. The statement in (9.14) follows from these considerations.

The influence of the magnetization on the dielectric tensor is a modest effect. We may proceed by expanding $\varepsilon_{\alpha\beta}(\omega; \boldsymbol{M}^{(0)})$ in powers of the cartesian components of the magnetization, retaining only the lowest order terms. Consider first diagonal terms, $\varepsilon_{\alpha\alpha}(\omega, \boldsymbol{M}^{(0)})$. The statement in (9.13) require these be real, for the transparent medium. Furthermore, (9.14) requires these be an even function of $\boldsymbol{M}^{(0)}$. Hence, the lowest-order terms in the expansion are quadratic in the components of the $\boldsymbol{M}^{(0)}$. Thus, if $\varepsilon^{(0)}_{\alpha\beta}$ is the dielectric tensor in the absence of spontaneous magnetization, we have

$$\varepsilon_{\alpha\alpha}(\omega, \boldsymbol{M}^{(0)}) = \varepsilon^{(0)}_{\alpha\alpha}(\omega) + \sum_{\nu r} K^{(2)}_{\alpha\alpha;\mu\nu}(\omega) M^{(0)}_\mu M^{(0)}_\nu + \dots \ . \tag{9.16}$$

Now consider off-diagonal matrix elements, $\varepsilon_{\alpha\beta}(\omega, \boldsymbol{M}^{(0)})$, where $\beta \neq \alpha$. If we separate $\varepsilon_{\alpha\beta}(\omega, \boldsymbol{M}^{(0)})$ into real and imaginary parts,

$$\varepsilon_{\alpha\beta}(\omega, \boldsymbol{M}^{(0)}) = \varepsilon_{\alpha\beta}^{(1)}(\omega, \boldsymbol{M}^{(0)}) + \mathrm{i}\varepsilon_{\alpha\beta}^{(2)}(\omega, \boldsymbol{M}^{(0)}) \,, \tag{9.17}$$

then by combining (9.13,14), we find

$$\varepsilon_{\alpha\beta}^{(1)}(\omega, \boldsymbol{M}^{(0)}) = +\varepsilon_{\alpha\beta}^{(1)}(\omega, -\boldsymbol{M}^{(0)}) \tag{9.18a}$$

and

$$\varepsilon_{\alpha\beta}^{(2)}(\omega, \boldsymbol{M}^{(0)}) = -\varepsilon_{\alpha\beta}^{(2)}(\omega, -\boldsymbol{M}^{(0)}) \,. \tag{9.18b}$$

When these statements are combined with (9.14), we see that for $\alpha \neq \beta$, $\varepsilon_{\alpha\beta}^{(2)}$ is an odd function of $\boldsymbol{M}^{(0)}$. The first term in its expansion is thus linear in $\boldsymbol{M}^{(0)}$. The real part, $\varepsilon_{\alpha\beta}^{(1)}$ is an even function of $\boldsymbol{M}^{(0)}$.

When these statements are combined, one then has an expansion of the form (for the transparent medium)

$$\varepsilon_{\alpha\beta}(\omega; \boldsymbol{M}^{(0)}) = \varepsilon_{\alpha\beta}^{(0)} + \mathrm{i}\sum_{\nu} K_{\alpha\beta\nu}^{(1)}(\omega)M_{\nu}^{(0)} + \sum_{\nu r} K_{\alpha\beta;\mu\nu}^{(2)}(\omega)M_{\mu}^{(0)}M_{\nu}^{(0)} + \dots \,, \tag{9.19}$$

where $K_{\alpha\beta\nu}^{(1)}(\omega)$ and $K_{\alpha\beta;\mu\nu}^{(2)}(\omega)$ are both real. Note $K_{\alpha\beta\mu}^{(1)}(\omega)$ is antisymmetric under interchange of α and β from (9.14), while $K_{\alpha\beta;\mu\nu}^{(2)}(\omega)$ is symmetric. Furthermore, one may see that for the transparent medium, $K_{\alpha\beta\nu}^{(1)}(\omega)$ is an odd function of frequency, and $K_{\alpha\beta;\mu\nu}^{(2)}(\omega)$ is an even function of frequency:

$$K_{\alpha\beta\nu}^{(1)}(\omega) = -K_{\alpha\beta\nu}^{(1)}(-\omega) \tag{9.20a}$$

and

$$K_{\alpha\beta;\mu\nu}^{(2)}(\omega) = -K_{\alpha\beta;\mu\nu}^{(2)}(-\omega) \,. \tag{9.20b}$$

One may verify the statements in (9.20) by using (A.27) to form expressions for the combinations $\varepsilon_{\alpha\beta}(\omega; \boldsymbol{M}^{(0)}) - \varepsilon_{\beta\alpha}(\omega; \boldsymbol{M}^{(0)})$ and $\varepsilon_{\alpha\beta}(\omega; \boldsymbol{M}^{(0)}) + \varepsilon_{\beta\alpha}(\omega; \boldsymbol{M}^{(0)})$, then noting for the case where the frequency ω lies in a region of transparency, one may discard contributions proportional to $\delta(\omega - \omega_{n0})$. By this means, one may see that off diagonal terms of the dielectric tensor are odd functions of frequency, and diagonal terms are even functions of frequency.

Our discussion of the propagation of electromagnetic waves of infrared or optical frequency will be based on the use of (9.19). We specialize to the case of a cubic material, with magnetization $\boldsymbol{M}^{(0)}$ aligned along the z direction, $\boldsymbol{M}^{(0)} = \hat{z}M^{(0)}$.

For this case, one may see that the only nonzero elements of $K_{\alpha\beta\nu}^{(1)}(\omega)$ are $K_{xyz}^{(1)}(\omega) = -K_{yxz}^{(1)}(\omega) \equiv K^{(1)}(\omega)$. One may see that $K_{yzz}^{(1)}(\omega)$ vanishes as follows. If this element were nonzero, application of an electric field parallel to the z axis would lead to an electric dipole moment/unit volume parallel to the y direction, $P_y = K_{yzz}^{(1)}M^{(0)}E_z$. Since $\boldsymbol{P}$ is a polar vector, reflection on the yz plane must leave P_y unchanged. But such a reflection changes the sign of $\boldsymbol{M}^{(0)}$ when it is aligned along $\hat{z}$, since this is an axial vector. Hence, we must have $K_{yzz}^{(1)} = 0$, and a similar argument applies to $K_{xzz}^{(1)}$.

The symmetry properties of $K^{(2)}_{\alpha\beta;\mu\nu}$ are the same as those of the elastic constants, so to use a notation borrowed from elasticity theory for a cubic material, the nonzero elements of this tensor are $K^{(2)}_{11} = K_{xx,xx}$, $K^{(2)}_{12} = K^{(2)}_{xx,yy}$, and $K^{(2)}_{44} = K^{(2)}_{xz,xz}$, and those equivalent by symmetry.

Hence, for a transparent cubic crystal with $\boldsymbol{M}^{(0)}$ along $\hat{z}$, the dielectric tensor may be written

$$\varepsilon_{\alpha\beta}(\omega; \boldsymbol{M}^{(0)}) = \varepsilon^{(0)}(\omega)\delta_{\alpha\beta} + \Delta\varepsilon_{\alpha\beta}(\omega; \boldsymbol{M}^{(0)}) , \tag{9.21}$$

where $\Delta\varepsilon_{\alpha\beta}(\omega; \boldsymbol{M}^{(0)})$ is the change in dielectric tensor induced by the presence of the spontaneous magnetization. We have

$$\Delta\varepsilon(\omega; \boldsymbol{M}^{(0)}) = \begin{pmatrix} K^{(2)}_{12}(\omega)M^{(0)2} & \mathrm{i}K^{(1)}(\omega)M^{(0)} & 0 \\ -\mathrm{i}K^{(1)}(\omega)M^{(0)} & K^{(2)}_{12}(\omega)M^{(0)2} & 0 \\ 0 & 0 & K^{(2)}_{44}(\omega)M^{(0)2} \end{pmatrix} . \tag{9.22}$$

Here $K^{(1)}(\omega)$ is real, and an odd function of frequency, while $K^{(2)}_{ij}(\omega)$ is real and an even function of frequency.

We have proceeded with the construction of (9.22) with some care, because the arguments we have used can be adapted to a variety of situations including the application of a magnetic field to a nonmagnetic medium. At first glance, one would suppose considerations of time reversal invariance would rule out any contribution linear in $\boldsymbol{M}^{(0)}$. We see this is not the case. The off diagonal terms linear in $\boldsymbol{M}^{(0)}$ will have most important consequences, as we will see in the next section.

We now turn our attention to aspects of electromagnetic wave propagation, in a medium described by the dielectric tensor in (9.22).

9.2 Electromagnetic Wave Propagation in Ferromagnetic Materials; Faraday Rotation and the Cotton-Mouton Effect

We now turn our attention to the nature of the electromagnetic waves which propagate in the material described by the dielectric tensor in (9.21). To make the notation simpler, we introduce the definitions

$$\varepsilon_{\perp}(M^{(0)}) = \varepsilon^{(0)}(\omega) + K^{(2)}_{12}(\omega)M^{(0)2} , \tag{9.23a}$$

$$\varepsilon_{\parallel}(M^{(0)}) = \varepsilon^{(0)}(\omega) + K^{(2)}_{44}(\omega)M^{(0)2} , \tag{9.23b}$$

and

$$\Delta(M^{(0)}) = K^{(1)}(\omega)M^{(0)} . \tag{9.23c}$$

In the interest of brevity, we suppress reference to the frequency variation of the quantities which enter (9.23) in what follows: The dielectric tensor is thus written

$$\varepsilon(M^{(0)}) = \begin{pmatrix} \varepsilon_{\perp}(M^{(0)}) & +\mathrm{i}\Delta(M^{(0)}) & 0 \\ -\mathrm{i}\Delta(M^{(0)}) & \varepsilon_{\perp}(M^{(0)}) & 0 \\ 0 & 0 & \varepsilon_{\parallel}(M^{(0)}) \end{pmatrix} . \tag{9.24}$$

Our task is to analyze the plane wave solutions to Maxwell's equations as stated in (9.2) with $\boldsymbol{H} = \boldsymbol{B}$, and for a medium with the dielectric tensor just quoted. Without loss of generality, we may let the wave vector lie in the xy plane, as we did in our earlier discussion of wave propagation in anisotropic dielectrics presented in Sect. 2.3. As in Fig. 2.7, we let $k_z = k\cos\theta$ and $k_x = k\sin\theta$. One may obtain a set of three homogeneous equations for the three electric field components E_x, E_y and E_z, by eliminating $\boldsymbol{B}$ from Maxwell's equations. These may be arranged to read

$$\left(\cos^2\theta - \frac{\omega^2\varepsilon_\perp(M^{(0)})}{c^2k^2}\right)E_x - \mathrm{i}\frac{\omega^2\Delta(M^{(0)})}{c^2k^2}E_y - \sin\theta\cos\theta E_z = 0\,, \tag{9.25a}$$

$$+\mathrm{i}\frac{\omega^2\Delta(M^{(0)})}{c^2k^2}E_x + \left(1 - \frac{\omega^2\varepsilon_\perp(M^{(0)})}{c^2k^2}\right)E_y = 0\,, \tag{9.25b}$$

and

$$-\sin\theta\cos\theta E_x + \left(\sin^2\theta - \frac{\omega^2\varepsilon_\parallel(M^{(0)})}{c^2k^2}\right)E_z = 0\,. \tag{9.25c}$$

If we arbitrarily set $\Delta(M^{(0)}) = 0$, then the dielectric tensor of the ferromagnet becomes identical in form to that of the anisotropic dielectric discussed in Sect. 2.3. In (9.25b), we have a description of the ordinary wave, while (9.25a,c) duplicate (2.64a,b), which describe the extraordinary wave.

Of course, in the ferromagnet, $\Delta(M^{(0)}) \neq 0$. For each electromagnetic mode which emerges from the solution of (9.25), all three field components are nonzero, for a general angle of propagation θ. We see from (9.25c) that E_x and E_z oscillate in phase, while from (9.25b) we see that E_y is 90° out of phase with both E_x and E_z. In these waves, then, the tip of the electric field vector traces out an ellipse. The waves are thus elliptically polarized. However, the plane of the ellipse is in general tipped out of the plane perpendicular to the wave vector. One may demonstrate, again for general values of θ, that the plane of the ellipse is not normal to the wave vector $\boldsymbol{k}$. As we see from (9.2a), the requirement $\boldsymbol{k}\cdot\boldsymbol{B} = 0$ means the $\boldsymbol{B}$ field always lies in the plane perpendicular to $\boldsymbol{k}$. Finally, the requirement $\boldsymbol{k}\times\boldsymbol{E} = (\omega/c)\boldsymbol{B}$ in (9.2b) makes $\boldsymbol{B}$ perpendicular to $\boldsymbol{E}$ always.

The dispersion relation of the waves is found by setting the appropriate 3×3 determinant to zero. We must solve

$$\left(\frac{c^2k^2}{\varepsilon_\perp(M^{(0)})} - \omega^2\right)\left(\frac{c^2k^2}{\varepsilon_\parallel(M^{(0)})}\sin^2\theta + \frac{c^2k^2}{\varepsilon_\perp(M^{(0)})}\cos^2\theta - \omega^2\right) + \omega^2\frac{\Delta(M^{(0)})^2}{\varepsilon_\perp(M^{(0)})^2}\left(\frac{c^2k^2}{\varepsilon_\perp(M^{(0)})}\sin^2\theta - \omega^2\right) = 0\,. \tag{9.26}$$

Since this is a quadratic equation for the quantity ω^2, for each choice of wave vector $\boldsymbol{k}$, we find two propagating modes, as expected. The solutions for general θ are readily written down, but do not prove enlightening. We thus turn to two special cases, $\theta = 0$ and $\theta = \frac{\pi}{2}$.

9.2.1 Propagation Parallel to the Magnetization; Faraday Rotation and the Kerr Effect

Consider the case where the wave vector $\boldsymbol{k}$ is directed parallel to the magnetization, so $\theta = 0$. From (9.26), we find two modes, with wave vectors $k_\pm$ given by

$$\frac{c^2k_\pm^2}{\omega^2} = \varepsilon_\perp(M^{(0)}) \pm \Delta(M^{(0)}) = \varepsilon_\perp(M^{(0)}) \pm K^{(1)}(\omega)M^{(0)} \tag{9.27a}$$

which we write as

$$\frac{c^2k_\pm^2}{\omega^2} = \varepsilon_\pm(M^{(0)}) \ . \tag{9.27b}$$

For both waves, (9.25c) requires $E_z = 0$, so for propagation parallel to the magnetization, the electric field is transverse. The two waves are each circularly polarized, with $|E_x| = |E_y|$. For one, as one looks up the direction of magnetization, the electric field vector rotates in the clockwise sense, and for the other, it rotates in the counterclockwise sense. The first is referred to often as a right-handed circularly polarized wave, and the second as a left-handed circularly polarized wave. When $k = k_+$, we have $E_y = -\mathrm{i}E_x$, while $E_y = +\mathrm{i}E_x$ when $k = k_-$.

Suppose we consider the propagation of a plane polarized electromagnetic wave in the medium, with the propagation direction parallel to the magnetization. Suppose in the xy plane, where $z = 0$, the electric field in this wave is parallel to the x axis. Thus, if we denote the electric field in the medium by $\boldsymbol{E}(z, t)$, then in the xy plane

$$\boldsymbol{E}(0, t) = \hat{x}E_0\cos(\omega t) = \mathrm{Re}\{\hat{x}E_0\mathrm{e}^{-\mathrm{i}\omega t}\} \ . \tag{9.28}$$

Such a plane polarized wave may be viewed as a linear superposition of a right- and left-hand circularly polarized disturbance:

$$\boldsymbol{E}(0, t) = \frac{E_0}{2}[\hat{x}\cos(\omega t) + \hat{y}\sin(\omega t)] + \frac{E_0}{2}[\hat{x}\cos(\omega t) - \hat{y}\sin(\omega t)] \tag{9.29a}$$

or

$$\boldsymbol{E}(0, t) = \frac{1}{2}\mathrm{Re}\{E_0(\hat{x} + \mathrm{i}\hat{y})\mathrm{e}^{-\mathrm{i}\omega t} + E_0(\hat{x} - \mathrm{i}\hat{y})\mathrm{e}^{-\mathrm{i}\omega t}\} \ . \tag{9.29b}$$

As the disturbance propagates along the z axis, the piece proportional to $(\hat{x}+\mathrm{i}\hat{y})$ does so with wave vector k_-, while the piece proportional to $(\hat{x} - \mathrm{i}\hat{y})$ does so with wave vector k_+. Hence

$$\boldsymbol{E}(z, t) = \frac{1}{2}\mathrm{Re}\{E_0(\hat{x} + \mathrm{i}\hat{y})\mathrm{e}^{\mathrm{i}(k_-z-\omega t)} + E_0(\hat{x} - \mathrm{i}\hat{y})\mathrm{e}^{\mathrm{i}(k_+z-\omega t)}\} \ ,$$

or

$$\begin{aligned}\boldsymbol{E}(z, t) = {} & \frac{E_0}{2}\hat{x}\,[\cos(k_+z - \omega t) + \cos(k_-z - \omega t)] \\ & + \frac{E_0}{2}\hat{y}\,[\sin(k_+z - \omega t) - \sin(k_-z - \omega t)] \ .\end{aligned} \tag{9.30}$$

Suppose we define an average wave vector

$$\bar{k} = \frac{1}{2}(k_+ + k_-) \tag{9.31a}$$

for the two waves, and also

$$\Delta k = k_+ - k_- \, . \tag{9.31b}$$

Then after some simple algebra, we find

$$E(z,t) = E_0 \cos(\bar{k}z - \omega t)\left[\hat{x}\cos\left(\frac{1}{2}\Delta kz\right) + \hat{y}\sin\left(\frac{1}{2}\Delta kz\right)\right] . \tag{9.32}$$

The electric field in (9.32) is plane polarized, within any plane z = constant. As one progresses up the z axis, however, the tip of the electric field vector executes a helical motion; the plane of polarization rotates continuously. In the plane $z = \pi/\Delta k$, the electric field vector is directed along $\hat{y}$, at $\frac{2\pi}{\Delta k}$ along $-\hat{x}$, at $\frac{3\pi}{\Delta k}$ along $-\hat{y}$, and then it is once again parallel to $+\hat{x}$ when $z = \frac{4\pi}{\Delta k}$. The rotation rate, in radians/cm is just $(\frac{\Delta k}{2})$. Since, in general, $K^{(1)}M^{(0)} \ll \varepsilon_\perp(M^{(0)})$, the rotation rate is

$$\frac{\Delta k}{2} = \frac{\omega}{c}\,\frac{K^{(1)}(\omega)M^{(0)}}{\varepsilon_\perp(M^{(0)})^{\frac{1}{2}}} \, . \tag{9.33}$$

The phenomenon just described is referred to as Faraday rotation. Notice that if the magnetization is reversed, the sense of rotation is as well.

Quite generally in ferromagnetic media, $K^{(1)}M^{(0)} \ll \varepsilon_\perp(M^{(0)})$. Consider then the description of the transmission of plane polarized light, which strikes a ferromagnet slab at normal incidence. Let the magnetization be normal to the surface. At the surface, as we have just done, we can decompose the incident plane wave into right- and left-circularly polarized waves, consider the transmission of each, then use the superposition principle to describe the transmitted wave. Since $\varepsilon_+(M^{(0)})$ and $\varepsilon_-(M^{(0)})$ differ little when $K^{(1)}M^{(0)}$ is small, the reflection and transmission amplitudes of the two circularly polarized waves at each surface differ very little. The effect of the magnetization is then primarily just to rotate the plane of polarization of the transmitted wave as it propagates through the slab. The transmitted wave will thus be plane polarized, but the plane of polarization will be rotated through the angle

$$\Delta\theta = \frac{\omega}{c}\,\frac{K^{(1)}(\omega)M^{(0)}D}{\varepsilon_\perp(M^{(0)})^{\frac{1}{2}}} \, , \tag{9.34}$$

where D is the thickness of the slab.

To see that the influence of the spontaneous magnetization on the dielectric constant is indeed a very modest effect, we illustrate with some numbers appropriate to a well-known ferromagnetic insulator used in many device applications. The material is yttrium iron garnet, referred to commonly in the literature by its acronym YIG. This material is cubic, ferromagnetic at and above room temperature, and transparent to electromagnetic radiation at near infrared frequencies. This material has been employed very widely in magnetic devices in the microwave frequency range, and also as we shall see later, in magneto-optic devices which require a long propagation length.

For YIG, $M^{(0)} \cong 150$ Gauss, and $K^{(1)} \cong 2.5 \times 10^{-6}$ (Gauss)$^{-1}$ in the near infrared. The product $K^{(1)}M^{(0)}$ is thus 4×10^{-4}, so the off diagonal terms in the dielectric tensor are of this magnitude. At a wavelength of one micron, using $\varepsilon_{\perp}(M^{(0)}) \cong 6$, one estimates the Faraday rotation rate is roughly 600°/cm of travel in the medium. Thus, despite the smallness of $K^{(1)}$, the Faraday rotations realized over macroscopic propagation lengths are very substantial indeed.

The presence of spontaneous magnetization, and its influence on the dielectric tensor through the term proportional to $K^{(1)}(\omega)$, influences the reflectivity of a ferromagnetic material most importantly. Suppose once again that the material is illuminated by plane polarized radiation at normal incidence. Let the z axis be normal to the surface, and suppose $\boldsymbol{E}$ is parallel to the x direction. The electric field inside the medium is also parallel to x, and this will induce an electric dipole moment $\boldsymbol{P}$ with a component parallel to y, proportional to $\mathrm{i}K^{(1)}(\omega)M^{(0)}E^{(0)}$, with $E^{(0)}$ the strength of the incident field, inside the material. A consequence is that the reflected light will contain a y component of electric field, 90° out of phase with the x component. The reflected radiation is thus elliptically polarized. The conversion of plane polarized incident radiation to elliptically polarized reflected radiation is referred to as the Kerr effect, or more completely the magneto-optic Kerr effect.

The reader may easily see that the phenomenon is found quite generally, except for selected geometries. For instance, if the magnetization is parallel to the surface, and we have s polarized radiation at non-normal incidence, the reflected light will be s polarized as well if the plane of incidence is aligned so the electric vector is parallel to the magnetization.

Remarkably, the Kerr effect can be employed to study monolayer quantities of magnetic material despite the smallness of $K^{(1)}(\omega)$. If one illuminates a monolayer of Fe with plane polarized radiation, the Kerr component of the electric field in the reflected light will be smaller than the incident electric field by roughly 10^{-3}. One can readily detect this component of the reflected field through use of a cross polarizer in the reflected beam which suppresses the much stronger signal from the dominant component of the electric field. This technique is referred to by the acronym SMOKE (Surface Magneto Optic Kerr Effect), and is now widely used to study the magnetism in ultrathin (few atomic layer) ferromagnetic films and structures. As an example of the excellent data one may obtain with this technique, in Fig. 9.1, we reproduce the hysteresis loop of a monolayer quantity of Fe on a Pd surface. The data is reproduced from [9.5]. For a more extensive discussion, and references to the literature, we refer the reader to the review paper by *Bader* [9.6].

9.2.2 Propagation Perpendicular to Magnetization; the Cotton–Mouton Effect

We once again have two electromagnetic modes, of course. We denote the frequency of these two modes by k_1, and k_2 respectively. We then have, from (9.26) with $\theta = \frac{\pi}{2}$,

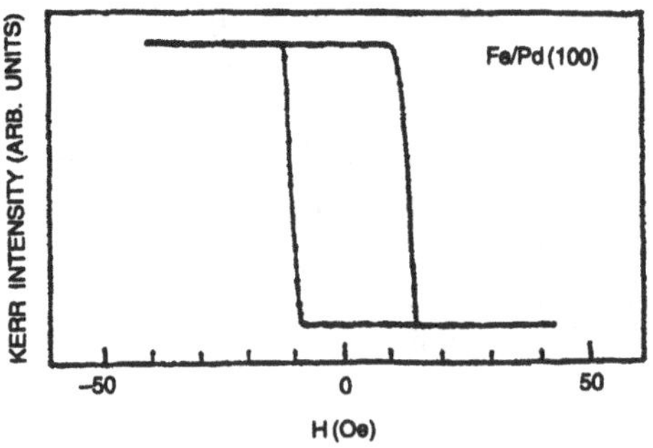

Fig. 9.1. A hysteresis loop, for one atomic layer of Fe deposited on a Pd surface. The loop has been measured by means of the surface magnetooptic Kerr effect (SMOKE) technique. The data has been reproduced from [9.5].

$$\frac{c^2k_1^2}{\omega^2} = \varepsilon_{\parallel}(M^{(0)}) = \varepsilon^{(0)}(\omega) + K_{44}^{(2)}(\omega)M^{(0)2} \tag{9.35a}$$

and

$$\frac{c^2k_2^2}{\omega^2} = \varepsilon_{\perp}(M^{(0)}) - \frac{\Delta^2(M^{(0)})}{\varepsilon_{\perp}(M^{(0)})}\,, \tag{9.35b}$$

or

$$\frac{c^2k_2^2}{\omega^2} = \varepsilon^{(0)}(\omega) + \left[K_{12}^{(2)}(\omega) - \frac{(K^{(1)}(\omega))^2}{\varepsilon^{(0)}(\omega)}\right]M^{(0)2}\,, \tag{9.35c}$$

where in (9.35c), we have replaced $\varepsilon_{\perp}(M^{(0)})$ in the denominator by $\varepsilon^{(0)}(\omega)$, since the corrections to the dielectric tensor associated with the presence of the magnetization are quite small.

In the presence of spontaneous magnetization, even though the crystal may remain cubic in structure, for propagation perpendicular to the magnetization, the material behaves as if it were a unaxial crystal, with an effective dielectric tensor similar to that studied in Sect. 2.3. We have two propagating modes, whose effective indices of refraction differ by terms proportional to the square of the magnetization $M^{(0)}$. This is referred to as the Cotton–Mouton effect. For the example considered earlier, the crystal YIG, the constants $K_{12}^{(2)}(\omega)$ and $K_{44}^{(2)}(\omega)$ fall in the range of 10^{-8} (Gauss)$^{-2}$. The difference in index of refraction of the two modes is thus quite small, in the range 10^{-4}.

For both modes, as we have seen quite generally, the magnetic field $\boldsymbol{B}$ is perpendicular to the wave vector. For the mode with wave vector k_1, the electric field $\boldsymbol{E}$ is parallel to the magnetization, with $\boldsymbol{B}$ parallel to the xy plane. For the mode with wave vector k_2, if we suppose the wave vector is parallel to the y axis, $\boldsymbol{B}$ is parallel to $\boldsymbol{M}^{(0)}$, while the electric field lies in the xy plane. Since, as we have seen, the magnitude of $K^{(1)}(\omega)M^{(0)}$ is quite small, the electric field is nearly parallel to the x axis, perpendicular to the wave vector, with a small component parallel to the y axis, 90° out of phase with the x component. This mode is thus elliptically polarized.

9.2.3 Final Remarks

We have considered the propagation of electromagnetic waves in a ferromagnetic medium, within which one has a spontaneous magnetization $\boldsymbol{M}^{(0)}$. If we have a nonmagnetic crystal, and subject it to a spatially uniform, dc magnetic field $\boldsymbol{H}^{(0)}$, then the influence of the magnetic field on the dielectric tensor may be found by simply substituting $\boldsymbol{H}^{(0)}$ for $\boldsymbol{M}^{(0)}$ everywhere in the discussion above. This follows from symmetry considerations only, after noting the behavior of $\boldsymbol{H}^{(0)}$ under rotations, inversions of the coordinate system or reflection of it through a plane, and also time reversal symmetry, are identical to $\boldsymbol{M}^{(0)}$. Thus, for example, plane polarized radiation which propagates parallel to $\boldsymbol{H}^{(0)}$ will experience Faraday rotation, very much as it does in ferromagnet.

9.3 Second-Harmonic Generation from Magnetic Materials; Surface Effects

We saw in Chap. 8 that in crystals which posseses an inversion center, so the second order susceptibility $\chi^{(2)}_{\alpha\beta\gamma}$ vanishes, atoms located within the crystal surface or molecules adsorbed on the surface possess nonzero second-order electric dipole moments, since their local environment lacks inversion symmetry. These entities are thus a source of second-harmonic radiation, and this radiation serves as a microscopic probe of the near-surface environment. The presence of spontaneous magnetization in the surface adds interesting and unique new aspects to this phenomenon. One sees second-harmonic radiation with origin in near-surface magnetic moments, with polarizations forbidden from a symmetry analysis based on atomic positions alone.

We first note once again that in a crystal with an inversion center, where $\chi^{(2)}_{\alpha\beta\gamma}$ vanishes, the presence of spontaneous magnetization $\boldsymbol{M}^{(0)}$ does not induce nonzero elements in this tensor. This is surprising at first glance, since in the presence of $\boldsymbol{M}^{(0)}$, the symmetry of the crystal is clearly lower than in its absence. There is now a preferred direction, for instance. However, the magnetization is an axial vector, which does not change sign upon carrying out an inversion operation. Thus, a crystal which has inversion symmetry above its Curie temperature where $\boldsymbol{M}^{(0)}$ vanishes, will continue to have inversion symmetry below, where $\boldsymbol{M}^{(0)} \neq 0$.

In the surface, of course $\chi^{(2)}_{\alpha\beta\gamma}$ is necessarily nonzero in any material, as we have seen in Chap. 8. If the surface atoms are magnetic, clearly this tensor will be influenced by the presence of the magnetization. For atoms in various surfaces of a cubic ferromagnetic crystal, *Pan* et al. [9.7] have presented an analysis of the influence of the surface magnetization on the structure of this tensor. Through symmetry arguments which may be viewed as an extension of those we have given earlier in the present chapter, *Pan* et al. found elements in $\chi^{(2)}_{\alpha\beta\gamma}(\boldsymbol{M}^{(0)})$ which are odd functions in $\boldsymbol{M}^{(0)}$, and which thus vanish in the absence of spontaneous magnetization. These give rise to second-harmonic radiation in polarization combinations forbidden when $\boldsymbol{M}^{(0)} = 0$.

Thus, one may use this phenomenon to probe the magnetism within microscopic distances from the surface. Spin polarized electron spectroscopies in various forms also serve as a probe of near surface magnetism, if one employs electron energies in the range where their mean free path is very short. However, such methods can be used only when the sample is in an ultra-high vacuum environment. Optical probes, including second-harmonic generation, allow study of surfaces in complex environments, and "hidden interfaces" as well.

At the time of this writing, experimental studies of second-harmonic generation from magnetic surfaces or interfaces are in their early stages. *Reif* et al. [9.8] have reported the influence of surface magnetization on the intensity of second-harmonic radiation from a ferromagnet. More recently, an elegant study of polarization selection rules and related phenomena have enabled *Stolle* et al. [9.9] to isolate signals from elements of $\chi^{(2)}_{\alpha\beta\gamma}(\boldsymbol{M}^{(0)})$ which change sign with reversal of $\boldsymbol{M}^{(0)}$, and those which are even in the Cartesian components of the magnetization. We may look forward to further studies in the near future.

9.4 Dynamic Response of the Magnetization and the Origin of Nonlinear Magneto-optic Interactions

9.4.1 General Remarks

The discussion above focused its attention on the influence of the magnetization $\boldsymbol{M}^{(0)}$, assumed spatially uniform, and independent of time, on the propagation characteristic of electromagnetic waves in ferromagnetic media. In Sect. 9.1, it was noted that in fact the magnetization may vary with position and time, as described by (9.1). There are two reasons why this may happen. In any material at finite temperature, thermal fluctuations are necessarily present. Just as there are thermal fluctuations in the density of any material, we will have thermal fluctuations in the magnitude and direction of the magnetization. It is also the case that one may drive the system through application of a time varying magnetic field, and thus induce a disturbance in the magnetization.

The discussion of electromagnetic wave propagation presented in Sect. 9.2 focused its attention on fields which oscillate with frequency high compared to the natural resonance frequency ω_H of the spin system. As mentioned, this condition is well-satisfied at near-infrared or visible frequencies, in virtually all microwave materials. If, however, we drive a ferromagnetic material with an externally applied magnetic field whose frequency is near ω_H, as discussed in Sect. 9.1, a time dependent component of the magnetization $\boldsymbol{m}(\boldsymbol{r}t)$ will be induced, as in (9.1).

An electromagnetic wave in the visible or near infrared frequency range will interact with $\boldsymbol{m}(\boldsymbol{r}, t)$, whether it has its origin in thermal fluctuations or whether it is produced by an external driving field in the microwave frequency range. A description of the interaction of the electromagnetic wave with magnetization fluctuations follows quite directly from the discussion presented in Sect. 9.1. The characteristic frequencies of the magnetization fluctuations are in the range of ω_H, once again. These frequencies are very low indeed, compared to those characteristic of the electronic degrees of freedom in the material. In this circumstance, we may just replace $\boldsymbol{M}^{(0)}$ in (9.18) by $\boldsymbol{M}^{(0)} + \boldsymbol{m}(\boldsymbol{r}, t)$.

If $\boldsymbol{m}(\boldsymbol{r}, t)$ is small compared to $\boldsymbol{M}^{(0)}$, we may retain only terms linear in $\boldsymbol{m}(\boldsymbol{r}, t)$. One may then write

$$\varepsilon_{\alpha\beta}(\omega; \boldsymbol{M}(\boldsymbol{r}, t)) = \varepsilon_{\alpha\beta}(\omega; \boldsymbol{M}^{(0)}) + \delta\varepsilon_{ab}(\omega; \boldsymbol{r}, t) \ , \tag{9.36}$$

where $\varepsilon_{\alpha\beta}(\omega; \boldsymbol{M}^{(0)})$ is defined in (9.18), and

$$\delta\varepsilon_{\alpha\beta}(\boldsymbol{r}, t) = \mathrm{i}\sum_{\mu} K^{(1)}_{\alpha\beta\mu}(\omega) m_{\mu}(\boldsymbol{r}, t) + 2\sum_{\mu, r} K^{(2)}_{\alpha\beta;\mu\nu}(\omega) M^{(0)}_{\mu} m_{\nu}(\boldsymbol{r}, t) \ . \tag{9.37}$$

Thus, fluctuations in the magnetization induce time and space varying fluctuations in the dielectric tensor. We may now return to earlier discussions, such as that in Sect. 5.4, where it is argued that light waves are diffracted or scattered from such fluctuations in the dielectric tensor. If $\boldsymbol{m}(\boldsymbol{r}, t)$ is produced by thermal fluctuations, we have inelastic scattering of light identical in nature to the Brillouin scattering described in Sect. 5.4. We shall explore Brillouin scattering by magnetic fluctuations later in this chapter, since there are striking aspects of this phenomenon.

If the spin system is driven by an externally applied magnetic field of wave vector $\boldsymbol{Q}$ and frequency Ω, then $\boldsymbol{m}(\boldsymbol{r}, t) = \boldsymbol{m}\ (\boldsymbol{Q}, \Omega)\cos(\boldsymbol{Q}\cdot\boldsymbol{r} - \Omega t)$. The magnetic disturbance creates a diffraction grating in the material. An incident light wave of wave vector $\boldsymbol{k}$ and frequency ω is diffracted to states with wave vector $\boldsymbol{k} \pm \boldsymbol{Q}$, while the diffracted light has frequency $\omega \pm \Omega$. We may view this as a nonlinear interaction between the light wave, and the "excitation wave" in the spin system, with a nature every similar to the nonlinear processes explored in Chap. 4.

To proceed, we need to understand more completely the nature of such "excitation waves" in magnetic media. We turn to this issue in the next section.

9.4.2 Collective Excitations (Spin Waves) in Magnetic Materials; Ferromagnets as an Example

Consider once again a ferromagnet with magnetization $\boldsymbol{M}^{(0)}$ parallel to the z-axis everywhere; an external magnetic field $\boldsymbol{H}^{(0)} = H^{(0)}\hat{z}$ is applied parallel to the z direction. It should be remarked that in practice, in experimental studies of ferromagnets or in devices which employ them, such a magnetic field is always present. This is because in its absence, the material will break up into small domains [9.1], as remarked earlier. Within each domain, $\boldsymbol{M}^{(0)}$ is uniform, but its direction varies from domain to domain. Application of a suitable external magnetic field will render $\boldsymbol{M}^{(0)}$ uniform throughout the sample.

Now tip the spins within a small volume ΔV away from the z direction, to make $\boldsymbol{m}(\boldsymbol{r}, t)$ nonzero in ΔV. The spins will precess around $\boldsymbol{H}^{(0)}$ at the Larmor frequency, as described by (9.4) with $\boldsymbol{H}_T$ replaced by $\hat{z}H^{(0)}$.

Each precessing spin bears a magnetic moment, and these precessing moments generate a time dependent magnetic field $\boldsymbol{H}^{(d)}(\boldsymbol{r}, t)$, referred to commonly as the dipole field. This field exerts a torque on the spins elsewhere in the system, and sets them into precession about the combination $\hat{z}H^{(0)} + \boldsymbol{H}^{(d)}$. If we make a localized disturbance as described in the previous paragraph, it propagates through the system by the mechanism just described.

Such disturbances, in fact, are described by the Maxwell equations, supplemented by the relation between $\boldsymbol{m}$ and the dipole field provided by (9.6). In (9.6), we replace H_β by $H_\beta^{(d)}$, and the use of this in Maxwell's equations describes the propagation of the dipolar field through the system, along with the electric field it necessarily generates through Faraday's law. We seek solutions of Maxwell's equations where, as in Sect. 9.1, all quantities vary in space and time as $\exp(i\boldsymbol{k}\cdot\boldsymbol{r} - i\omega t)$. When this is done, we are discussing wave-like collective excitations of the spin system, which generate electric and magnetic fields. These are, of course, just electromagnetic waves whose frequency lies in a regime where the approximation in (9.10) breaks down badly, and the resonant character of $\mu_{\alpha\beta}(\omega)$ in (9.8) enters importantly. These waves are then magnetic analogues of the polaritons discussed in Sect. 2.1. Indeed, they are referred to frequently as magnetic polaritons.

The collective excitations of interest in ferromagnets will have frequencies in the range of 10 GHz, as we have discussed. Hence, $\omega \sim 10^{11}$ radians/s, typically. Here our interest will be in collective excitations of the spin system which interact with light. Thus, we consider wave vectors k in the range 10^5 cm^{-1}. Under these conditions, the full Maxwell equations in (9.2) may be simplified greatly. Notice that $\omega/c \sim 3$ cm^{-1} for the numbers above, while once again $k \sim 10^5$ cm^{-1}. The terms proportional to ω/c in Maxwell's equations may then be discarded, since ω/c is orders of magnitude smaller than k.

This decouples the magnetic and electric fields, so we have $\boldsymbol{E} = \boldsymbol{D} = 0$. The fluctuating magnetic moments thus generate the magnetic field $\boldsymbol{H}^{(d)}$ only in this limit, and we examine the set of equations

$$\boldsymbol{k}\cdot\boldsymbol{B} = 0\ , \tag{9.38a}$$

$$\boldsymbol{k}\times\boldsymbol{H}^{(d)} = 0\ , \tag{9.38b}$$

where, as we see from (9.8)

$$B_\alpha = \sum_\beta \mu_{\alpha\beta}(\omega) H_\beta^{(d)}\ , \tag{9.39}$$

and $\mu_{\alpha\beta}(\omega)$ is defined in (9.9).

The approximation contained in (9.38,39), used widely in magnetism, is referred to as the magnetostatic limit. The frequencies of the collective modes, called spin waves in this limit, are found by seeking values of ω for which (9.38,39) can be satisfied. While our attention here is directed toward ferromagnetic materials, the method is readily extended to more complex magnetic crystals, such as antiferromagnets [9.10].

With no loss of generality, we may allow the wave vector $\boldsymbol{k}$ to be in the xz plane, as illustrated in Fig. 9.2a. Then (9.38b) reduces to the two statements

$$H_y^{(d)} = 0 \tag{9.40a}$$

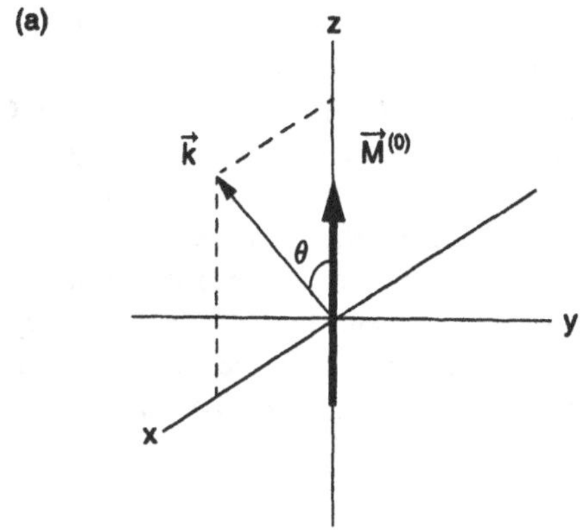

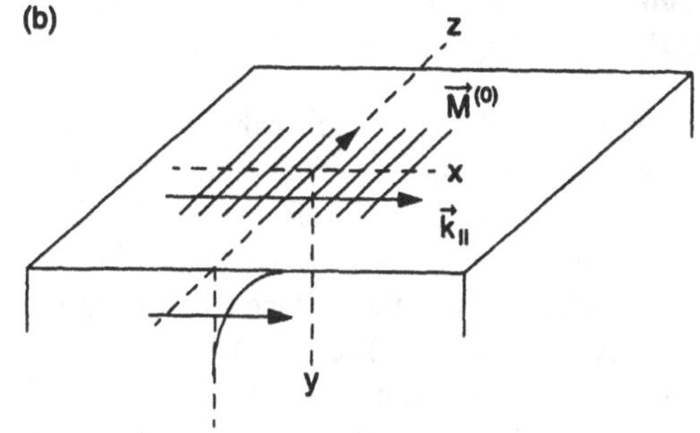

Fig. 9.2. **a** The geometry considered in the discussion of bulk spin-wave propagation. The magnetization $M^{(0)}$ is aligned along the z direction. **b** The geometry considered in the discussion of the Damon–Eshbach surface spin wave. The magnetization is parallel to the surface, and the propagation direction is perpendicular to the magnetization.

$$\sin\theta H_z^{(d)} = \cos\theta H_x^{(d)} , \tag{9.40b}$$

while (9.38) requires, noting (9.40a),

$$\sin\theta\mu_1(\omega)H_x^{(d)} + \cos\theta H_z^{(d)} = 0 . \tag{9.41}$$

The spin wave frequency is thus found from

$$\sin^2\theta\mu_1(\omega) + \cos^2\theta = 0 , \tag{9.42}$$

or with

$$\mu_1(\omega) = 1 + \frac{4\pi\omega_M\omega_H}{\omega_H^2 - \omega^2} , \tag{9.43}$$

we have for the frequency

$$\omega(\boldsymbol{k}) = \gamma[H^{(0)}(H^{(0)} + 4\pi M_s \sin^2\theta)]^{\frac{1}{2}} . \tag{9.44}$$

with γ the gyromagnetic ratio introduced earlier.

In the magnetostatic limit, the frequencies of the spin waves depend on the direction of the wave vector, but are independent of its magnitude. For $\theta = 0$ (propagation parallel to $M^{(0)}$), the frequency is just the Larmor frequency $\gamma H^{(0)}$. No dipolar field is generated for this direction of propagation. For $\theta = \pi/2$, propagation in the xy plane, the frequency rises to $[H^{(0)}(H^{(0)} + 4\pi M_s)]^{\frac{1}{2}}$.

The dipolar fields generated by the spin motion "stiffen" the response of the spin system, and raise the frequency of the spin waves. A measure of the magnitude of this effect is provided by $4\pi M_s$, as one sees from (9.44). As we shall see later in this chapter, microwave devices exploit and utilize spin waves in ferromagnetic materials. If one is interested in high-frequency devices, one means of achieving this goal is to see a material with a large value of $4\pi M_s$. Suppose we consider Fe as an example (Fe has been used little in devices to date, since it is a metal and introduces Ohmic dissipation). In a 2 kiloGauss field, the Larmor frequency $\gamma H^{(0)}$ is roughly 6 Ghz. In Fe, $4\pi M_s = 21$ kGauss. For a device based on spin waves with the frequency $\gamma[H^{(0)}(H^{(0)} + 4\pi M_s)]^{\frac{1}{2}}$, the "internal stiffening" raises the frequency to over 20 Ghz.

9.4.3 Surface Spin Waves on Ferromagnetic Surface; the Damon–Eshbach Mode and Non Reciprocal Propagation on Magnetic Surfaces

In Chap. 8, we saw that on the surface of a dielectric which exhibits a resonant response to an electromagnetic field, one realizes surface electromagnetic waves, or surface polaritons which are "bound" to the surface. The ferromagnet exhibits a resonant response to time varying magnetic fields, as we have seen. We have surface spin waves in this case, similar qualitatively to the surface electromagnetic waves (or surface polaritons) of the previous chapter. There is a new and most interesting feature of these modes unique to the magnetic material, as we shall see.

The geometry of interest is illustrated in Fig. 9.2b. We have a ferromagnetic medium, whose surface coincides with the xz plane. The magnetization $\boldsymbol{M}^{(0)}$ is parallel to the surface, which is the usual configuration for a magnetic film. (If, for example, the magnetization is normal to the surface, elementary magnetostatics informs one that there is an internal demagnetizing field $\boldsymbol{H}^{(I)} = -4\pi \boldsymbol{M}^{(0)}$. This is antiparallel to the magnetization, and the configuration is unstable unless an external magnetic field $\boldsymbol{H}^{(0)}$ greater in magnitude than $4\pi \boldsymbol{M}^{(0)}$ is applied normal to the surface.) We suppose a dc magnetic field of strength $\boldsymbol{H}^{(0)}$ is applied parallel to $\boldsymbol{M}^{(0)}$.

The fields generated by the spin system may be described by generalizing (9.38). In the magnetostatic approximation, we may write

$$\nabla \cdot \boldsymbol{B} = 0 \tag{9.45a}$$

and

$$\nabla \times \boldsymbol{H}^{(d)} = 0 \,. \tag{9.45b}$$

The statement in (9.45b) allows us to introduce a magnetic scalar potential by writing

$$\boldsymbol{H}^{(d)} = -\nabla \Phi^{(M)} \tag{9.46}$$

with $\Phi^{(M)}(x, y)$ a magnetic potential, which generates the demagnetizing field $\boldsymbol{H}^{(d)}$ through (9.46). We seek wavelike solutions, where

$$\Phi^{(M)}(x, y) = \phi^{(M)}(y)e^{ik_{\|}x} , \tag{9.47}$$

so that

$$H_x^{(d)} = -ik_{\|}\phi^{(M)}(y)e^{ik_{\|}x} \tag{9.48a}$$

and

$$H_y^{(d)} = -\frac{d\phi^{(M)}}{dy}e^{ik_{\|}x} . \tag{9.48b}$$

The reader should keep in mind the fact that all quantities display the time dependence $\exp(-i\omega t)$.

Both in the medium and in the vacuum above the ferromagnet, (9.45a) requires $\phi_M(y)$ to obey the simple Laplace equation

$$\frac{d^2\phi^{(M)}}{dy^2} + k_{\|}^2\phi^{(M)} = 0 , \tag{9.49}$$

so we have, if we seek disturbances localized on the surface,

$$\phi^{(M)}(y) = \begin{cases} \phi_+^{(M)} \exp(-|k_{\|}|y), & y \geq 0 \\ \phi_-^{(M)} \exp(+|k_{\|}|y), & y \leq 0 . \end{cases} \tag{9.50}$$

One requires boundary conditions to be satisfied at the surface of the ferromagnet, just as in the discussion of electromagnetic waves. One sees easily that conservation of tangential $\boldsymbol{H}^{(d)}$ is ensured by the choice $\phi_+^{(M)} = \phi_-^{(M)} = \phi^{(M)}$. For the normal component of $\boldsymbol{B}$ inside and outside the ferromagnet, we have

$$B_y = |k_{\|}|\phi^{(M)}e^{ik_{\|}x} \begin{cases} [\mu_1(\omega) + \sigma\mu_2(\omega)]e^{|k_{\|}|y}, & y \geq 0 \\ -e^{+|k_{\|}|y}, & y \leq 0 \end{cases} \tag{9.51}$$

where

$$\sigma = \frac{k_{\|}}{|k_{\|}|} . \tag{9.52}$$

The parameter $\sigma = +1$ if $k_{\|} > 0$, and $\sigma = -1$, if $k_{\|} < 0$. The presence of the factor of σ in (9.51) will have a most dramatic effect, as we shall see.

The conservation of normal $\boldsymbol{B}$ at the surface leads us to require

$$1 + \mu_1(\omega) + \sigma\mu_2(\omega) = 0 , \tag{9.53}$$

a statement which fixes the frequency ω. It is simple to show from (9.53) that

$$\omega_{ssw} = \frac{\sigma}{2}(\omega_H + \omega_B) \tag{9.54}$$

is the frequency for surface spin wave propagation in the direction perpendicular to the magnetization.

Suppose that we choose $k_{\|}$ to be positive in the above discussion. Then since all quantities vary with time and the coordinate x in the manner $\exp(ik_{\|}x - i\omega t)$, we have a disturbance which propagates from left to right across the magnetization. The frequency ω is positive, as we see from (9.55) with $\sigma = +1$.

By changing the sign of $k_\parallel$, one would suppose we obtain a disturbance which propagates from right to left across the magnetization. But if we let $k_\parallel \to -k_\parallel$, σ becomes -1, and the frequency ω_{ssw} in (9.55) changes sign as well. The factor $\exp(ik_\parallel x - i\omega t)$, with both $k_\parallel$ and ω negative, describes a disturbance that still propagates in the original direction, from left to right!

The surface spin waves just discussed are thus "one way modes," which may traverse the magnetizations in one sense only! As the reader will appreciate from Problem 9.2, if we consider a film with thickness D, and suppose $k_\parallel D \gg 1$, then on the top surface we have a mode which propagates only from left to right, while on the bottom surface of the film we have a mode which propagates in the oppose sense.

These remarkable waves are referred to often as Damon–Eshbach modes, after the researchers who first discussed them [9.11]. When we have a circumstance where there is a left/right asymmetry in wave propagation characteristics, one says the medium exhibits nonreciprocal propagation. The example just given is the most dramatic possible, where one realizes the "one way" characteristic just described.

Nonreciprocal propagation is most useful in device applications. Suppose, for example, we have a device in the form of a planar structure. We wish to transmit information in the form of a pulse from A to B. If we have a transducer at A and a receiver at B, at a time subsequent to the arrival of the signal pulse of interest, one will have a second unwanted "false signal" from the pulse backreflected from the sample boundary. If the information is carried by a Damon–Eshbach wave, there is no backreflected pulse. There is a generation of ferrite devices that exploit these "one-way" modes.

It is striking that one finds nonreciprocal propagation characteristics for spin waves localized onto magnetic surfaces, while there is no evidence for such behavior for spin wave propagation in the bulk, even for ferromagnetic materials. This last point is illustrated by the results of the previous section. The following argument shows this behavior to be a combined effect of spontaneous magnetization (or an applied magnetic field) and the presence of a surface [9.12].

Consider first the propagation of spin waves in an infinitely extended crystal, with magnetic field $\boldsymbol{H}^{(0)}$ and magnetization $\boldsymbol{M}^{(0)}$ along the z axis. The wave vector $\boldsymbol{k}$ lies in the xz plane. The geometry is that in Fig. 9.2a. We refer to the dispersion relation of the spin waves as $\omega(k_x, k_z)$, and we use R_{xz} and R_{yz} to denote the operation of reflection in the xz and yz plane. We assume R_{xz} and R_{yz} are good symmetry operations, when $\boldsymbol{H}^{(0)}$ and $\boldsymbol{M}^{(0)}$ are both zero. The operation R_{yz} lets $k_x \to -k_x$, and leaves k_z unchanged. When this is a "good" symmetry operation that leaves the crystal unchanged, we may then conclude $\omega(+k_x, k_z) = \omega(-k_x, k_z)$. However, the presence of either the magnetic field or magnetization destroys R_{yz} as a good symmetry operation, since both are axial vectors, and change sign under reflection through any plane to which they are parallel. However, when $\boldsymbol{H}^{(0)}$ and $\boldsymbol{M}^{(0)}$ are nonzero, the product $R_{xz}R_{yz}$ remains a symmetry operation in their presence, and the net effect of application of both is to change k_x to $-k_x$. Since the product leaves the underlying crystal structure invariant (by assumption) we may prove $\omega(+k_x, k_z) = \omega(-k_x, k_z)$. Reflection in the xy plane remains a "good" symmetry operation when $\boldsymbol{M}^{(0)}$ and $\boldsymbol{H}^{(0)}$ are nonzero, since any axial vector is unchanged by reflection in the plane to which it is perpendicular. But R_{xy} changes k_z to $-k_z$, so $\omega(k_x, k_z) \equiv \omega(k_x, -k_z)$.

Hence, any bulk wave in a ferromagnetic medium such as that just described, has reciprocal propagation characteristics. Now consider the semi-infinite crystal depicted in Fig. 9.2b. Let the direction of the wave vector $\boldsymbol{k}_{\parallel}$ be canted away from the x direction, and let $\omega_s(k_x, k_z)$ be the dispersion relation of a surface wave. In the presence of the surface, R_{xz} takes the crystal from the upper half space $y > 0$ and maps it into the lower half space $y < 0$. Hence the product of reflections $R_{yz}R_{xz}$ is no longer a symmetry operation for the semi-infinite crystal, and we cannot prove $\omega_s(+k_x, k_z) = \omega_s(-k_x, k_z)$. Nonreciprocal propagation is allowed on the surface. Since $\boldsymbol{M}^{(0)}$ and $\boldsymbol{H}^{(0)}$ are unchanged by a reflection in a plane perpendicular to themselves, it is the case that $\omega_s(k_x, k_z)$ is an even function of k_z. We would see this if we extend our discussion of surface spin wave propagation to general propagation directions [9.11].

Note that the argument just given, when applied to a film, requires a wave on the lower surface which propagates from right to left, if one propagates from left to right on the upper surface. This is illustrated by Problem 9.4. For the film, in the absence of magnetization R_{xz} applied to the mid plane is restored as a good symmetry operation. The operation $R_{yz}R_{xz}$ changes the sign of k_x and maps a wave localized to the upper surface onto the lower surface.

The argument above applies to any surface wave on a medium with a net magnetic moment $\boldsymbol{M}^{(0)}$, or a nonmagnetic medium to which a magnetic field is applied to the surface. The surface of a crystal supports surface acoustic waves, known since the time of Lord Rayleigh, see [9.13]. They are referred to commonly as Rayleigh waves, or in the engineering literature by the acronym SAW. When such waves are launched on the surface of a ferromagnet, with propagation direction perpendicular to the magnetization, their propagation characteristics become nonreciprocal [9.12]. We may consider surface electromagnetic waves, or surface polaritons on the surface of a doped semiconductor with magnetic field parallel to the surface. The free carriers form a plasma whose response is affected strongly by the magnetic field. This gives rise into dramatic nonreciprocal propagation characteristics for surface polaritons on such materials [9.14]. Antiferromagnets are of interest from this point of view. In the absence of an external magnetic field $\boldsymbol{H}^{(0)}$, the macroscopic magnetization vanishes. One realizes surface electromagnetic waves or surface polaritons of magnetic character in the far infrared, whose propagation characteristics are fully reciprocal [9.15]. Application of an external magnetic field introduces very interesting nonreciprocal behavior [9.15]. Such modes have been explored in detail in a recent experiment [9.16].

The argument above applies not only to surfaces waves, but to the interaction of bulk waves with the surface environment. For example, if a bulk acoustic wave or sound wave reflects off the surface of a ferromagnetic crystal, and its plane of incidence is perpendicular to the magnetization, there is a substantial left/right asymmetry in the reflection coefficient, if its frequency lies in the vicinity of the spin wave frequencies of the material [9.17]. Similar remarks apply to the reflection of electromagnetic waves from such a surface.

As remarked earlier, nonreciprocal propagation characteristics of Damon–Eshbach surface spin waves have played an important role in microwave technology, through ferrite-based devices. A three port device referred to as a circulator is used very widely in many microwave circuits, most particularly in phase-array antennas [9.18]. The discussion just presented shows nonreciprocal wave propagation near and on surfaces can be realized in diverse conditions. It is the view of the present author that in the near future, we shall see diverse, new applications to this most interesting phenomenon.

9.5 Nonlinear Interaction of Light with Spin Waves in Ferromagnets

In Sect. 9.4, we discussed how fluctuations in the magnetization modify the dielectric tensor of a ferromagnetic material. As we see from (9.36,37), the presence of a spin wave, wherein the magnetization components $m_\mu(\boldsymbol{r}, t)$ vary with space and time as $\exp(\mathrm{i}\boldsymbol{k}\cdot\boldsymbol{r} - \mathrm{i}\omega t)$, will induce a moving diffraction grating in the material. Light waves are diffracted from such a moving grating. We may view this process as a nonlinear interactions of the light wave with the spin wave. If the light wave has the wave vector $\boldsymbol{k}_l$ and frequency ω_l, then considerations of phase matching lead to output waves of frequency $\omega_l \pm \omega$, and wave vectors $\boldsymbol{k}_l \pm \boldsymbol{k}$. We have here a magnetic analogue of the phenomenon of Brillouin scattering discussed in Sect. 9.4.

In this section, we discuss two examples of the nonlinear interaction of light with spin waves. First we examine the Brillouin scattering of laser light by spin waves present as thermal excitations on the surface of ferromagnetic materials, or in films. Then we explore the physics of a most interesting device based on the interaction of laser radiation with spin waves generated in a ferromagnetic film by a transducer. This is the magneto-optic Bragg cell, developed in recent years.

9.5.1 Brillouin Scattering of Light by Thermally Excited Spin Waves

As noted several times above, in a ferromagnet, thermal fluctuations produce time and space varying components to the magnetization; we may describe these through the function $\boldsymbol{m}(\boldsymbol{r}, t)$ introduced above. At any instant of time, we may Fourier analyze this function, just as we did for the thermal fluctuations in displacement discussed in Sect. 5.4. Thus, at any instant of time, in a sample of volume V we may write

$$\boldsymbol{m}(\boldsymbol{r}, t) = \frac{1}{\sqrt{v}} \sum_{\boldsymbol{k}} \boldsymbol{m}(\boldsymbol{k}, t) \mathrm{e}^{\mathrm{i}\boldsymbol{k}\cdot\boldsymbol{r}} . \tag{9.55}$$

The time dependence of the amplitude $\boldsymbol{m}(\boldsymbol{k}, t)$ is controlled by precisely the same equation of motion we used to analyze spin waves of wave vector $\boldsymbol{k}$. Hence, if

$\omega(\boldsymbol{k})$ is the frequency of a spin wave with wave vector $\boldsymbol{k}$, we have

$$\boldsymbol{m}(\boldsymbol{r},t) = \frac{1}{\sqrt{V}} \sum_{\boldsymbol{k}} \boldsymbol{m}(\boldsymbol{k}) \mathrm{e}^{\mathrm{i}[\boldsymbol{k}\cdot\boldsymbol{r}-\omega(\boldsymbol{k})t]} \,. \tag{9.56}$$

When (9.56) is combined with (9.36,37), we see that each thermally excited spin wave mode creates a moving grating in the material. Each such grating diffracts laser light; this is the view of the Brillouin scattering even provided by classical physics.

Quantum mechanics provides another physical picture of this scattering event. Just as is the case with the acoustic waves discussed in Sect. 5.4, the spin waves are quantized [9.19]. Their energy thus comes in packets of size $\hbar\omega(\boldsymbol{k})$, and in the nonlinear diffraction process, $\hbar\boldsymbol{k}$ plays the role of the momentum. The diffraction process, at the quantum level, is viewed as the emission (Stokes process) or absorption (anti-Stokes process) of a spin wave quantum by the laser photon, and its consequent scattering into its final state. If ω_I and $\boldsymbol{k}_I$ are the frequency and wave vector of the incident laser photon, and ω_s and $\boldsymbol{k}_s$ that of the scattered proton, then energy conservation requires $\hbar\omega_s = \hbar\omega_I \mp \hbar\omega(\boldsymbol{k})$, and momentum conservation requires $\hbar\boldsymbol{k}_s = \hbar\boldsymbol{k}_I \mp \hbar\boldsymbol{k}$, where the upper sign refers to the Stokes process, and the lower sign the anti-Stokes event. With factors of $\hbar$ cancelled, these relations are identical to those provided by the classical theory of nonlinear wave interactions, with the conservation of wave vector a consequence of the interaction be phase matched for the output to be appreciable.

An impressive sequence of experiments have explored backscattered light from the surface of ferromagnetic metals, and thin films of such metals. In such measurements, the penetration of the light into the material is limited to the optical skin depth, which is typically in the range of 150–200 Ångstroms. Thus, such experiments probe the dynamical response of the material very close to the surface. Two features appear in such Brillouin spectra. The first is produced by bulk spin waves, which propagate up to the surface and reflect off of it. The second arises from the thermally excited Damon–Eshbach surface spin waves, which skitter along the surface.

In order to interpret the spectra, one needs to discuss the experimental arrangement for a moment. We illustrate this schematically in Fig. 9.3. The incident light makes some angle θ_I with respect to the surface normal, as indicated. The detector is set to examine scattered light which makes a particular angle θ_s with respect to the normal. In a Stokes process, viewed in the quantum theoretic picture, the laser photon emits a spin wave quantum. If $\theta_s > \theta_I$, as illustrated in Fig. 9.3a, the wave vector $\boldsymbol{k}$ of the spin wave is directed toward the left.

In an anti-Stokes process, the spin wave is absorbed by the photon. For the scattered photon to enter the detector, the wave vector of the spin wave must be directed toward the right. Thus, if the scattering geometry is arranged as illustrated in the figure, the Stokes portion of the Brillouin spectrum explores spin waves which propagate from right to left, and the anti-Stokes side explores those which propagate from left to right.

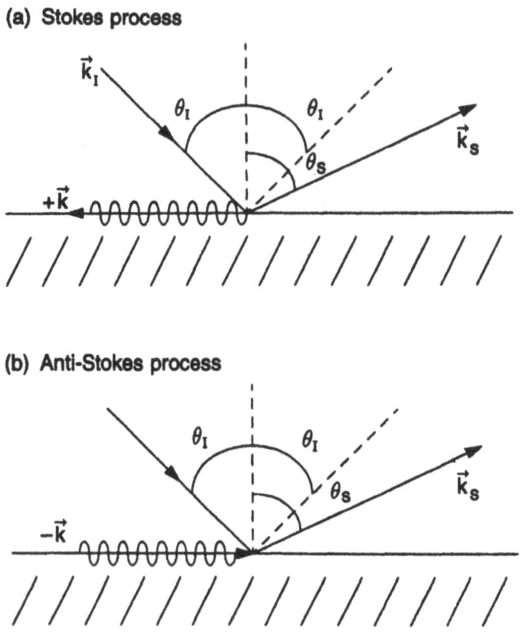

Fig. 9.3. An illustration of a Brillouin scattering experiment, carried out in a geometry where the light is backscattered from the surface. We show, for a given experimental geometry, **a** a Stokes scattering event, and **b** an anti-Stokes scattering event.

We show an experimental Brillouin spectrum in Fig. 9.4, for backscattering from the surfaces of ferromagnetic Fe. The data (open squares) is that reported by *Sandercock* [9.20], and the solid line is generated from the theory developed by *Camley* and *Mills* [9.21]. The theory has been extended to ferromagnetic films in [9.22]. In the experiment, the magnetization is parallel to the surface, and perpendicular to the scattering plane. Thus, all of the spin waves probed in the experiment propagate in the plane perpendicular to the magnetization.

The most dramatic feature in the spectrum is the sharp peak labeled SM, on the Stokes side of the spectrum. This feature is produced by scattering from the Damon–Eshbach surface spin wave described in the previous section. (The label "SM" stands for surface magnon. The elementary quanta of spin waves are referred

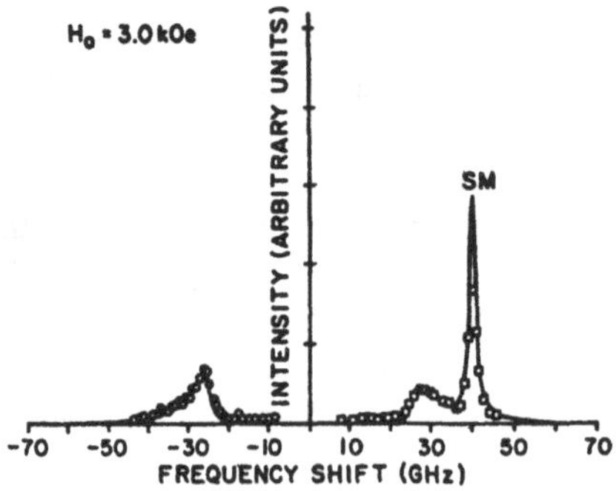

Fig. 9.4. Experimental Brillouin spectra produced by spin waves in ferromagnetic Fe (open squares), along with a theoretically generated spectrum. The experimental geometry is as illustrated in Fig. 9.3. The magnetization is parallel to the surface, and normal to the scattering plane. The data is that reported by *Sandercock* [9.20], and the figure is taken from a paper by *Camley* et al. [9.22]

to as magnons in the literature on magnetism). Notice that there is no Damon–Eshbach loss feature on the anti-Stokes side of the spectrum. This is dramatic confirmation of the "one way" propagation characteristics of these most remarkable waves. If we have, as in this case, a geometry where the spin wave propagates from right to left across the magnetization, there are no Damon–Eshbach waves which propagate in the opposite sense. Thus, if we see a feature on the Stokes side of the spectrum from these waves, as in Fig. 9.4, the line on the anti-Stokes side is missing. It is verified experimentally that if the direction of the externally applied magnetic field is reversed (and along with this the direction of the magnetization is necessarily reversed), then the loss feature associated with the Damon–Eshbach wave moves from the Stokes to the anti-Stokes side of the laser line. Its sense of propagation has been reversed.

At the time of this writing, the Brillouin scattering method has proved to be a powerful means of probing the nature of spin waves in diverse materials. Ferromagnetic films as thin as a few atomic layers may be explored by this method.

9.5.2 Nonlinear Mixing of Light with Macroscopic Spin Waves; the Magneto-optic Bragg Cell as an Example

In the previous section we discussed the Brillouin scattering of light by spin waves present as thermal excitations in ferromagnetic media. Within the framework of classical physics, as we have seen, one may view this process as a nonlinear interaction of the incident electromagnetic wave with a spin wave, present as a thermal excitation.

As noted above, one may also use microstrips which launch spin waves in ferromagnetic materials. This may be accomplished as follows: Consider a thin ferromagnetic film of some thickness D. One may deposit a thin metallic microstrip on the surface, whose width w is very small. A current of frequency ω is then passed through the microstrip. The microstrip then has in its near vicinity a magnetic field which oscillates with frequency ω, qualitatively similar to that which surrounds a thin conducting wire. This magnetic field may drive the spins within the ferromagnetic film, and their response may be described by (9.4), provided the field from the strip is incorporated into the total magnetic field $\boldsymbol{H}_T$. If the frequency ω lies in the frequency band where spin wave excitations exist, then spin waves of frequency ω are launched from the near vicinity of the microstrip. Similarly, a microstrip serves as a detector of spin waves. When a spin wave passes by such a structure, currents with the frequency ω of the spin wave are induced with the microstrip. As we shall see shortly, such externally generated spin waves may interact with light which propagates in the ferromagnetic film, through the modulation of the dielectric tensor such as that described in (9.37).

We saw in Subsect. 9.4.2 that for spin waves with the wavelength of interest for such interactions, the frequency of the spin wave is independent of the magnitude of its wave vector $\boldsymbol{k}$, but depends on its direction relative to the magnetization, as one sees from (9.44). This is true when the spin wave propagates in the infinitely extended crystal, as considered in Subsect. 9.4.2. If the discussion of the

Damon–Eshbach wave were extended to a general direction of propagation on the surface, as considered in Problem 9.2, one would see that a similar result follows for this case as well. However, if we consider spin waves which propagate within a film of finite thickness D, then the modes have a frequency which depends on the magnitude of the wave vector. The spin wave frequencies lie within a frequency band, whose limits are controlled by the strength of the externally applied dc field, and its orientation relative to the film surface. Since the location of this band may be "tuned" by variations in either the strength or direction of the applied field, spin wave modes are utilized as the basis for numerous microwave devices. In Problem 9.4, the propagation of Damon–Eshbach waves in a film of thickness D is considered, for the case where the propagation is perpendicular to the magnetization. From the solution to this problem, one sees that as the product kD ranges from 0 to ∞, where k is the wave vector of the mode, the frequency of the mode ranges from $(\omega_H \omega_B)^{\frac{1}{2}}$ to $(\omega_H + \omega_B)/2$, in notation similar to Subsect. 9.4.3. We define $\omega_B = \gamma(H^{(0)} + 4\pi M_S)$. We turn to another geometry for the moment.

A geometry used in a device considered below is that illustrated in Fig. 9.5a. We have a ferromagnetic film of thickness D placed in an external magnetic field of strength $H^{(0)}$ perpendicular to its surfaces. As mentioned earlier, if the magnetization $\mathbf{M}^{(0)}$ is perpendicular to the surface as well, as illustrated, elementary magnetostatics informs one there is an internal field of strength $H^{(0)} - 4\pi M^{(0)}$. If

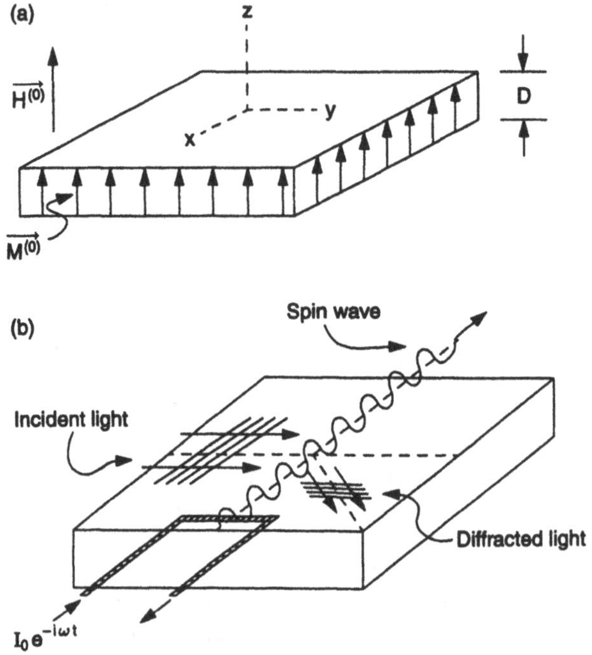

Fig. 9.5. a The magnetostatic forward volume wave configuration discussed in Subsect. 9.5.2. An external magnetic field is applied normal to the surface, with strength sufficient to orient the magnetization as shown. **b** A schematic drawing of the magnetooptic Bragg cell

$H^{(0)} < 4\pi M^{(0)}$, the internal field is anti-parallel to the magnetization, so the perpendicularly magnetized state is unstable. However, if $H^{(0)} > 4\pi M^{(0)}$, the internal field and the magnetization are parallel, and we realize a stable state. In the material YIG used earlier as an example, $4\pi M^{(0)} = 1750$ Gauss so the perpendicular state is stable even for modest external magnetic fields. We explore the spin waves which propagate in such a film, for this state.

We may describe the magnetic response of the film by the permeability tensor described in (9.9), except the magnetic field $H^{(0)}$ is to be replaced by the internal field $H^{(i)} = H^{(0)} - 4\pi M^{(0)}$ appropriate to the present problem. Thus in the formulae given in (9.9), in place of $\omega_H = \gamma H^{(0)}$, we have $\omega_H = \gamma(H^{(0)} - 4\pi M^{(0)})$ everywhere for the present geometry.

We may use the magnetostatic theory to describe the spin waves once again. Thus, as in (9.46), once again we write

$$\boldsymbol{H}^{(d)} = -\nabla \Phi^{(M)} , \tag{9.57}$$

where now, noting the coordinate axes in Fig. 9.5a, we have

$$\Phi^{(M)}(x, z) = \phi^{(M)}(z)\mathrm{e}^{ikx} \tag{9.58}$$

for spin waves which propagate parallel to the x-axis. Within the magnetic film, the requirement $\nabla \cdot \boldsymbol{B} = 0$ leads us to the condition

$$\frac{d^2\phi^{(M)}}{dz^2} - k^2\mu_1(\omega)\phi^{(M)} = 0 . \tag{9.59}$$

The quantity $\mu_1(\omega)$ is positive for frequencies less than $\gamma(H^{(0)} - 4\pi M^{(0)})$, negative in the region where $\gamma(H^{(0)} - 4\pi M^{(0)}) < \omega < \gamma\ [(H^{(0)} - 4\pi M^{(0)})H^{(0)}]^{\frac{1}{2}}$, and is positive for all frequencies above this band. If one consults Subsect. 9.4.2 and realizes the effective field to which the spins are exposed in $H^{(i)} = H^{(0)} - 4\pi M^{(0)}$, then the regime where $\mu_1(\omega)$ is negative is the frequency band where bulk spin waves may propagate. One may see this from (9.42), notice. In the present geometry, the frequency region of interest is that where $\mu_1(\omega) < 0$.

The film illustrated in Fig. 9.5a is left unchanged by the operation of reflection through the midplane. This is perhaps surprising, since both $\boldsymbol{H}^{(0)}$ and $\boldsymbol{M}^{(0)}$ are perpendicular to this plane. But each is an axial vector, and remains unchanged by a reflection through a plane to which it is perpendicular.

It follows that $\phi^{(M)}(z)$ in (9.58) has well defined parity; we may seek solutions either even under reflection through the mid plane, or odd under this reflection. If the film occupies the region $-D/2 < z < +D/2$, for the even case we may suppose

$$\phi^{(M)}(z) = \phi_i^{(M)} \cos(Qz), \quad -\frac{D}{2} < z < +\frac{D}{2} , \tag{9.60}$$

where, when $\mu_1(\omega) < 0$,

$$Q = k|\mu_1(\omega)|^{\frac{1}{2}} . \tag{9.61}$$

In the vacuum above the film,

$$\phi^{(M)}(z) = \phi_>^{(M)} \exp(-kz) , \tag{9.62}$$

with a related form below the film.

It is a straightforward matter to generate expressions for the fields $\boldsymbol{H}$ and $\boldsymbol{B}$ associated with the spin wave mode, and impose boundary conditions at the film surface. As in various examples considered earlier, this leads us to an implicit dispersion relation, which in the present case reads

$$\cot\left(\frac{1}{2}kD|\mu_1(\omega)|^{\frac{1}{2}}\right) = |\mu_1(\omega)|^{\frac{1}{2}} . \tag{9.63}$$

In the thin film limit $kD \ll 1$, we may obtain an analytic expression for the dispersion relation of the spin wave by using the small argument form for the cotangent. We have in this limit

$$\frac{1}{|\mu_1(\omega)|} = \frac{1}{2}kD . \tag{9.64}$$

The quantity $1/|\mu_1(\omega)|$ has a zero at the frequency $\gamma H^{(i)} = \gamma(H^{(0)} - 4\pi M^{(0)})$, so when $kD \ll 1$, the frequency of the mode lies near this value. A bit of algebra leads to the dispersion relation

$$\omega^2 \cong \gamma^2 H^{(i)}[H^{(i)} + 2\pi M^{(0)} kD] \tag{9.65}$$

in the long-wavelength limit.

The magnetic film thus acts as a waveguide for spin waves; the modes acquire a frequency that depends on the magnitude of the wave vector k of the spin waves, by virtue of the finite thickness of the film. As kD is increased, we find additional modes from (9.63), in a manner reminiscent of the mode structure of the optical fiber, discussed in Sect. 7.1. One also finds a set of modes for which the magnetic potential $\phi^{(M)}(z)$ is odd under reflection through the film. Here $\phi^{(M)}(z)$ is proportional to $\sin(Qz)$. The mode described by (9.65) is the only mode which survives in the limit $kD \ll 1$, and is the wave excited in the usual experimental geometry which employs the microstrip generator discussed above. The various remaining modes have cut off wave vectors k_n the order of $n\pi/D$, where $n = 1, 2, \ldots$. In frequency regimes where $\mu_1(\omega) > 0$, one finds no solutions compatible with the boundary conditions.

The geometry just explored is used frequently in device applications, since the magnetic field perpendicular to the film is produced easily by placing the film between poles of a magnet. There are three common geometries, and in the literature one encounters acronyms for such. The mode described by (9.65) is the "MagnetoStatic Forward Volume Wave," or the MSFVW. They are "volume waves," since they have standing wave character in the direction perpendicular to the film surface. The word "forward" enters because the phase velocity (ω/k) and the group velocity $(\partial\omega/\partial k)$ are parallel to each other. A second geometry has the external magnetic field $\boldsymbol{H}^{(0)}$ parallel to both the film surface and the propagation direction. Of course, the magnetization is then parallel to the surface and $\boldsymbol{H}^{(0)}$. For this case, in the limit $kD \ll 1$, one finds the dispersion relation becomes

$$\omega^2 = \gamma^2 H^{(0)}(H^{(0)} + 4\pi M^{(0)} - 2\pi M^{(0)} kD) . \tag{9.66}$$

For this mode, the phase velocity and the group velocity are antiparallel. In the literature, this is referred to as the "MagnetoStatic Backward Volume Wave," or MSBVW for this reason. The third configuration has the applied magnetic field applied parallel to the surface, but perpendicular to the direction of propagation. Here the mode launched by the microstrip is the Damon–Eshbach mode explored in Problem 9.3. This is referred to as the MagnetoStatic Surface Wave, or MSSW.

The discussion above shows that a thin ferromagnetic film supports a rich spectrum of guided spin-wave modes, which can be launched through use of a microstrip, and detected similarly. The frequency of a mode with given wavelength can be tuned by varying the external field $H^{(0)}$, for any of the three geometries just described. If the microstrip is driven at a selected frequency, then variation of the magnetic field can be used to "tune" the wavelength. We have seen that the frequency of these modes lie in the microwave regime. The spin waves are thus used as the "working waves" for a number of devices which operate in the microwave frequency regime.

We conclude with a description of a particular device which is based on the nonlinear interaction of light with spin waves such as those this discussed. This the magneto-optic Bragg cell [9.23]. We show a schematic illustration of the device in Fig. 9.5b. One has a YIG film, with a magnetic field $\boldsymbol{H}^{(0)}$ perpendicular to the surface, as illustrated in Fig. 9.5a. The magnetization is perpendicular to the surface as well. This material is transparent to electromagnetic radiation in the near infra-red, in the wavelength region longer than 1.5 μm. Laser light in this wavelength region can thus propagate for centimeters. A thin film of YIG can serve as an optical waveguide, through principles very similar to those discussed in Chap. 7, when we explored optical fibers. Thus, laser radiation in the 1–1.5 μm can be introduced into the YIG film, and be guided by it. A microstrip transducer is affixed to the film as indicated, and a magnetostatic forward volume wave is launched into the film. The spin wave propagates perpendicular to the direction of the incident light beam. There is a nonlinear interaction between the incident light and the spin wave, through (9.37). Diffracted light beams are produced, with wave vectors $\boldsymbol{k}_s = \boldsymbol{k}_l \pm \boldsymbol{k}$, and frequency $\omega_s = \omega_l \pm \omega$, where ω and $\boldsymbol{k}$ are the frequency and wave vector of the spin wave. By introducing Bi as an impurity in YIG, one can greatly increase the strength of the Faraday term in (9.37). In such films, diffraction efficiencies in the range of 14–20% are realized. Through tricks which lead to focusing of the spin wave, one can increase the diffraction efficiency well beyond these values.

One can use this device for several purposes. By modulating the amplitude of the spin wave, one can produce modulated diffracted beams. If the incident light is a shaped pulse, one can produce an output light beam with an altered shape by mixing the incident pulse with a suitably shaped spin wave pulse. It serves also as a fast optical switch. With spin wave off, there is no diffracted beam, of course, and one may switch a fraction of the energy in the incident pulse to a desired direction by launching a spin wave. Note that if the microscopic strip that launches the spin wave is operated at fixed frequency ω, the wave vector of the spin wave may be varied by changing the strength of the external magnetic field $\boldsymbol{H}^{(0)}$. Thus, one can direct the diffracted radiation through a range of angles. Finally, one can use the device as a spectrum analyzer in the microwave regime. Suppose a signal is introduced into the microstrip with two frequencies ω_1 and ω_2. One then generates two spin waves, of wave vector $\boldsymbol{k}_1$ and $\boldsymbol{k}_2$, respectively. One then has two diffracted beams, whose angular direction allows one to deduce ω_1 and ω_2.

We thus have a very compact device that can perform many tasks. To understand its operation, we must invoke many concepts developed earlier in this text.

Problems

9.1 In Chap. 2, we considered contributions to the dielectric tensor from conduction electrons. In their presence, the dielectric constant of the medium is given by (2.46a). Suppose the conductivity has tensor character, so $J_\alpha = \sum_\beta \sigma_{\alpha\beta}(\omega)E_\beta$. Then, upon ignoring other contributions to the dielectric tensor, we have

$$\varepsilon_{\alpha\beta}(\omega) = \delta_{\alpha\beta} + \frac{4\pi}{\omega}\sigma_{\alpha\beta}(\omega) \ .$$

(a) Consider conduction electrons placed in an external magnetic field $\boldsymbol{H}^{(0)} = \hat{z}H_0$. Obtain the form of the conductivity tensor $\sigma_{\alpha\beta}(\omega)$. For simplicity, let the relaxation time $\tau \to \infty$.

(b) Find the dispersion relation which describes propagation parallel to the magnetic field. Describe their polarization. Consider the special case $\omega \ll \omega_c$, where $\omega_c = eH^{(0)}/mc$ is the cyclotron frequency. You should find the curious dispersion relation $\omega = c^2k^2(\omega_c/\omega_p^2)$, with ω_p the plasma frequency. Such highly dispersive modes are referred to as helions in condensed matter physics, and whistlers in atmospheric physics.

9.2 A plane polarized electromagnetic wave strikes a semi-infinite ferromagnetic material, at normal incidence. The material is described by the dielectric tensor in (9.24). Derive the amplitude of the reflected wave for three cases: (i) the magnetization is perpendicular to the surface, (ii) the magnetization is parallel to the surface and to the electric field in the incident wave, and (iii) the magnetization is parallel to the surface, but perpendicular to the electric field in the incident wave.

9.3 In the text, the discussion of the semi-infinite ferromagnet was confined to the case where the wave propagated in the x direction illustrated in Fig. 9.2b. Extend the discussion to the case where the wave vector makes an angle ψ with the x axis. You should find a surface-wave solution exists only in the angular regime $-\psi_c \leq \psi + \psi_c$, where $\cos(\psi_c) = H^{(0)}/B^{(0)}$, where $B^{(0)} = H^{(0)} + 4\pi M^{(0)}$.

9.4 Consider a ferromagnetic film of thickness D with magnetization parallel to the surface, and to the z axis, which lies in the surface. The x axis lies in the surface as well, as in Fig. 9.2b. Consider an extension of the Damon–Eshbach spin-wave discussion to this slab geometry, for the case where the mode propagates parallel to the x axis. In the film, the function $\phi^{(M)}(y)$ is now a superposition of $\exp(+|k_\parallel|y)$, and $\exp(-|k_\parallel|y)$. Note that in this geometry, reflection through the midplane of the slab is not a "good symmetry operation," since the magnetization is an axial vector. You should find that the dispersion relation is now an even function of $k_\parallel$, so waves propagate in both directions across the magnetization. Convince yourself that in the limit $k_\parallel D \gg 1$, the eigenvector becomes localized on the upper surface for one direction of propagation, and the lower surface of the other.

10. Chaos

Our discussion of nonlinear optical phenomena began with a perturbation theoretic treatment of the consequences of the various nonlinear susceptibilities. Our first example was the response of the anharmonic oscillator described by (3.15). This led us to understand, on the basis of this simple model, the qualitative aspects of the frequency variation of the nonlinear susceptibilities, and also the influence of the anharmonic terms on the frequency spectrum of the motion $x(t)$ of the oscillator.

Suppose we recall the results for the symmetric oscillator, where the $Max^3/3$ term in (3.14) vanishes. If the driving field $E(t)$ has the form $E\cos(\omega t)$, characterized by the single frequency ω, then the bx^3 term generates output at the frequency 3ω, according to the perturbation theory development in Chap. 3. If the perturbation theory is continued beyond lowest order, then we generate various other harmonics at the frequencies $n\omega$, where n ranges over the positive integers. For example, a contribution from the second order of perturbation theory describes the interaction of the third harmonic with the fundamental at frequency ω, to generate responses at $3\omega - \omega = 2\omega$, and $3\omega + \omega = 4\omega$. The resulting displacement $x(t)$ still has the period $T = 2\pi/\omega$. The higher harmonics distort the shape of $x(t)$, from the simple harmonic oscillator form, but leave the period unchanged.

As we continued with our development of the propagation of electromagnetic waves in nonlinear media, we encounter remarkable features with no counterpart in perturbation theory. These are the solitons, which differ in nature qualitatively from any solution of linear theory, which describes only spatially extended plane waves. We can, of course, superimpose such plane waves to form wave packets, but in general these lack the striking shape stability we found in the solitons, as we have seen.

There are additional very fascinating features of the response of nonlinear systems that we have yet to discuss. For example, the simple anharmonic oscillator driven by an external field has response characteristics remarkably more complex and richer than one can envision from the perturbation theory picture of its response. Ultimately, when the amplitude of the driving field $E(t)$ becomes sufficiently large, the motion $x(t)$ of the oscillator acquires a random character; one says the motion of the oscillator is chaotic. The transition from the orderly regime described by perturbation theory to the chaotic regime takes place in a fascinating series of steps. This chapter summarizes the various means by which chaotic motion is achieved in nonlinear systems, along with examples provided by optical systems. Once again optical physics has become an arena within which fundamental new theoretical predictions can be set alongside data.

10.1 Duffing Oscillator: Transition to Chaos

We begin with a somewhat modified version of the anharmonic oscillator described by (3.15). First, we suppose $a = 0$, as in the previous subsection, we let b be negative, $b = -|b|$, and then we add a damping term, so we have

$$\frac{d^2x}{dt^2} + \gamma\frac{dx}{dt} + \omega_0^2 x - |b|x^3 = \frac{e}{M}E_0\cos(\omega t)\ . \tag{10.1}$$

A particle with this equation of motion is referred to as a Duffing oscillator. As indicated, we are interested in the response of such an oscillator to a driving field with the frequency ω, and amplitude E_0. This describes a particle of mass M, moving to the potential

$$V(x) = \frac{1}{2}M\left(\omega_0^2x^2 - \frac{1}{2}|b|x^4\right)\ , \tag{10.2}$$

which becomes "softer" as x increases, as illustrated in Fig. 8.1. Indeed $V(x)$ has maxima at $x_0 = \pm(\omega_0^2/|b|)^{1/2}$, so if the driving force is so large the particle is "lifted" over the maximum, it will move off to $x = +\infty$ or $x = -\infty$, without returning to the origin. Throughout this discussion, we shall be concerned with motions whose amplitude is always less than $\pm x_0$.

The damping term is essential, in that it allows the system to settle down into steady motion, at least for small amplitude fields. If the driving field is switched on, with the oscillator in some initial state, the motion will contain initial transients that die out, and then (for E_0 very small) we shall have $x(t) = A\cos(\omega t + \phi)$, so that the velocity $\dot{x}(t) = -\omega A\ \sin(\omega t + \phi)$. One may readily obtain formulae for A and ϕ by submitting the solution to (10.1), at least in the limit where E_0 is so small the anharmonic term $|b|x^3$ in (10.1) may be ignored. The motion of the particle can be pictured in a phase space plot, where its position and velocity form a trajectory in a plane whose horizontal axis is labeled by position, and vertical axis by velocity. If the axes are labeled v/ω, and x, as in Fig. 10.2 a, we then have a simple circle of radius A in such a plot.

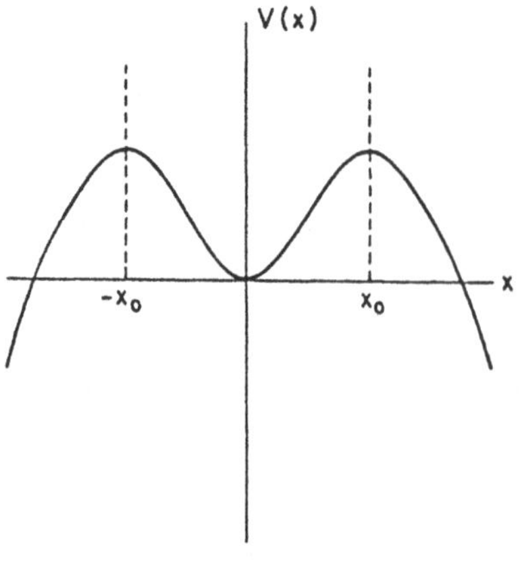

Fig. 10.1. The potential energy as a function of position for the Duffing oscillator described by (10.1)

As the amplitude of the driving field is increased, the phase space of the steady state orbit is no longer a circle, but becomes a curve of more complex shape, according to perturbation theory. It remains topologically equivalent to that in (10.1) in that it remains a simple closed curve which the particle traverses in the period $2\pi/\omega$.

Further increase in the driving field leads one to a transition to a regime whose description lies outside perturbation theory. There is a critical value of E_0, $E_0^{(1)}$ above which the period of the particle motion doubles.

One sees this through numerical studies of the power spectrum, $|x(\omega)|^2$ of the motion defined as follows:

$$|x(\omega)|^2 = \int_{-\infty}^{+\infty} d\tau e^{i\omega\tau} \langle x(t)x(t+\tau)\rangle \, , \tag{10.3}$$

where the angular brackets denote an average over a large number of times t. The power spectrum in the regime $E_0 < E_0^{(1)}$ consists of a sequence of delta functions at $\omega_n = n\omega$, and as soon as E_0 is increased and exceeds $E_0^{(1)}$, i.e., $E_0 > E_0^{(1)}$, subharmonics at all possible multiples of $(\omega/2)$ appear. One refers to the field $E_0^{(1)}$ as a bifurcation point of the solution. Above $E_0^{(1)}$, a schematic illustration of the particle's phase space trajectory is given in Fig. 10.2 b.

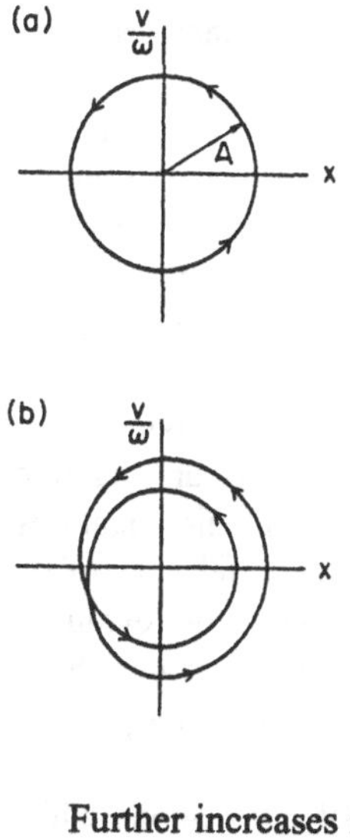

Fig. 10.2. The phase space plot of the particle's motion, **a** when the amplitude of the applied field is small, and linear theory is applicable, and **b** when the amplitude of the applied field is above the threshold for period doubling

Further increases in E_0 lead one to a second bifurcation field $E_0^{(2)}$, above which the period is $2^2T = 4T$ where $T = 2\pi/\omega$. One encounters a sequence of bifurcations, where above $E_0^{(n)}$, the period of the particle motion is 2^nT. The sequence of bifurcation fields approach a limiting field $E_0^{(\infty)}$,

$$\lim_{n\to\infty} E_0^{(n)} = E_0^{(\infty)} \, . \tag{10.4}$$

Above $E_0^{(\infty)}$, the particle's motion is no longer periodic at all. Its power spectrum has the character of broad band noise! One refers to the particle's motion in this regime as chaotic. The particle's trajectory wanders through phase space in a seemingly random manner, in response to the perfectly periodic driving field!

The motion of the particle thus has the character of a one particle analogue of turbulence. The notion that such random motion could emerge from a simple deterministic equation such as (10.1) came as a great surprise in theoretical physics. The concept appeared first in a paper by *Lorenz* [10.1], who explored the behavior of a very different set of nonlinear equations, which arise in a schematic picture of nonlinear hydrodynamics. We are schooled to view random phenomena as present only in physical systems which contain an enormous number of particles ($\sim 10^{23}$, as in a macroscopic sample of dense matter). The motion of any one particle is viewed as rather simple in nature, but macroscopic variables such as the pressure of a gas will display fluctuations with a broad frequency spectrum as a consequence of the actions of a very large number of particles.

There is another remarkable aspect of a chaotic motion, in nonlinear systems. This is an extreme sensitivity to initial conditions.

Of course, if we wish to find the value of $x(t)$ at times greater than some time t_0, assuming for the moment the driving field is turned on at the time t_0, then we must specify the initial values $x(t_0)$ and $\dot{x}(t_0)$. Suppose E_0 is so small anharmonicity can be ignored completely. After the initial transient has died out, in a time the order of γ^{-1}, we have $x(t) = A\cos(\omega t + \phi)$ and $\dot{x}(t) = -\omega A\cos(\omega t + \phi)$ in our earlier notation, where in fact A and ϕ are independent of $x(t_0)$, and $\dot{x}(t_0)$. More generally, we expect small changes in the initial conditions to lead to small changes in the subsequent time evolution of the system.

In the chaotic regime, very tiny differences in the initial conditions send the particle off onto vastly different trajectories. Let $\delta x(t_0)$ and $\delta\dot{x}(t_0)$ denote infinitesimal changes in the boundary conditions, and $\delta x(t)$, $\delta\dot{x}(t)$ the difference in particle trajectories at times subsequent to t_0. Then initially $\delta x(t) \sim \exp(\Lambda t)$ where $\Lambda > 0$, in the chaotic regime. Thus, to predict the future course of the particle's trajectory, we require the initial conditions with infinite precision!

This has profound implications for areas of science where the underlying mathematics involves highly nonlinear equations. Weather prediction is an example. Atmospheric motions are controlled by highly nonlinear hydrodynamic equations. We never have enough information in hand to specify the initial conditions of the atmosphere with precision sufficient to allow us to determine the future evolution of the earth's atmosphere. There may thus be fundamental limitations to our ability to make long term weather predictions, even if the physics of the atmosphere is understood fully and completely.

While we have phrased our discussion of the sequence of bifurcations which lead to chaos in language appropriate to the much analyzed Duffing oscillator, as the remarks above suggest, transitions to chaos occur for a very large variety of nonlinear systems. We may have a sequence of variables $x_1, x_2, \ldots, x_N$, which obey a set of differential equations of the form

$$\dot{x}_i(t) = F_i^{(\lambda)}(x_1, \ldots, x_N)\,, \tag{10.5}$$

where λ stands symbolically for a parameter or set of parameters that enter the functions $F_i^{(\lambda)}$ which link $\dot{x}_i(t)$ to the set $\{x_i\}$. If the functional relations are nonlinear, then as λ is increased with λ arranged such that $\lambda = 0$ describes a simple linear regime, we may encounter a sequence of bifurcations leading to chaos. Our Duffing oscillator fits this scheme if we choose $x_1 = x$, and $x_2 = \dot{x}$. We then have

$$F_1^{(\lambda)} = x_2 , \tag{10.6 a}$$

$$F_2^{(\lambda)} = -\omega_0^2 x_1 + |b| x_1^3 - \gamma x_2 + \frac{eE_0}{M} \cos(\omega t) , \tag{10.6 b}$$

where λ stands for the parameter pair (E_0, ω). In a classic paper, *Crutchfield* and *Huberman* [10.2] have explored, for large E_0, the sequence of period doublings and the subsequent chaotic behavior, as the frequency ω is varied.

In this chapter, we present only a brief summary of the chaotic behavior of nonlinear systems of interest to our discussions of nonlinear optics. We refer the reader to the very excellent review article by *Ackerhalt* et al. [10.3] for an introduction to general aspects of nonlinear phenomena framed in the modern viewpoint, and for numerous applications to optics. This review provides a complete survey of the field as it stood in the recent past.

So far, we have spoken as if the motion of the particle is completely random, in the chaotic regime. In fact, there is a certain underlying structure, in a sense that we now describe. In general, a nonlinear equation which describes a dissipative system has associated with it a feature in phase space called an attractor. Consider (10.1), for the case where E_0 is so small, the anharmonic term $|b|x^3$ may be ignored, Then at long times, no matter what the initial condition, the particle will execute the circular motion depicted in Fig. 10.2 a. We may start the particle out at an initial position $x(t_0)$, and normalized velocity $v(t_0)/\omega$ which lies off the circle, and the phase space trajectory will spiral toward the circle, until at times large compared to λ^{-1}, the phase space motion is that depicted in Fig. 10.2 a. This periodic motion on a closed trajectory in phase space is an example of a limit cycle, and the circle in Fig. 10.2 a is an attractor.

Any attractor has a basin of attraction, with the property that any initial value inside the basin will lead to subsequent motions wherein the particle is "driven" to the attractor. In the example just given, where the $|b|x^3$ term is set to zero, the basin is the entire plane.

With $|b| \neq 0$, the attractor remains the circle illustrated in Fig. 10.2 a, but the basin is now finite in area, because if $x(t_0) > (\omega_0^2/|b|)^{1/2}$, and the initial velocity is sufficiently large, the particle will simply fly off to infinity, when it moves in the full potential given in (10.2).

In the chaotic state, while the power spectrum associated with the particle motion has the character of broad band noise (but not white noise, i.e., $|x(\omega)|^2$ has clear dependence on the frequency), its phase space trajectory lies on an object called a strange attractor. Properties of strange attractors are discussed in the review of *Ackerhalt* et al. [10.3] cited above, and precise descriptions of attractors, including the strange attractor, are given by *Eckerman* [10.4], in a review that explores chaos and the approach to chaos in a manner more formal than much of the discussion in the literature of theoretical physics.

There is universal behavior in the approach to chaos by means of the sequence of period doublings described above, as discussed in classic papers by *Feigenbaum* [10.5]. Let λ be a parameter which is varied, to drive the system through the sequence of period doubling bifurcations, into the chaotic state. In our earlier discussion of the Duffing oscillator, we could identify $\lambda^{(1)}, \lambda^{(2)}, \ldots$ with the sequence of critical fields $E_0^{(1)}, E_0^{(2)}, \ldots$, and in the paper by Crutchfield and Huberman with the frequencies $\omega^{(1)}, \omega^{(2)}, \ldots$ of the driving field at which the various bifurcations occur. The analysis of Feigenbaum shows that

$$\lim_{n\to\infty} \frac{(\lambda_{n+1} - \lambda_n)}{(\lambda_{n+2} - \lambda_{n+1})} = \delta , \tag{10.7}$$

where $\delta = 4.669\ldots$ is a universal number. The numerical simulations reported by Crutchfield and Huberman are compatible with this rule.

Also, Feigenbaum argues that when n is large, there are universal relations between the intensities associated with the frequencies in the power spectrum of the particle.

Linsay [10.6] has carried out a very simple, but elegant experiment which explores the behavior of simple nonlinear oscillator similar to that described by (10.1). He constructed a simple LRC circuit, in which the capacitor is a semiconducting device with capacitance that varies in a nonlinear manner with voltage. The circuit is driven by an oscillatory voltage whose frequency is fixed at 1.78

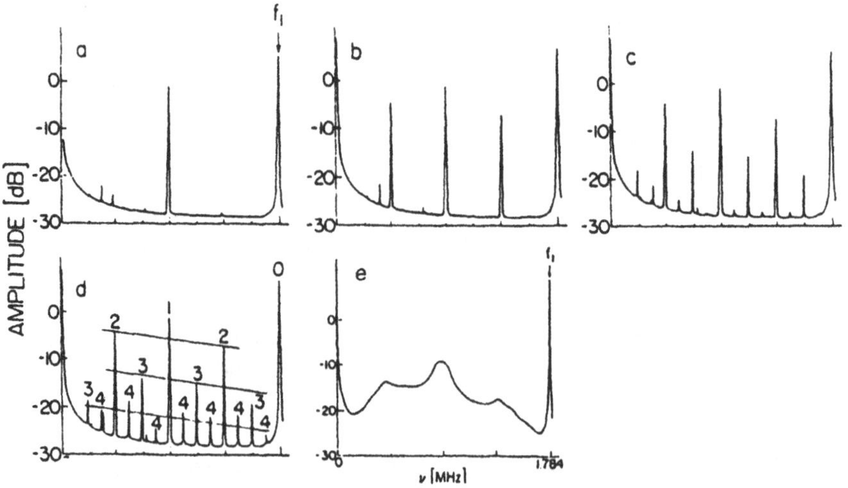

Fig. 10.3. The data of Linsay, who explored the power spectrum of current induced in a nonlinear LRC circuit by an applied voltage $V\cos(2\pi f_1 t)$ with f_1 the linear resonance frequency of the circuit. We show **a** the subharmonic at the frequency $f_1/2$ above the first bifurcation threshold, **b** the spectrum above the second, **c**, the third, **d** the fourth, and finally **e** a spectrum when the circuit exhibits chaotic response. The data is taken from *Linsay* [10.6]. The various lines in **d** relate to predictions by *Feigenbaum* [10.5], as discussed by Linsay

Mhz, which is in fact the small amplitude resonance frequency of this particular circuit. A sequence of period doublings was found at a sequence of critical voltages $V^{(1)}, V^{(2)}, V^{(3)} \dots$.

In Fig. 10.3, we show a sequence of Linsay's measure power spectra. In Fig. 10.3 a, we see the fundamental frequency, labeled f_1 and the first subharmonic at $f_1/2$. The applied voltage here lies in the range of $V^{(1)} < V < V^{(2)}$. In Fig. 10.3 b, we see $f_1/4$ and its various multiples, structures at $f_1/2^3 = f_1/8$ make their appearance in Fig. 10.3 c, and finally $f_1/2^4$ in Fig. 10.3 d. In Fig. 10.3 e, we show a chaotic spectrum. The various lines in Fig. 10.3 d are the predictions of Feigenbaum for the relative intensity of the various spectral components. We refer the reader to Linsay's paper for a complete discussion of comparison between theory and experiment.

Quite clearly, the behavior of the driven anharmonic oscillator is immeasurably richer than one would expect from the discussion of Chap. 3!

10.2 Routes to Chaos

The mathematical analysis of nonlinear equations is less complete than the theory of linear differential equations in a number of very important ways. There are, in the theory of linear differential equations, powerful and general theorems which outline the conditions required for solutions to exist for broad classes of differential equations, and enumerate their properties. This body of general theory is lacking for nonlinear equations. Our knowledge is obtained from study of the structure of selected equations, and often this follows from numerical studies such as those reported in the paper by *Huberman* and *Crutchfield* [10.2].

Nonetheless, systematic features emerge by collecting such analyses together. We saw that the Duffing oscillator approached the chaotic region of its parameter space in a sequence of period doubling bifurcations, which lead to a critical value of the parameter varied beyond which the chaos is realized. One refers to this scenario as the period doubling route to chaos. This approach to chaos is not unique to the Duffing oscillator, as the remarks of the previous section indicate, but is encountered rather frequently in studies of nonlinear dynamic systems.

There are two other scenarios that have been discussed or proposed in the literature, and experience to date suggests that one always encounters one of the three routes to chaos, in analyses of nonlinear systems. There is no guarantee or proof that the three schemes exhaust all possibilities, but this is viewed as the case by numerous authors.

One scenario was put forward by *Ruelle* and *Takens* [10.7]. This applies to a system of equations which, for a range of the parameter λ introduced in (10.5), has a stable fixed point, wherein the set of equations admit a solution with $x_1 = x_1^*, \dots, x_N = x_N^*$ independent of time. If one considers small amplitude motions around the fixed point, the solution $\{x_i(t)\}$ spirals into the fixed point by virtue of the dissipation.

In the Ruelle-Takens scenario, as λ is varied, a bifurcation point is reached where the system executes a closed trajectory in phase space (a limit cycle) with period $T_1 = 2\pi/f_1$. The power spectrum of the particle motion thus contains various integral multiples of f_1. Further changes in λ lead one to a second bifurcation point, above which the motion involves two incommensurate frequencies f_1 and f_2. There is then a third bifurcation, which leads one to the chaotic state, with the particle motion on the strange attractor.

The third scenario has its origin in the work of *Pomeau* and *Manneville* [10.8]. One begins with a parameter regime where the particle motion is periodic. An increase of λ leads one to a bifurcation point above which the motion is chaotic. There is thus a sudden transition from the initially periodic to the chaotic motion. If λ is increased further, one leaves the chaotic regime, and enters a new region of parameter space wherein the motion is again periodic. Then increasing λ further leads to a new chaotic regime. One proceeds into the final chaotic state by a sequence of transitions from periodic to chaotic motion and back again.

It is remarkable that these three scenarios occur again and again, for the diverse array of nonlinear equations that exhibit chaotic motion. We turn next to examples of optical systems wherein chaotic behavior has been observed and classified.

10.3 Experimental Observations of Chaos in Optical Systems

As we now know well, various optical systems respond in a highly nonlinear manner, and display behavior not described by perturbation theoretic analyses of the underlying equations. Optical systems also exhibit chaotic behavior, as we shall see.

The output of lasers show unstable behavior under certain conditions which, when analyzed carefully, can be understood within the framework developed here. We must begin with brief comments on the nature of lasing systems.

We have encountered the phenomenon of stimulated emission, in our earlier discussions. Consider an excited state 2 of system, and a lower energy level 1 separated by the energy $\hbar\omega_L$. If the system is placed in level 2, it will decay to level 1 with emission of a photon of frequency ω_L. The rate is proportional to $(1 + n_L)$, with n_L the number of photons in the particular mode that forms the final state. A lasing device consists of atoms or molecules with such levels, placed in a cavity or between mirrors that "trap" photons in one particular normal mode (or perhaps a small number of modes) with frequency ω_L. As the atoms radiate, the photons are trapped in the cavity, and n_L builds up in time. The build up in the photon population then enhances the decay rate through influence of the stimulated emission factor $(1 + n_L)$. If there is a large number of atoms in state 2, there is an avalanche, and a very large number of photons of frequency ω_L is produced. They are allowed to leak slowly out of the cavity.

For a laser to operate, we must have a large number of atoms in state 2. If, in addition, there are atoms in state 1 (which there will be after the emission process), these atoms will absorb the laser photons, thus degrading their number. An analysis

[10.9], based on a simple rate equation approach, shows that the requirement for the avalanche instability to occur is $N_2 > N_1$, where N_2 and N_1 are the populations of the levels responsible for the laser photons. If this condition can be maintained in a steady manner, then a very large number of laser photons can be produced. When one has established a configuration with $N_2 > N_1$, one says there is a population inversion in the system.

From our earlier discussions, it is clear that one cannot create a population inversion by exposing the system to radiation at the frequency ω_L, thus promoting any atoms in state 1 (which might be the ground state) into state 2. We saw earlier that under such conditions, at high intensity, the absorption rate saturates at the point where $N_1 \cong N_2$. Through use of intense resonant radiation, we can never achieve a population inversion.

Lasers operate through use of three levels, and not just two. Typically the two laser levels, state 2 and state 1, are excited states which lie above the ground state 0. Resonant radiation, referred to as the pump radiation, promotes atoms from the ground level to level 2. Once in level 2 (if level 1 and level 0 are the only two levels of lower energy), an atom may decay to level 1 thus emitting a laser photon, then subsequently decay to the ground state at a rate γ_{10}. They may also decay to the ground state directly from the excited level 2 at the rate γ_{20}. If the decay rate γ_{10} is very large, then state 1 is depleted as soon as the atom arrives there after emitting its laser photon. State 1 is depleted, while the pump beam continues to refresh the population of state 2. Under these conditions, one can establish a population inversion, and realize laser action. We refer the reader elsewhere for a detailed discussion [10.9].

For many years, it had been realized that at high pump powers, the output of laser systems can behave in an erratic manner. Several years ago, this behavior was studied carefully, noting the underlying rate equations are nonlinear in nature. There are now numerous examples in the literature where this erratic behavior has been shown to illustrate fundamental aspects of the theory of chaos.

Early papers of *Casperson* [10.10] provided impressive quantitative contact between theory and experimental observations. In this work, the output at 3.15 μm of a xenon laser was analyzed. At low pump intensities, the output is steady and time independent. A threshold was crossed with increasing pump intensity after which the output consists of periodically repeated pulses, each of which contains four or five beats. We may understand this in terms of the first step in the Ruelle-Takens route to chaos. At low pump intensities, there is a stable fixed point, and the output is time independent. There is a bifurcation at higher pump intensities above which limit cycle behavior is realized; this is the regime where the laser output consists of a repeated sequence of pulses. In his theoretical analysis, Casperson reproduces the observed pulse shapes in a most impressive manner. He emphasizes that the rate equations used commonly to describe laser action do not contain a description of this phenomenon. It is essential to base the theory on the use of the field amplitudes.

In an experimental study of the output of a He-Ne laser at 3.39 μm, *Weiss* et al. [10.11] have reported observation of each of the three routes to chaos described in the previous section. This was achieved within one single laser, by continuously tilting one particular cavity mirror. With the mirror set so the period doubling route to chaos is followed, a sequence of four period doublings was reported, before the onset of chaos was seen. There is no detailed theory which relates the particular route to chaos pursued by the laser, to the mirror orientation, for the apparatus used in this study.

There is another arena in which the onset of chaos has been reported. There are numerous examples of systems which exhibit the phenomenon of optical bistability [10.12]. These are typically resonant media, where in the resonant frequency has a dependence on the intensity of the field that illuminates the material [10.13]. The resonance may have its origin in the matching of the input frequency to an excitation of the system, whose energy is affected by the field. We saw, for instance, in Chap. 2 that semiconductors have exciton levels, whose absorption frequency lies very close to that of the absorption edge of the material. If such a material is illuminated with light very close in frequency to that of an exciton absorption, some electrons will also be placed in the conduction band, since the band has a "tail" that extends below the nominal absorption edge. These free cariers will affect the binding energy of the exciton, and shift the exciton absorption band. The denominator in the exciton contribution to $\varepsilon(\omega)$, the resonant term in [2.38], thus depends on power.

There also are systems with a geometrical resonance; a Fabry-Perot cavity is an example. If L is the length of the cavity, there will be a transmission resonance when the wave length λ of the radiation is such that an integral number of half wavelengths fit into the cavity. The condition is thus $\lambda = 2nL$. The frequency of the nth resonance is $\omega_n = \pi c/nL\sqrt{\varepsilon}$, with ε the dielectric constant of the medium. If ε depends on field intensity, there will be a power dependent shift in the resonance position that depends on field intensity. The gap soliton induced transmission resonances are an example of an internal resonance, dependent on nonlinear optical response for its existence, which leads to bistability [10.14].

Observation of a transition to chaos by the period doubling route in a nonlinear optical fiber was reported by *Nakatsuka* et al. [10.15]. In a device which employed a bistable element, Gibbs and co-workers reported a transition to chaos consistent with the period doubling scenario [10.16]. The experimental observations are in fact compatible with theoretical descriptions appropriate to this class of systems. A review of the relevant theory, and a more complete list of references, is found in the review article by Akerhalt et al. cited earlier.

The experimental study of chaos, and the various routes to chaos was not initiated in the field of optics. The early experimental research centered on the study of hydrodynamic instabilities in various simple geometries. However, the full set of equations describing the hydrodynamic studies are complex nonlinear differential equations in space and time. As a consequence, full and quantitative contact between theory and experiment such as that found in the work of Casperson is difficult to achieve. One or a few electromagnetic modes, isolated by a cavity arrangement, are described by differential equations for the mode amplitudes that depend on a single variable, the time. Similarly, an ensemble of atoms, viewed as two level systems, are described by small set of time-dependent variables, as in our discussion of the Bloch equations. Thus, nonlinear optics in an excellent arena within which the predictions of theory can be placed alongside data in a quantitative manner.

Problems

10.1 Consider the Duffing oscillator, described by (10.1) of the text.

We might try to explore solutions when E_0 is large by seeking a solution in the form

$$x(t) = a(t)e^{-i\omega t} + a^*(t)e^{+i\omega t} ,$$

where, following the spirit of the slowly varying envelope approximation used in earlier chapters, we assume the time variation of $a(t)$ is slow compared to that of $\exp(\pm i\omega t)$.

(a) Show that if terms in d^2a/dt^2 are neglected, the envelope function obeys

$$(\gamma - 2i\omega)\frac{da}{dt} + (\omega_0^2 - \omega^2 - 2i\gamma\omega - 3b|a|^2)a = \frac{eE_0}{2M} .$$

For the case where we have a steady state oscillation, $a(t)$ is independent of time. Derive a cubic equation for $|a|^2$, when this is assumed.

(b) From the structure of the cubic equation derived in (a), demonstrate that

(i) when $\omega > \omega_0$, there is only one physical solution for $|a|^2$.

(ii) when $\omega < \omega_0$, and also when $\omega_0^2 - \omega^2 \gg 4\gamma\omega$, show the change of variables

$$x = \frac{b|a|^2}{\omega_0^2 - \omega^2}$$

maps the equation for $|a|^2$ into the form

$$x^3 - \frac{2}{3}x^2 + \frac{1}{9} - \frac{1}{9}\lambda = 0 ,$$

where $\lambda = b(eE_0/2M)^2(\omega_0^2 - \omega^2)^3$. Study the solutions as a function of λ, to find a bifurcation point when λ exceeds a vertain critical value.

With this approach, while interesting behavior emerges from the analysis, the approximation scheme provides no direct contact with the period doubling route to chaos. Nontheless, as discussed in [10.3], aspects of the present solution. suitably interpreted, can be placed alongside information extracted from the full numerical analysis of the Duffing oscillator.

10.2 A very interesting nonlinear mapping that is easily studied with either a programmable hand calculator or very small computer is the logistic map,

$$x_{n+1} = \alpha x_n(1 - x_n) ,$$

where $0 < \alpha < 4$, and x lies in the range [0, 1].

(a) Find the two fixed points of this mapping, i.e., the special values of x where the "input" equals the "output". Investigate the range of α within which each fixed point is stable by a linear stability analysis, i.e., if x^* is the value x at a fixed point, let $x_n = x^* + \delta x_n$, and for small δx_n inquire (by analytic means) when successive iterations drive x_n back to x^*.

(b) Program the logistic map on a hand calculator, and explore the output as a function of α. You should realize chaotic output when $\alpha > \alpha_c = 2.77\ldots$, and as α is increased to α_c from below, the chaotic regime is approached by the period doubling route.

Appendix A: Structure of the Wave Vector and Frequency Dependent Dielectric Tensor

We shall examine the response of a general physical system to an electric field with the space and time variation

$$E(r,t) = E(k)e^{ik\cdot r}e^{-i\omega t} + \text{c.c.}\,. \tag{A.1}$$

The electric field may be derived from the vector potential

$$A(r,t) = A(k)e^{ik\cdot r}e^{-i\omega t} + \text{c.c.} \tag{A.2}$$

by using $E = -(1/c)\partial A/\partial t$. Hence, $A(k) = -icE(k)/\omega$.

We proceed by calculating the expectation value of the current density, $\langle J_\alpha(r)\rangle$. For any array of particles described by a Hamiltonian invariant under rigid translation of the particles, we will have

$$\langle J_\alpha(r)\rangle = J_\alpha(k)e^{ik\cdot r}e^{-i\omega t} + \text{c.c.}\,, \tag{A.3}$$

where we define the wave vector and frequency dependent conductivity tensor $\sigma_{\alpha\beta}(k,\omega)$ by calculating $\langle J_\alpha(r)\rangle$ to first order in $E(k)$, then writing

$$J_\alpha(k) = \sum_\beta \sigma_{\alpha\beta}(k,\omega)E_\beta(k)\,. \tag{A.4}$$

If all charges in the system, including bound charges, are allowed to contribute to $J_\alpha(k)$, the dielectric tensor $\varepsilon_{\alpha\beta}(k,\omega)$ is related to $\sigma_{\alpha\beta}(k,\omega)$ by the relation [see the discussion which follows (A.49)]

$$\varepsilon_{\alpha\beta}(k,\omega) = \delta_{\alpha\beta} + \frac{4\pi i}{\omega}\sigma_{\alpha\beta}(k,\omega)\,. \tag{A.5}$$

To begin, recall that for a single particle of mass m and charge e, the current density in quantum mechanics is

$$j(\mathrm{r}) = \frac{e\hbar}{2im}[\psi^*\nabla\psi - (\nabla\psi^*)\psi] - \frac{e^2}{mc}A(r)|\psi(r)|^2\,, \tag{A.6}$$

when the particle is in the state $\psi(r)$. This is the expectation value of an operator j_{op}

$$j(r) = \int d^3r'\psi^*(r')j_{\mathrm{op}}\psi(r')\,, \tag{A.7}$$

where

$$j_{\text{op}} = \frac{e\hbar}{2\text{i}m}[\delta(\boldsymbol{r}-\boldsymbol{r}')\nabla_{r'} + \nabla_{r'}\,\delta(\boldsymbol{r}-\boldsymbol{r}')] - \frac{e^2}{mc}\delta(\boldsymbol{r}-\boldsymbol{r}')\boldsymbol{A}(\boldsymbol{r}')\,. \tag{A.8}$$

For a system of many particles, generalizing this gives the current operator

$$\boldsymbol{J}_{\text{op}} = \frac{e\hbar}{2i}\sum_j \left[\frac{1}{m_j}\delta(\boldsymbol{r}-\boldsymbol{r}_j)\nabla_{r_j} + \frac{1}{m_j}\nabla_{r_j}\delta(\boldsymbol{r}-\boldsymbol{r}_j)\right] - \frac{e^2}{c}\sum_j \frac{\delta(\boldsymbol{r}-\boldsymbol{r}_j)}{m_j}\boldsymbol{A}(\boldsymbol{r}_j)\,. \tag{A.9}$$

Recall that the momentum operator is given by $\boldsymbol{p}_j = (\hbar/i)\nabla_{r_j}$. We write

$$\boldsymbol{J}_{\text{op}} = \boldsymbol{J}_{\text{op}}^{(\text{P})} + \boldsymbol{J}_{\text{op}}^{(\text{D})}\,, \tag{A.10}$$

where $\boldsymbol{J}_{\text{op}}^{(\text{D})}$, often called the diamagnetic contribution to the current density, is the term proportional to $\boldsymbol{A}$. The remainder, $\boldsymbol{j}_{\text{op}}^{(\text{P})}$ is referred to as the paramagnetic contribution.

We suppose all particles are confined to a large volume V, and write for the total current density operator and its components,

$$\boldsymbol{J}_{\text{op}} = \frac{1}{V}\sum_{\boldsymbol{k}} \boldsymbol{J}(\boldsymbol{k})\text{e}^{\text{i}\boldsymbol{k}\cdot\boldsymbol{r}}\,. \tag{A.11}$$

One has

$$\boldsymbol{J}_\alpha^{(\text{P})}(\boldsymbol{k}) = \sum_j \left(\text{e}^{-\text{i}\boldsymbol{k}\cdot\boldsymbol{r}_j}\frac{e_j\boldsymbol{p}_j}{2m_j} + \frac{e_j\boldsymbol{p}_j}{2m_j}\text{e}^{-\text{i}\boldsymbol{k}\cdot\boldsymbol{r}_j}\right) \tag{A.12}$$

and

$$\boldsymbol{J}_\alpha^{(\text{D})}(\boldsymbol{k}) = -\sum_j \frac{e_j^2}{m_j c}\text{e}^{-\text{i}\boldsymbol{k}\cdot\boldsymbol{r}_j}\boldsymbol{A}(\boldsymbol{r}_j)\,. \tag{A.13}$$

Let our system of particles occupy the state $|\psi_0\rangle$, which we assume to be the ground state, and also an eigenstate of total momentum

$$\boldsymbol{p}^{(\text{T})} = \sum_j \boldsymbol{p}_j \tag{A.14}$$

with eigenvalue zero.

The calculation of $\langle\psi_0|\boldsymbol{J}_\alpha^{(\text{D})}(\boldsymbol{k})|\psi_0\rangle$ to first order in the applied field is straightforward, since $J_\alpha^{(\text{D})}$ is itself proportional to $\boldsymbol{A}$. One simply takes the expectation value of the operator:

$$\langle\psi_0|\boldsymbol{J}_\alpha^{(\text{D})}(\boldsymbol{k}')|\psi_0\rangle = -\frac{1}{c}\boldsymbol{A}(\boldsymbol{k})\text{e}^{-\text{i}\omega t}\sum_j \frac{e_j^2}{m_j}\langle\psi_0|\text{e}^{\text{i}(\boldsymbol{k}-\boldsymbol{k}')\cdot\boldsymbol{r}_j}|\psi_0\rangle - \frac{1}{c}\boldsymbol{A}^*(\boldsymbol{k})\text{e}^{+\text{i}\omega t}\sum_j \frac{e_j^2}{m_j}\langle\psi_0|\text{e}^{-\text{i}(\boldsymbol{k}+\boldsymbol{k}')\cdot\boldsymbol{r}_j}|\psi_0\rangle\,. \tag{A.15}$$

Suppose we define the operator

$$S(\boldsymbol{Q}) = \sum_j \frac{e_j^2}{m_j} e^{i\boldsymbol{Q}\cdot\boldsymbol{R}_j} . \tag{A.16}$$

Upon noting $[\boldsymbol{p}^{(T)}, S(\boldsymbol{Q})] = \hbar\boldsymbol{Q}S(\boldsymbol{Q})$, one sees easily that $S(\boldsymbol{Q})|\psi_0\rangle$ is an eigenfunction of $\boldsymbol{p}^{(T)}$ with eigenvalue $\hbar\boldsymbol{Q}$. It follows that $\langle\psi_0|S(\boldsymbol{Q})|\psi_0\rangle$ is proportional to the Kronecker delta $\delta_{\boldsymbol{Q},0}$, which is unity when $\boldsymbol{Q} \equiv 0$, and zero otherwise. Then the only nonzero contributions to $\langle\psi_0|J_\alpha^{(D)}(\boldsymbol{k}')|\psi_0\rangle$ are those for which $\boldsymbol{k}' = \pm\boldsymbol{k}$. A bit of algebra leads to

$$\langle\psi_0|\boldsymbol{J}_{op}^{(D)}|\psi_0\rangle = \frac{i}{\omega V}\left(\sum_j \frac{e_j^2}{m_j}\right)[\boldsymbol{E}(\boldsymbol{k})e^{i\boldsymbol{k}\cdot\boldsymbol{r}}e^{-i\omega t} - \boldsymbol{E}^*(\boldsymbol{k})e^{-i\boldsymbol{k}\cdot\boldsymbol{r}}e^{+i\omega t}] . \tag{A.17}$$

To calculate the expectation value of the paramagnetic current density, which is by itself zero order in the electric field or vector potential, we need to expand the wave function $|\psi_0\rangle$ in powers of the field. If this is done, and $|\psi_0^{(1)}\rangle$ is the piece of the wave function first order in $\boldsymbol{A}$, then

$$\langle\psi|\boldsymbol{J}_{op}^{(P)}|\psi\rangle = \langle\psi_0^{(1)}|\boldsymbol{J}_{op}^{(P)}|\psi_0\rangle + \langle\psi_0|\boldsymbol{J}_{op}^{(P)}|\psi_0^{(1)}\rangle . \tag{A.18}$$

To find $|\psi_0^{(1)}\rangle$, we must examine the manner in which $\boldsymbol{A}$ enters the Hamiltonian of the system. The Hamiltonian is written, in the presence of the vector potential,

$$H = \sum_j \frac{1}{2m_j}\left[\boldsymbol{p}_j - \frac{e_j}{c}\boldsymbol{A}(\boldsymbol{r}_j)\right]^2 + V , \tag{A.19}$$

where V contains the interaction between the particles; we shall not require its precise form. To first order in $\boldsymbol{A}$, with the vector potential given in (A.2), one has

$$H = H_0 - \frac{1}{c}\boldsymbol{J}^{(P)}(-\boldsymbol{k})\cdot\boldsymbol{A}(\boldsymbol{k})e^{-i\omega t}e^{\eta t} - \frac{1}{c}\boldsymbol{J}^{(P)}(\boldsymbol{k})\cdot\boldsymbol{A}^*(\boldsymbol{k})e^{+i\omega t}e^{\eta t} , \tag{A.20}$$

where we append the factor of $e^{\eta t}$ to the time dependence of the vector potential, so the perturbation is turned on adiabatically.

Application of time dependent perturbation theory to (A.20) yields an expression for $|\psi_0^{(1)}\rangle$. Given the Hamiltonian

$$H = H_0 + v e^{-i\omega t}e^{\eta t} + v^+ e^{i\omega t}e^{\eta t} , \tag{A.21}$$

by this means the wave function to order in v is

$$|\psi\rangle = e^{-iE_0/\hbar t}\left[|\psi_0\rangle - \frac{1}{\hbar}\sum_n \frac{\langle\psi_n|v|\psi_0\rangle}{\omega_{n0} - \omega - i\eta} e^{-i(\omega+i\eta)t}|\psi_n\rangle \right.$$
$$\left. - \frac{1}{\hbar}\sum_n \frac{\langle\psi_n|v^+|\psi_0\rangle}{\omega_{n0} + \omega + i\eta} e^{i(\omega-i\eta)t}|\psi_n\rangle\right], \tag{A.22}$$

where E_0 is the energy of state $|\psi_0\rangle$, E_n that of state $|\psi_n\rangle$, and $\omega_{n0} = (E_n - E_0)/\hbar$. Given any operator O with the property $\langle\psi_0|O|\psi_0\rangle = 0$, to first order in v we find, after a short calculation

$$\langle\psi|O|\psi\rangle = -\frac{1}{\hbar}\sum_n\left[\frac{\langle\psi_0|O|\psi_n\rangle\langle\psi_n|v|\psi_0\rangle}{\omega_{n0}+\omega+\mathrm{i}\eta} + \frac{\langle\psi_0|O|\psi_n\rangle\langle\psi_n|v|\psi_0\rangle}{\omega_{n0}-\omega-\mathrm{i}\eta}\right] \times \mathrm{e}^{-\mathrm{i}(\omega+\mathrm{i}\eta)t} + \text{c.c.} \,. \tag{A.23}$$

With these results in hand, for $\langle\psi|\boldsymbol{J}^{(\mathrm{P})}_{\mathrm{op}}|\psi\rangle$ one finds, with the subscript α referring to the αth Cartesian component of the current density,

$$\langle\psi|\boldsymbol{J}^{(\mathrm{P})}_{\mathrm{op}_\alpha}|\psi\rangle = \frac{1}{V\hbar c}\sum_\beta\sum_n\left[\frac{\langle\psi_0|J^{(\mathrm{P})}_\beta(\boldsymbol{k})^+|\psi_n\rangle\langle\psi_n|J^{(\mathrm{P})}_\alpha(\boldsymbol{k})|\psi_0\rangle}{\omega_{n0}+\omega+\mathrm{i}\eta} + \frac{\langle\psi_0|J^{(\mathrm{P})}_\alpha(\boldsymbol{k})|\psi_n\rangle\langle\psi_n|J^{(\mathrm{P})}_\beta(\boldsymbol{k})^+|\psi_0\rangle}{\omega_{n0}-\omega-\mathrm{i}\eta}\right]A_\beta(\boldsymbol{k})\mathrm{e}^{\mathrm{i}\boldsymbol{k}\cdot\boldsymbol{r}}\mathrm{e}^{-\mathrm{i}\omega t} + \text{c.c.} \tag{A.24}$$

We have used the identity $\boldsymbol{J}^{(\mathrm{P})}_\alpha(-\boldsymbol{k}) = \boldsymbol{J}^{(\mathrm{P})}_\alpha(\boldsymbol{k})^+$.

We pause to look more closely at the properties of the states which enter (A.24). We have said that the unperturbed Hamiltonian H_0 is invariant under a rigid body translation of the whole system. This implies that all eigenstates of H_0 are eigenstates of the total momentum $\boldsymbol{p}^{(\mathrm{T})}$ defined in (A.14); we have assumed that $|\psi_0\rangle$, the ground state, is in eigenstate with zero total momentum.

One may show that the state $J^{(\mathrm{P})}_\alpha(\boldsymbol{k})|\psi_0\rangle$ is an eigenstate of $\boldsymbol{p}^{(\mathrm{T})}$ with momentum $-\hbar\boldsymbol{k}$. To see this, one notes the commutation relation

$$[p^{(\mathrm{T})}_\gamma, J^{(\mathrm{P})}_\alpha(\boldsymbol{k})] = -\hbar k_\gamma J^{(\mathrm{P})}_\alpha(\boldsymbol{k}) \,. \tag{A.25}$$

Then

$$\begin{aligned} p^{(\mathrm{T})}_\gamma(J^{(\mathrm{P})}_\alpha(\boldsymbol{k})|\psi_0\rangle) &= (J^{(\mathrm{P})}_\alpha(\boldsymbol{k})p^{(\mathrm{T})}_\gamma - \hbar k_\gamma J^{(\mathrm{P})}_\alpha(\boldsymbol{k}))|\psi_0\rangle \\ &= -\hbar k_\gamma(J^{(\mathrm{P})}_\alpha(\boldsymbol{k}))|\psi_0\rangle \,. \end{aligned} \tag{A.26}$$

Similarly, $J^{(\mathrm{P})}_\beta(\boldsymbol{k})^+$ operating on $|\psi_0\rangle$ creates a state with total momentum $+\hbar\boldsymbol{k}$.

From this it follows that the intermediate state $|\psi_n\rangle$ in the first term of (A.24) must be an eigenstate of total momentum with eigenvalue $-\hbar\boldsymbol{k}$, and the state $|\psi_n\rangle$ in the second term must be an eigenstate with momentum $+\hbar\boldsymbol{k}$. We should label the various energies and eigenstates which enter (A.24) with momentum indices. Thus, in the first term of (A.24), the matrix element $\langle\psi_n|J^{(\mathrm{P})}_\alpha(\boldsymbol{k})|\psi_0\rangle$ will be replaced by the more explicit form $\langle\psi_n(-\boldsymbol{k})|J^{(\mathrm{P})}_\alpha(\boldsymbol{k})|\psi_0\rangle$. In the energy denominator ω_{n0} is replaced by $\omega_{n0}(-\boldsymbol{k}) = [E_n(-\boldsymbol{k}) - E_0(0)]/\hbar$.

Suppose the physical system of interest is a collection of atoms, a case considered below. The energy of an atom consists of two parts, the kinetic energy associated with the center of mass motion, and the internal energy, independent of center of mass motion, that we may call ε_n. Then for a single atom $E_n(\boldsymbol{k}) = \varepsilon_n + \hbar^2k^2/2M$, with M the mass of the atom. This provides an example of a means through which the energy denominators acquire a wave vector dependence.

In physical terms, the paramagnetic current density is nonzero within the atom, because in the presence of the electromagnetic field, the atom has absorbed a (virtual) photon; it recoils as it does so, and thus acquires the center of mass momentum $\hbar \boldsymbol{k}$.

It is now a matter of collecting various terms together to generate an expression for $\varepsilon_{\alpha\beta}(\boldsymbol{k}, \omega)$, using its relation to the frequency and wave vector dependent conductivity given in (A.5). Using the explicit notation allows us to write this in the form

$$\varepsilon_{\alpha\beta}(\boldsymbol{k}, \omega) = \delta_{\alpha\beta} + \frac{4\pi}{\omega^2 \hbar V} \sum_n \left[\frac{\langle \psi_0(0) | J_\beta^{(\mathrm{P})}(\boldsymbol{k})^+ | \psi_n(-\boldsymbol{k}) \rangle \langle \psi_n(-\boldsymbol{k}) | J_\alpha^{(\mathrm{P})}(\boldsymbol{k}) | \psi_0(0) \rangle}{\omega_{n0}(-\boldsymbol{k}) + \omega + \mathrm{i}\eta} \right.$$
$$\left. + \frac{\langle \psi_0(0) | J_\alpha^{(\mathrm{P})}(\boldsymbol{k}) | \psi_n(\boldsymbol{k}) \rangle \langle \psi_n(\boldsymbol{k}) | J_\beta^{(\mathrm{P})}(\boldsymbol{k})^+ | \psi_0(0) \rangle}{\omega_{n0}(\boldsymbol{k}) - \omega - \mathrm{i}\eta} \right]$$
$$- \frac{4\pi\delta_{\alpha\beta}}{\omega^2 V} \left(\sum_j \frac{e_j^2}{m_j} \right) . \tag{A.27}$$

From the identity $J_\alpha^{(\mathrm{P})}(-\boldsymbol{k}) = J_\alpha^{(\mathrm{P})}(\boldsymbol{k})^+$, one may establish that $\varepsilon_{\alpha\beta}^*(\boldsymbol{k}, \omega) = \varepsilon_{\alpha\beta}(-\boldsymbol{k}, -\omega)$.

The expression in (A.27) is very general, and may be used for explicit calculations for a wide range of physical systems. It has the awkward feature that the quantization volume V appears explicitly. However, as the following example illustrates, in any explicit calculation, this disappears and is replaced by a physically meaningful quantity.

Consider, for example, that the system of interest consists of N atoms, each of which contains n particles, $n - 1$ electrons and a nucleus. Suppose the atoms are viewed as non-interacting, and distinguishable. The terms in the paramagnetic current density may be arranged so $J_\alpha^{(\mathrm{P})}(\boldsymbol{k})$ is written

$$J_\alpha^{(\mathrm{P})}(\boldsymbol{k}) = j_\alpha^1(\boldsymbol{k}) + j_\alpha^2(\boldsymbol{k}) + \cdots + j_\alpha^N(\boldsymbol{k}) , \tag{A.28}$$

where

$$j_\alpha^1(\boldsymbol{k}) = \sum_{i=1}^{n} \left[\mathrm{e}^{\mathrm{i}\boldsymbol{k}\cdot\boldsymbol{r}_i} \frac{e_i \boldsymbol{p}_i}{2m_i} + \frac{e_i \boldsymbol{p}_i}{2m_i} \mathrm{e}^{-\mathrm{i}\boldsymbol{k}\cdot\boldsymbol{r}_i} \right] \tag{A.29}$$

contains only the coordinates and momenta of the constituents of atom 1, $j_\alpha^2(\boldsymbol{k})$ those associated with atom 2, and so on.

For this picture $|\psi_0\rangle = \Pi_{m=1}^N |\phi_0(m)\rangle$, with $|\phi_0(m)\rangle$ the ground state of atom m. The excited states $|\psi_n\rangle$ of the array which contribute to the dielectric tensor describe a configuration with $N - 1$ atoms in the ground state, and one of them, say atom ℓ, in the excited state ϕ_n. Thus,

$$|\psi_n\rangle = |\phi_n(\ell)\rangle \prod_{\substack{m=1 \\ m\neq\ell}}^{N} |\phi_0(m)\rangle \tag{A.30}$$

and

$$\langle\psi_n|J_\alpha^{(\mathrm{P})}(\boldsymbol{k})|\psi_0\rangle = \langle\phi_n(\ell)|j_\alpha^\ell(\boldsymbol{k})|\phi_0(\ell)\rangle \equiv \langle n|j_\alpha(\boldsymbol{k})|0\rangle\,, \tag{A.31}$$

where in the last step, the notation acknowledges that the value of the matrix element is independent of which atom is excited.

There are N different such excited states $|\psi_n\rangle$ corresponding to the various choices of ℓ. When the contribution of each of these is included we have, reverting to a notation which explicitly indicates the center of mass momentum of the various atoms,

$$\begin{aligned}\varepsilon_{\alpha\beta}(\boldsymbol{k},\omega) = \delta_{\alpha\beta} + \frac{4\pi N}{\omega^2\hbar V}\sum_n \Bigg[&\frac{\langle\phi_0(0)|j_\beta(\boldsymbol{k})^+|\phi_n(-\boldsymbol{k})\rangle\langle\phi_n(-\boldsymbol{k})|j_\alpha(\boldsymbol{k})|\phi_0(0)\rangle}{\omega_{n0}(-\boldsymbol{k}) + \omega + \mathrm{i}\eta} \\ &+ \frac{\langle\phi_0(0)|j_\alpha(\boldsymbol{k})|\phi_n(\boldsymbol{k})\rangle\langle\phi_n(\boldsymbol{k})|j_\beta(\boldsymbol{k})^+|\phi_0(0)\rangle}{\omega_{n0}(\boldsymbol{k}) - \omega - \mathrm{i}\eta}\Bigg] \\ &- \delta_{\alpha\beta}\frac{4\pi N}{\omega^2 V}\sum_{i=1}^{n}\frac{e_i^2}{m_i}\,.\end{aligned} \tag{A.32}$$

We now see that the electric susceptibility is proportional to N/V, the density of atoms. While this simple dependence on density is a consequence of the special assumption that the atoms are a non-interacting ensemble of atoms, in any application of the general formula in (A.27), the quantization volume V will disappear, and only physically meaningful parameters will appear in the final expression.

We now turn to special limits of the general formula in (A.27). It is argued in Chap. 2, that since the wavelength of light is very long compared to the microscopic length scales that characterize numerous systems of interest, one may replace the numerator in (A.27) by its limiting form as $\boldsymbol{k}\to 0$, in a number of circumstances. The denominators, however, are sensitive to $\boldsymbol{k}$ near a resonance where $\omega \cong \omega_{n0}(0)$, so for the moment we retain the denominators in full. We examine the limit just described, and in the process arrange (A.27) into an alternate form that is seen frequently. Upon noting that the interaction term V in (A.19) depends on the coordinates of the various particles, so the momentum operators enter only the kinetic energy, one may establish the identity

$$J_\alpha^{(\mathrm{P})}(\boldsymbol{k}) = \frac{1}{\mathrm{i}\hbar}\sum_j e_j[r_\alpha^j\,\mathrm{e}^{-\mathrm{i}\boldsymbol{k}\cdot\boldsymbol{r}_j}, H_0] + \frac{i}{2}k_\gamma\sum_j\sum_\gamma\frac{e_j}{m_j}Q_{\alpha\gamma}^j \tag{A.33}$$

where $Q_{\alpha\gamma}^j = r_\alpha^j p_\gamma^j + p_\gamma^j r_\alpha^j$. One then has

$$\begin{aligned}\langle\psi_0(0)|J_\alpha^{(\mathrm{P})}(\boldsymbol{k})|\psi_n(\boldsymbol{k})\rangle = -\mathrm{i}\,\omega_{n0}(\boldsymbol{k})&\left\langle\psi_0(0)\left|\sum_j e_j r_\alpha^j\mathrm{e}^{-\mathrm{i}\boldsymbol{k}\cdot\boldsymbol{r}_j}\right|\psi_n(\boldsymbol{k})\right\rangle \\ &+\frac{\mathrm{i}}{2}\sum_\gamma k_\gamma\left\langle\psi_0(0)\left|\frac{e_j}{m_j}Q_{\alpha\gamma}^j\right|\psi_n(\boldsymbol{k})\right\rangle\,.\end{aligned} \tag{A.34}$$

In the limit of small $\boldsymbol{k}$, we approximate this by

$$\langle\psi_0(0)|J_\alpha^{(\mathrm{P})}(\boldsymbol{k})|\psi_n(\boldsymbol{k})\rangle \cong -\mathrm{i}\omega_{n0}(\boldsymbol{k})\left\langle \psi_0\left|\sum_j e_j r_\alpha^j\right|\psi_n\right\rangle, \tag{A.35}$$

where on the right-hand side, both $|\psi_0\rangle$ and $|\psi_n\rangle$ describe states with $\boldsymbol{k} = 0$. Then (A.27) becomes, noting we will have $\omega_{n0}(\boldsymbol{k}) = \omega_{n0}(-\boldsymbol{k})$ on general grounds,

$$\varepsilon_{\alpha\beta}(\boldsymbol{k},\omega) = \delta_{\alpha\beta} + \frac{4\pi}{\omega^2\hbar V}\sum_n \omega_{n0}^2(\boldsymbol{k})\left[\frac{\left\langle\psi_0\left|\sum_j e_j r_\beta^j\right|\psi_0^n\right\rangle\left\langle\psi_n\left|\sum_j e_j r_\alpha^j\right|\psi_0\right\rangle}{\omega_{n0}(\boldsymbol{k}) + \omega + \mathrm{i}\eta} + \frac{\left\langle\psi_0\left|\sum_j e_j r_\alpha^j\right|\psi_n\right\rangle\left\langle\psi_n\left|\sum_j e_j r_\beta^j\right|\psi_0\right\rangle}{\omega_{n0}(\boldsymbol{k}) - \omega - \mathrm{i}\eta}\right] - \frac{4\pi\delta_{\alpha\beta}}{\omega^2 V}\left(\sum_j \frac{e_j^2}{m_j}\right). \tag{A.36}$$

The diagonal element $\varepsilon_{\alpha\alpha}(\boldsymbol{k}\omega)$ becomes simpler:

$$\varepsilon_{\alpha\alpha}(\boldsymbol{k},\omega) = 1 + \frac{8\pi}{\omega^2\hbar V}\sum_n \frac{\omega_{n0}^3(\boldsymbol{k})}{\omega_{n0}^2(\boldsymbol{k}) - (\omega + \mathrm{i}\eta)^2}\left|\left\langle\psi_n\left|\sum_j e_j r_\alpha^j\right|\psi_0\right\rangle\right|^2 - \frac{4\pi}{\omega^2 V}\sum_j \frac{e_j^2}{m_j}. \tag{A.37}$$

This expression may be arranged so the behavior of $\varepsilon_{\alpha\alpha}$ as $\omega \to 0$ is clear and evident. Notice that

$$\frac{\omega_{n0}^2(k)}{\omega_{n0}^2(k) - (\omega + \mathrm{i}\eta)^2} = 1 + \frac{\omega^2}{\omega_{n0}^2(k) - (\omega + \mathrm{i}\eta)^2}. \tag{A.38}$$

One may then replace the remaining factors of $\omega_{n0}(\boldsymbol{k})$ in the numerator by $\omega_{n0}(0)$, and make use of the identity

$$\sum_n \omega_{n0}\left|\left\langle\psi_n\left|\sum_j e_j r_\alpha^j\right|\psi_0\right\rangle\right|^2 = \frac{1}{2}\hbar\sum_j \frac{e_j}{m_j}, \tag{A.39}$$

which is often referred to as the f sum rule. We then find

$$\varepsilon_{\alpha\alpha}(\boldsymbol{k},\omega) = 1 + \frac{8\pi}{\hbar V}\sum_n \frac{\omega_{n0}(0)\left|\left\langle\psi_n\left|\sum_j e_j r_\alpha^j\right|\psi_0\right\rangle\right|^2}{\omega_{n0}^2(\boldsymbol{k}) - (\omega + \mathrm{i}\eta)^2}. \tag{A.40}$$

In (A.40), the states $|\psi_0\rangle$ and $|\psi_n\rangle$ describe our many particle system, with particles distributed throughout the quantization volume V. The sum on j ranges over all particles. If we apply this to a set of N non-interacting atoms or molecules, each of which contains n particles, one finds, following arguments similar to those which precede (A.32),

$$\varepsilon_{\alpha\alpha}(\boldsymbol{k}, \omega) = 1 + \frac{8\pi N}{\hbar V} \sum_n \frac{\omega_{n0}(0) \left| \left\langle \phi_n \left| \sum_{j=1}^{n} e_j r_\alpha^j \right| \phi_0 \right\rangle \right|^2}{\omega_{n0}^2(\boldsymbol{k}) - (\omega + i\eta)^2}, \tag{A.41}$$

where $|\phi_0\rangle$, $|\phi_n\rangle$ are states for one atom or molecule, and $\omega_{n0}(\boldsymbol{k})$ is now the excitation spectrum of such an object, with center of mass motion included.

Quite clearly, this analysis can be extended to higher order in the vector potential, to generate microscopic expressions for the nonlinear susceptibilities in the text. The reader should consult [A.1] for such an analysis.

Appendix B: Aspects of the Sine-Gordon Equation

We discuss the sine-Gordon equation within the framework of the continuum theory of a nonlinear one-dimensional line, with amplitude θ that depends on position x and the time, which we refer to as τ, to follow the notation of Chap. 6.

One describes such a system by a Lagrangian L, composed of a Lagrangian density that is a function of $\dot{\theta} = \partial\theta/\partial\tau$, θ itself, and $\partial\theta/\partial x$:

$$L = \int dx\, \mathcal{L}(\dot{\theta}, \theta, \partial\theta/\partial x) . \tag{B.1}$$

The equation of motion is generated by requiring the action $S = \int_{\tau_a}^{\tau_b} d\tau\, L$ be an extremum with respect to variations $\delta\theta(x, \tau)$ in $\theta(x, \tau)$, subject to the constraint $\delta\theta(x, \tau_a) = \delta\theta(x, \tau_b) = 0$. This yields

$$\frac{\partial}{\partial\tau}\left(\frac{\partial\mathcal{L}}{\partial\dot{\theta}}\right) = -\frac{\partial\mathcal{L}}{\partial\theta} + \frac{\partial\mathcal{L}}{\partial(\partial\theta/\partial x)} . \tag{B.2}$$

The sine-Gordon equation, (6.76) is generated by

$$\mathcal{L} = \frac{1}{2}\left(\frac{\partial\theta}{\partial\tau}\right)^2 - \frac{1}{2}\bar{c}^2\left(\frac{\partial\theta}{\partial x}\right)^2 + \mu^2\cos\theta . \tag{B.3}$$

The Hamiltonian H is given by

$$H = \int dx(\dot{\theta}\pi - \mathcal{L}) , \tag{B.4}$$

where $\pi = \partial\mathcal{L}/\partial\dot{\theta}$ is the momentum density. Thus, upon subtracting a constant to render the quiescent state $\theta = 0$ the state of zero energy, we have

$$H = \int dx\left\{\frac{1}{2}\left(\frac{\partial\theta}{\partial\tau}\right)^2 + \frac{1}{2}\bar{c}^2\left(\frac{\partial\theta}{\partial x}\right)^2 - \mu^2[\cos(\theta) - 1]\right\} \tag{B.5 a}$$

or

$$H = \int dx\left\{\frac{1}{2}\left(\frac{\partial\theta}{\partial\tau}\right)^2 + \frac{1}{2}\bar{c}^2\left(\frac{\partial\theta}{\partial x}\right)^2 + 2\mu^2\sin^2\left(\frac{\theta}{2}\right)\right\} . \tag{B.5 b}$$

The soliton emerges by seeking a solution of (6.76) with $\theta(x, \tau)$ in the special form $\theta = \theta(\tau - x/v)$, where $v < \bar{c}$. Hence

$$\frac{\partial^2\theta}{\partial\tau^2} = \frac{\mu^2 v^2}{\bar{c}^2 - v^2}\sin\theta . \tag{B.7}$$

The remainder of the discussion proceeds along the lines given in Sect. 6.3. One has

$$\theta(x, \tau) = \pm 4 \tan^{-1}\left\{\exp\left[\pm\frac{\mu v}{(\bar{c}^2 - v^2)^{1/2}}\left(\tau - \frac{1}{v}x\right)\right]\right\}. \tag{B.8}$$

To calculate the energy, note $\partial\theta/\partial x = -(\partial\theta/\partial\tau)/v$, and the analogue of (6.59) reads $(\partial\theta/\partial\tau) = \pm[2\mu v \sin(\theta/2)]/(\bar{c}^2 - v^2)^{1/2}$. Hence, calling $E(v)$ the energy of a soliton with velocity v,

$$E(v) = \frac{4\mu^2\bar{c}^2}{\bar{c}^2 - v^2}\int_{-\infty}^{+\infty} dx \sin^2\left(\frac{\theta}{2}\right), \tag{B.9}$$

or changing variables to $\xi = \mu x/(\bar{c}^2 - v^2)^{1/2}$,

$$E(v) = \frac{4\mu\bar{c}^2}{(\bar{c}^2 - v^2)^{1/2}}\int_{-\infty}^{+\infty}\frac{d\xi}{\cosh^2(\xi)}, \tag{B.10}$$

or

$$E(v) = \frac{E(0)}{[1 - (v^2/\bar{c}^2)]^{1/2}}, \tag{B.11}$$

the relation quoted in Sect. 6.5. One has, upon evaluating the integral in (B.10), $E(0) = 8\mu$ as the energy of a soliton at rest.

In the text, we discussed the fact that the solitons which emerge from the sine-Gordon equation behave as noninteracting particles, in that they pass through each other, to emerge from the interaction unchanged in shape or amplitude. One may construct N soliton solutions of the sine-Gordon equation with this property [8.1,2]. There are two soliton states, which display this property and which may be expressed in terms of elementary functions.

For example, the function

$$\theta(x, \tau) = 4 \tan^{-1}\left\{\frac{v}{\bar{c}}\,\frac{\sinh\left[\frac{\mu}{\bar{c}}\gamma(v)x\right]}{\cosh\left[\frac{\mu}{\bar{c}}\gamma(v)v\tau\right]}\right\} \tag{B.12}$$

is a solution, for any value of v in the range $0 < v < \bar{c}$. Here $\gamma(v) = [1 - (v^2/\bar{c}^2)]^{-1/2}$. As $\tau \to -\infty$, this solution has the asymptotic form

$$\lim_{\tau\to-\infty} \tan\left(\frac{\theta}{4}\right) = \exp\left[\frac{\mu}{\bar{c}}\gamma(x - x_0 + v\tau)\right] - \exp\left[-\frac{\mu}{\bar{c}}\gamma(x + x_0 - v\tau)\right], \tag{B.13}$$

which describes a pair of solitons approaching each other with equal and opposite velocity. Here $x_0 = (\bar{c}/\mu\gamma)\ \ln(\bar{c}/v)$.

As $\tau \rightarrow +\infty$, we have two solitons which are receding from each other:

$$\lim_{\tau \rightarrow +\infty} \tan\left(\frac{\theta}{4}\right) = \exp\left[\frac{\mu\gamma}{\bar{c}}(x - x_0 - v\tau)\right] - \exp\left[-\frac{\mu\gamma}{\bar{c}}(x + x_0 + v\tau)\right]. \tag{B.14}$$

An excellent discussion of the rich range of solutions of the sine-Gordon equation has been given by *Curie* [B.3]. A splendid general discussion of soliton theory may be found in the text by *Rajaraman* [B.4].

Appendix C: Structure of the Electromagnetic Green's Functions

In Sect. 8.1, we addressed the task of solving for the electric fields generated by an electric dipole source term placed near the boundary of a dielectric medium. This is a special case of a more general problem: the description of electromagnetic wave generation by sources in spatially inhomogeneous media. For a source with frequency Ω, and for a medium whose (complex) dielectric constant depends on only z, the coordinate normal to the surface, we wish to solve a generalized version of (4.2b):

$$\nabla^2 \boldsymbol{E}(\boldsymbol{r},\Omega) - \nabla[\nabla \cdot \boldsymbol{E}(\boldsymbol{r},\Omega)] + \left(\frac{\Omega}{c}\right)^2 \varepsilon(\Omega,z)\boldsymbol{E}(\boldsymbol{r},\Omega)$$

$$= -\frac{4\pi\Omega^2}{c^2}\boldsymbol{P}^{(s)}(\boldsymbol{r},\Omega). \tag{C.1}$$

(Here we use Ω rather than ω, because at some points in the main text, we use 2ω in explicit application to second harmonic generation.) Note that any solution of (C.1) also satisfies the generalized version of (4.2.a), as one sees upon taking the divergence of (C.1). We remark that problems such as those described in (C.1) occur not only in the nonlinear optical response of the near surface environment, but in diverse problems in linear optics as well. For example, see [8.3]. Thus, we use a notation for the source term in (C.1) that is more general than that found in the main text.

We shall suppose that $\boldsymbol{P}^{(s)}(\boldsymbol{r},\Omega)$ has a simple sinusoidal dependence on the coordinates parallel to the surface. The coordinate system may then by oriented so that

$$\boldsymbol{P}^{(s)}(\boldsymbol{r},\Omega) = \wp(z,\Omega)\mathrm{e}^{\mathrm{i}k_\parallel x} . \tag{C.2}$$

The response to more complex source structures may be analyzed using the approach described here, combined with Fourier synthesis, as in [8.1]. For the source term in (C.2), the electric field has the form

$$\boldsymbol{E}(\boldsymbol{r},\Omega) = \mathcal{E}(z,\Omega)\mathrm{e}^{\mathrm{i}k_\parallel x} . \tag{C.3}$$

The various components of the electric field thus satisfy

$$\left[\frac{d^2}{dz^2} + \left\{\frac{\Omega^2}{c^2}\varepsilon(\Omega,z) - k_\parallel^2\right\}\right]\mathcal{E}_y(z,\Omega) = -\frac{4\pi\Omega^2}{c^2}\wp_y(z,\Omega) \tag{C.4a}$$

and

$$\left[\frac{d^2}{dz^2}+\left(\frac{\Omega}{c}\right)^2\varepsilon(\Omega,z)\right]\mathcal{E}_x(z,\Omega)-ik_\parallel\frac{d\mathcal{E}_z}{dz}(z,\Omega)=-\frac{4\pi\Omega^2}{c^2}\wp_x(z,\Omega) \quad \text{(C.4b)}$$

with also

$$-ik_\parallel\frac{d\mathcal{E}_x(z,\Omega)}{dz}+\left[\left(\frac{\Omega}{c}\right)^2\varepsilon(\Omega,z)-k_\parallel^2\right]\mathcal{E}_z(z,\Omega)=-\frac{4\pi\Omega^2}{c^2}\wp_z(z,\Omega)\,. \quad \text{(C.4c)}$$

In (C.4a), we have an s polarized electromagnetic disturbance, generated by a source polarized normal to the xz plane, while the combination (C.4b,c) describe p polarized fields, in response to a source parallel to the xz plane.

In what follows, we shall assume $\varepsilon(\Omega,z)\to\varepsilon_>(\Omega)$ as $z\to\infty$, where $\varepsilon_>(\Omega)$ is complex, with imaginary part that is necessarily positive. Also $\varepsilon(\Omega,z)\to\varepsilon_<(\Omega)$ as $z\to-\infty$, where $\varepsilon_<(\Omega)$ is complex as well. We assume the transition from $\varepsilon_<$ to $\varepsilon_>$ takes place in a selvage region of width D, as illustrated in Fig. C.1. We do not need to specify the precise behavior of $\varepsilon(\Omega,z)$ in the selvage region.

A solution of (C.4) is achieved readily by adopting the well-known Green's function method from mathematical physics [8.2]. We introduce a Green's function referred to here as $\mathcal{G}_{yy}(k_\parallel\Omega|z,z')$ which satisfies

$$\left[\frac{d^2}{dz^2}+\left\{\frac{\Omega^2}{c^2}\varepsilon(\Omega,z)-k_\parallel^2\right\}\right]\mathcal{G}_{yy}(k_\parallel\Omega|z,z')=4\pi\delta(z-z') \quad \text{(C.5)}$$

and boundary conditions noted below. Then our solution of (C.4a) is

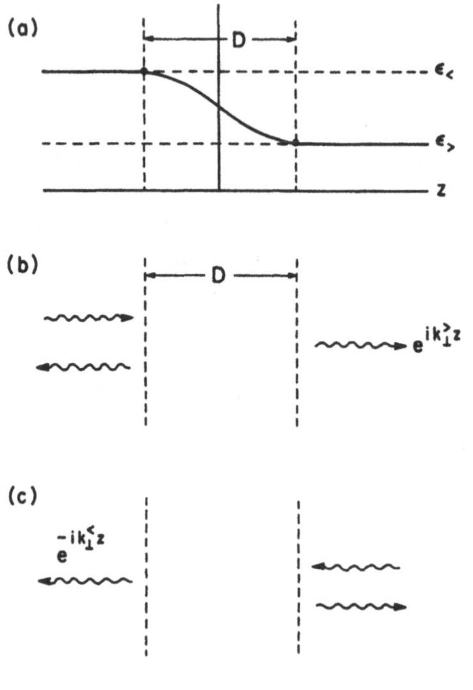

Fig. C.1. **a** The spatial variation assumed for the complex dielectric constant $\varepsilon(\Omega,z)$. Its variation in the selvedge region can be arbitrary. **b** A schematic representation of the field $\mathcal{E}_y^>(z,\Omega)$ which enters the Green's function $\mathcal{G}_{yy}(k_\parallel\omega|z,z')$. **c** A schematic illustration of the field $\mathcal{E}_y^<(z,\Omega)$ which enters the Green's function $\mathcal{G}_{yy}(k_\parallel\omega|zz')$

$$\mathcal{E}_y(z,\Omega) = -\left(\frac{\Omega}{c}\right)^2 \int_{-\infty}^{+\infty} \mathcal{G}_{yy}(k_\| \Omega|z,z') \wp_y(z',\Omega) dz' . \tag{C.6}$$

The boundary conditions on $\mathcal{G}_{yy}(k_\|\Omega|z,z')$ are found by noting that the Green's function describes the response to a source localized in the plane $z = z'$. Since $\varepsilon(\Omega,z)$ describes an absorptive medium, we require that for fixed z',

$$\lim_{z\to\infty} \mathcal{G}_{yy}(k_\|\Omega|z,z') = 0 \tag{C.7a}$$

and

$$\lim_{z\to-\infty} \mathcal{G}_{yy}(k_\|\Omega|z,z') = 0 . \tag{C.7b}$$

The method for constructing this Green's function is discussed in standard texts on mathematical physics [8.2]. One constructs this function as follows:

There are two linearly independent solutions to the homogeneous version of (C.5). We refer to these, chosen as described below, as $\mathcal{E}_y^>(k_\|\Omega|z)$ and $\mathcal{E}_y^<(k_\|\Omega|z)$. Each thus satisfies

$$\left[\frac{d^2}{dz^2} + \left\{\frac{\Omega^2}{c^2}\varepsilon(\Omega,z) - k_\|^2\right\}\right] \mathcal{E}_y^{>,<}(k_\|\Omega|z) = 0 . \tag{C.8}$$

As $z \to +\infty$, $\varepsilon(\Omega,z) \to \varepsilon_>$, and $\{\frac{\Omega^2}{c^2}\varepsilon(\Omega,z) - k_\|^2\} \to \{\frac{\Omega^2}{c^2}\varepsilon_> - k_\|^2\}$, independent of z. We then define $\mathcal{E}_y^>(k_\|\Omega|z)$ to be that solution for which

$$\lim_{z\to\infty} \mathcal{E}_y^>(k_\|\Omega|z) = \mathrm{e}^{ik_\perp^> z} , \tag{C.9}$$

where

$$k_\perp^> = \left(\frac{\Omega^2}{c^2}\varepsilon_> - k_\|^2\right)^{\frac{1}{2}} , \tag{C.10}$$

and we are to choose the complex square root in (C.10) so that $\mathrm{Im}(k_\perp^>) > 0$. Then $\mathcal{E}_y^>(k_\|\Omega|z) \to 0$ as $z \to \infty$.

Similarly, $\mathcal{E}_y^<(k_\|\Omega|z)$ is defined to be that solution of (C.8) for which

$$\lim_{z\to\infty} \mathcal{E}_y^<(k_\|\Omega|z) = e^{-ik_\perp^< z} , \tag{C.11}$$

where

$$k_\perp^< = \left(\frac{\Omega^2}{c^2}\varepsilon_< - k_\|^2\right)^{\frac{1}{2}} \tag{C.12}$$

and again $\mathrm{Im}(k_\perp^<) > 0$. Thus, by construction, $\mathcal{E}_y^<(k_\|\Omega|z) \to 0$ as $z \to -\infty$.

The physical nature of the two solutions of (C.8) just introduced is illustrated in Fig. C.1b,c. The field $\mathcal{E}_y^>(k_\|\Omega|z)$ describes a wave transmitted through the selvedge region in response to a wave incident from the left. A consequence is that we have a reflected wave on the left. Similarly, $\mathcal{E}_y^<(k_\|\Omega|z)$ describes a wave incident from the right.

Once these solutions are constructed, then [8.2]

$$\mathcal{G}_{yy}(k_\parallel\omega|zz') = \frac{4\pi}{W_s(k_\parallel\Omega)}\left[\mathcal{E}_y^>(k_\parallel\omega|z)\mathcal{E}_y^<(k_\parallel\omega|z')\theta(z-z')\right.$$
$$\left. + \mathcal{E}_y^<(k_\parallel\omega|z')\mathcal{E}_y^>(k_\parallel\omega|z)\theta(z-z')\right] , \tag{C.13}$$

where

$$W_s(k_\parallel\Omega) = \left(\frac{d\mathcal{E}_y^>}{dz}\right)\mathcal{E}_y^< - \left(\frac{d\mathcal{E}_y^<}{dz}\right)\mathcal{E}_y^> . \tag{C.14}$$

Here once again $\theta(x)$ is the Heaviside step function, equal to unity for positive values of its argument, and zero for negative values. Differentiation of $W_s(k_\parallel, \Omega)$ with respect to z, then use of (C.8) demonstrates this object (the Wronskian formed from $\mathcal{E}_y^>$ and $\mathcal{E}_y^<$) is independent of z.

We conclude with one technical comment. It is crucial to choose the proper square roots in both (C.10,12) to obtain physically meaningful results from the formalism. Often one wishes to apply the formalism to the case where one medium, say that in the upper half space, is vacuum. Then $\varepsilon_> = 1$, a purely real number, and $k_\perp^>$ is purely real when $(\Omega/c) > k_\parallel$. It is useful in this case to let $\varepsilon_> \to 1+i\eta$, where η is regarded as a positive infinitesimal. Then when $(\Omega/c) > k_\parallel$, the complex root for which $\text{Im}(k_\perp^>) > 0$ describes a plane wave propagating away from the interface. This is the physically meaningful solution, in the limit $\varepsilon_> \to 1$. Similar remarks apply to the treatment of $\varepsilon_<$, if one wishes the medium for large negative z to be the vacuum. In numerical studies based on use of this formalism, the author will keep η very small but finite, possibly in the range of 10^{-3} or 10^{-4}.

We may solve the coupled pair of equations, (C.4b,c), by a suitable extension of the Green's function formalism to multicomponent fields. We first introduce two objects $\mathcal{G}_{xx}(k_\parallel\Omega|zz')$ and $\mathcal{G}_{zx}(k_\parallel\Omega|zz')$ which satisfy the pair of coupled equations

$$\left[\frac{d^2}{dz^2} + \left(\frac{\Omega}{c}\right)^2 \varepsilon(\Omega, z)\right]\mathcal{G}_{xx}(k_\parallel\Omega|zz') - ik_\parallel\frac{d}{dz}\mathcal{G}_{zx}(k_\parallel\Omega|zz') = 4\pi\delta(z-z') , \tag{C.15a}$$

$$-ik_\parallel\frac{d\mathcal{G}_{xx}}{dz}(k_\parallel\Omega|zz') + \left[\frac{\Omega^2}{c^2}\varepsilon(\Omega, z) - k_\parallel^2\right]\mathcal{G}_{zx}(k_\parallel\Omega|zz') = 0 , \tag{C.15b}$$

and then two additional objects $\mathcal{G}_{xz}(k_\parallel\Omega|zz')$ and $\mathcal{G}_{zz}(k_\parallel\Omega|zz')$ generated from

$$\left[\frac{d^2}{dz^2} + \left(\frac{\Omega}{c}\right)^2 \varepsilon(\Omega, z)\right]\mathcal{G}_{xz}(k_\parallel\Omega|zz') - ik_\parallel\frac{d}{dz}\mathcal{G}_{zz}(k_\parallel\Omega|zz') = 0 \tag{C.16a}$$

and

$$-ik_\parallel\frac{d}{dz}\mathcal{G}_{xz}(k_\parallel\Omega|zz') + \left[\frac{\Omega^2}{c^2}\varepsilon(\Omega, z) - k_\parallel^2\right]\mathcal{G}_{zz}(k_\parallel\Omega|zz') = 4\pi\delta(z-z') . \tag{C.16b}$$

The set of four equations, (C.15,16), are the explicit forms of (8.26) of the main text, written out in detail. All four of the functions just introduced vanish as $z \to +\infty$, and as $z \to -\infty$, as in the boundary conditions stated in (C.7).

Once these functions are in hand, it is a straightforward matter to demonstrate that the solution of (C.4b,c) has the form

$$\mathcal{E}_\alpha(z,\Omega) = -\left(\frac{\Omega}{c}\right)^2 \sum_\beta \int_{-\infty}^{+\infty} \mathcal{G}_{\alpha\beta}(k_\parallel \Omega|zz')\wp_\beta(z',\Omega)dz' \; , \tag{C.17}$$

where α and β range over x and z.

We may write the four functions just introduced in terms of two p polarized electric fields $[\mathcal{E}_x^>(k_\parallel\Omega|z), \mathcal{E}_z^>(k_\parallel\Omega|z)]$, and $[\mathcal{E}_x^<(k_\parallel\Omega|z), \mathcal{E}_z^<(k_\parallel\Omega|z)]$ which are solutions of the homogeneous versions of (C.15,16). We have

$$-ik_\parallel \frac{d}{dz}\mathcal{E}_x^{>,<}(k_\parallel\Omega|z) + \left[\frac{\Omega^2}{c^2}\varepsilon(\Omega,z) - k_\parallel^2\right]\mathcal{E}_z^{>,<}(k_\parallel\Omega|z) = 0 \tag{C.18a}$$

and

$$\left[\frac{d^2}{dz^2} + \frac{\Omega^2}{c^2}\varepsilon(\Omega,z)\right]\mathcal{E}_x^{>,<}(k_\parallel\Omega|z) - ik_\parallel\frac{d}{dz}\mathcal{E}_z^{>,<}(k_\parallel\Omega|z) = 0 \; . \tag{C.18b}$$

We append the following boundary conditions to the fields:

$$\lim_{z\to\infty} \mathcal{E}_z^>(k_\parallel\Omega|z) = e^{ik_\perp^> z} \tag{C.19a}$$

and

$$\lim_{z\to\infty} \mathcal{E}_z^<(k_\parallel\Omega|z) = e^{-ik_\perp^< z} \; . \tag{C.19b}$$

Once we impose the conditions in (C.19), then the behavior of $\mathcal{E}_x^>$ and $\mathcal{E}_x^<$ as $z \to +\infty$, and $z \to -\infty$, respectively, is fixed by (C.18a) and/or (C.18b). One has

$$\lim_{z\to\infty} \mathcal{E}_x^>(k_\parallel\Omega|z) = -\frac{k_\perp^>}{k_\parallel} e^{+ik_\perp^> z} \tag{C.20a}$$

and

$$\lim_{z\to\infty} \mathcal{E}_x^<(k_\parallel\Omega|z) = +\frac{k_\perp^<}{k_\parallel} e^{-ik_\perp^< z} \; . \tag{C.20b}$$

When these fields are constructed, one may verify that the Green's functions can be written as follows:

$$\begin{aligned}\mathcal{G}_{xx}(k_\parallel\Omega|z,z') &= \frac{4\pi}{W_p(k_\parallel\Omega)}\big[\mathcal{E}_x^>(k_\parallel\Omega|z)\mathcal{E}_x^<(k_\parallel\Omega|z')\theta(z-z') \\ &\quad + \mathcal{E}_x^<(k_\parallel\Omega|z)\mathcal{E}_x^>(k_\parallel\Omega|z')\theta(z'-z)\big] \; , \end{aligned} \tag{C.21a}$$

$$\begin{aligned}\mathcal{G}_{zx}(k_\parallel\Omega|z,z') &= \frac{4\pi}{W_p(k_\parallel,\Omega)}\big[\mathcal{E}_z^>(k_\parallel\Omega|z)\mathcal{E}_x^<(k_\parallel\Omega|z')\theta(z-z') \\ &\quad + \mathcal{E}_z^<(k_\parallel\Omega|z)\mathcal{E}_x^>(k_\parallel\Omega|z')\theta(z'-z)\big] \; , \end{aligned} \tag{C.21b}$$

$$\begin{aligned}\mathcal{G}_{xz}(k_\parallel\Omega|z,z') &= -\frac{4\pi}{W_p(k_\parallel,\Omega)}\big[\mathcal{E}_x^>(k_\parallel\Omega|z)\mathcal{E}_z^<(k_\parallel\Omega|z')\theta(z-z') \\ &\quad + \mathcal{E}_x^<(k_\parallel\Omega|z)\mathcal{E}_z^>(k_\parallel\Omega|z')\theta(z'-z)\big] \; , \end{aligned} \tag{C.21c}$$

and finally

$$\begin{aligned}\mathcal{G}_{zz}(k_\parallel \Omega|z,z') &= \frac{4\pi c^2}{\Omega^2\varepsilon(z)}\delta(z-z') \\ &\quad - \frac{4\pi}{W_p(k_\parallel,\Omega)}\left[\mathcal{E}_z^>(k_\parallel\Omega|z)\mathcal{E}_z^<(k_\parallel\Omega|z')\theta(z-z')\right. \\ &\quad \left.+\mathcal{E}_z^<(k_\parallel\Omega|z)\mathcal{E}_z^>(k_\parallel\Omega|z')\theta(z'-z)\right] .\end{aligned} \tag{C.21d}$$

There are several equivalent forms for $W_p(k_\parallel, \Omega)$, which as is the case with its counterpart in (C.14), may be demonstrated to be independent of z. Direct substitution of (C.21a,b) into (C.15a,b) gives

$$\begin{aligned}W_p(k_\parallel,\Omega) &= \left[\frac{d\mathcal{E}_x^>(k_\parallel\Omega|z)}{dz} - ik_\parallel\mathcal{E}_z^>(k_\parallel\Omega|z)\right]\mathcal{E}_x^<(k_\parallel\Omega|z) \\ &\quad - \left[\frac{d\mathcal{E}_x^<(k_\parallel\Omega|z)}{dz} - ik_\parallel\mathcal{E}_z^<(k_\parallel\Omega|z)\right]\mathcal{E}_x^>(k_\parallel\Omega|z) ,\end{aligned} \tag{C.22}$$

which is equal to the more compact form

$$W_p(k_\parallel,\Omega) = \mathrm{i}\frac{\Omega^2\varepsilon(\Omega,z)}{k_\parallel c^2}\left[\mathcal{E}_z^<(k_\parallel\Omega|z)\mathcal{E}_x^>(k_\parallel\Omega|z) - \mathcal{E}_z^>(k_\parallel\Omega|z)\mathcal{E}_x^<(k_\parallel\Omega|z)\right] . \tag{C.23}$$

One has also a form quoted earlier [8.2]

$$W_p(k_\parallel,\Omega) = W_{xx}(k_\parallel,\Omega) - W_{zz}(k_\parallel,\Omega) , \tag{C.24}$$

where $W_{xx}(k_\parallel, \Omega)$ and $W_{zz}(k_\parallel, \Omega)$ are calculated as in (C.14), but with y replaced by x and z, respectively.

As remarked at the outset of this discussion, we have made no specific assumption about $\varepsilon(\Omega, z)$, save for the asymptotic behaviors illustrated in Fig. C.1. Thus, we can apply the above formulae to the study of the electromagnetic response of dielectric multilayers, surfaces or interfaces where we may model the dielectric function as a continuous function of z. Our primary task is to construct the fields $\boldsymbol{E}^{>,<}(k_\parallel\Omega|z)$, and we can then form the complete response functions from the procedures just outlined.

For the example in Sect. 8.1 of the text, we require the Green's functions near the surface of a semi-infinite dielectric, where we have vacuum in the region $z > 0$, and a dielectric with dielectric constant $\varepsilon(\Omega)$ when $z < 0$. We conclude by constructing the various fields for this example. We have the following expressions for this particular case. One has

$$k_\perp^< = \left(\frac{\Omega^2}{c^2}\varepsilon(\Omega) - k_\parallel^2\right)^{\frac{1}{2}} \tag{C.25a}$$

and with our earlier remarks on handling the vacuum region in mind,

$$k_\perp^> = \left(\frac{\Omega^2}{c^2}[1+i\eta] - k_\parallel^2\right)^{\frac{1}{2}} . \tag{C.25b}$$

Then

$$\mathcal{E}_y^>(k_\parallel \Omega|z) = \begin{cases} e^{ik_\perp^> z} & , \quad z > 0 \\ A_s^{(+)} e^{ik_\perp^< z} + A_s^{(-)} e^{ik_\perp^< z} & , \quad z < 0, \end{cases} \tag{C.26a}$$

$$\mathcal{E}_y^<(k_\parallel \Omega|z) = \begin{cases} B_s^{(+)} e^{ik_\perp^> z} + B_s^{(-)} e^{-ik_\perp^> z} & , \quad z > 0 \\ e^{-ik_\perp^< z} & , \quad z < 0 \end{cases} \tag{C.26b}$$

where

$$A_s^{(\pm)} = \frac{1}{2k_\perp^<}\left(k_\perp^< \pm k_\perp^>\right) , \tag{C.27a}$$

$$B_s^{(\pm)} = \frac{1}{2k_\perp^>}\left(k_\perp^> \mp k_\perp^<\right) , \text{ and} \tag{C.27b}$$

$$W_s(k_\parallel, \Omega) = i\left(k_\perp^> + k_\perp^<\right) . \tag{C.28}$$

Then for the p polarized fields,

$$\mathcal{E}_z^>(k_\parallel \Omega|z) = \begin{cases} e^{ik_\perp^> z} & , \quad z > 0 \\ A_p^{(+)} e^{ik_\perp^< z} + A_p^{(-)} e^{-ik_\perp^< z} & , \quad z < 0 , \end{cases} \tag{C.29a}$$

and

$$\mathcal{E}_x^>(k_\parallel \Omega|z) = \begin{cases} -\frac{k_\perp^>}{k_\parallel} e^{ik_\perp^> z} & , \quad z > 0 \\ -\frac{k_\perp^<}{k_\parallel}\left(A_p^{(+)} e^{ik_\perp^< z} - A_p^{(-)} e^{-ik_\perp^< z}\right) & , \quad z < 0 , \end{cases} \tag{C.29b}$$

where

$$A_p^{(\pm)} = \frac{1}{2k_\perp^< \varepsilon(\Omega)}\left[k_\perp^< \pm \varepsilon(\Omega) k_\perp^>\right] \tag{C.30}$$

while

$$\mathcal{E}_z^<(k_\parallel \Omega|z) = \begin{cases} B_p^{(+)} e^{ik_\perp^> z} + B_p^{(-)} e^{-ik_\perp^> z} & , \quad z > 0 \\ e^{-ik_\perp^< z} & , \quad z < 0 , \end{cases} \tag{C.31a}$$

and

$$\mathcal{E}_x^<(k_\parallel \Omega|z) = \begin{cases} -\frac{k_\perp^>}{k_\parallel}\left(B_p^{(+)} e^{ik_\perp^> z} - B_p^{(-)} e^{-ik_\perp^> z}\right) & , \quad z > 0 \\ +\frac{k_\perp^<}{k_\parallel} e^{-ik_\perp^< z} & , \quad z < 0 , \end{cases} \tag{C.31b}$$

where

$$B_p^{(\pm)} = \frac{1}{2k_\perp^>}\left[\varepsilon(\Omega) k_\perp^> \mp k_\perp^<\right] \tag{C.32}$$

and finally

$$W_p(k_\parallel, \Omega) = \frac{\mathrm{i}\Omega^2}{c^2 k_\parallel^2} \left[\varepsilon(\Omega) k_\perp^> + k_\perp^<\right] . \tag{C.33}$$

We have discussed the use of the electromagnetic Green's functions as a means of generating descriptions of the linear and nonlinear optical response of materials, near surfaces and interfaces. The fluctuation-dissipation theorem of statistical mechanics [C.1] allows one to use these functions to describe thermodynamic fluctuations of the electric field, in the near vicinity of surfaces and interfaces. Elegant applications of this scheme can be found in the classic text by *Abrikosov* et al. [C.2]. As these authors demonstrate, even though these Green's functions have been obtained by solving the Maxwell equations of classical physics, they can be used to describe the thermodynamic fluctuations in a fully quantum theoretic manner, so long as one can sensibly describe the media of interest by dielectric response tensors. The author and his colleagues have used this formalism to obtain complete and fully quantitative accounts of Raman scattering from semiconducting films and surfaces [C.3], and also Brillouin scattering from magnetic media [C.4, 5].

References

Chapter 1

1.1 J.D. Jackson: *Classical Electrodynamics*, 1st edn. (Wiley, New York 1962) p. 103ff.
1.2 Y.R. Shen: *The Principles of Nonlinear Optics* (Wiley, New York 1984)
1.3 C. Kittel: *Introduction to Solid State Physics* 5th edn. (Wiley, New York 1976) p. 413
1.4 J.D. Jackson: *Classical Electrodynamics*, 2nd edn. (Wiley, New York 1975) Chaps. 1–4
1.5 L.D. Landau, E.M. Lifshitz: *Theory of Elasticity* (Pergamon, Oxford 1959) Sect. 10, p. 36
1.6 L.D. Landau, E.M. Lifshitz: *Electrodynamics of Continuous Media* (Pergamon, Oxford 1960) Chap. 9
1.7 J.A. Armstrong, N. Bloembergen, J. Ducuing, P.S. Pershan: Phys. Rev. **127**, 1918 (1962)

Chapter 2

2.1 V.M. Agranovich, V.L. Ginzburg: *Crystal Optics with Spatial Dispersion and Excitons*, 2nd edn., Springer Ser. Solid-State Sci., Vol. 42 (Springer, Berlin, Heidelberg 1984)
2.2 J.D. Jackson: *Classical Electrodynamics*, 2nd edn. (Wiley, New York 1975) Sect. 6.8, p. 236
2.3 L.D. Landau, E.M. Lifshitz: *Electrodynamics of Continuous Media* (Pergamon, Oxford 1960)
2.4 J.M. Stone; *Radiation and Optics* (McGraw-Hill, New York 1963)
2.5 E. Merzbacher: *Quantum Mechanics*, 2nd edn. (Wiley, New York 1970) p. 85
2.6 [Ref. 2.5, Chap. 2]
2.7 L. Brillouin: *Wave Propagation and Group Velocity* (Academic, New York 1960)
2.8 E. Burstein: Introductory Remarks, in *Polaritons*, ed. by E. Burstein, F. de Martini (Pergamon, New York 1974)
2.9 D.L.Mills, E. Burstein: Repts. Prog. Phys. **37**, 817 (1972)
2.10 C. Kittel: *Introduction to Solid State Physics*, 2nd edn. (Wiley, New York 1953) Chap. 10
2.11 [Ref. 2.10, p. 112ff.]
2.12 See the review article by D.L. Mills, C.J. Duthler, M. Sparks: In *Dynamical Properties of Solids*, Vol 4, ed. by G.K. Horton, A.A. Maradudin (North-Holland, Amsterdam 1980) Chap. 4
2.13 [Ref. 2.5, p. 458ff.]
2.14 C. Kittel: *Quantum Theory of Solids* (Wiley, New York 1963) Chap. 16

Chapter 3

3.1 L.D. Landau, E.M. Lifshitz: *Electrodynamics of Continuous Media* (Pergamon, Oxford 1960) p. 259ff.
3.2 N. Bloembergen: *Nonlinear Optics* (Benjamin, New York 1965) p. 5
3.3 J.A. Armstrong, N. Bloembergen, J. Ducuing, P.S. Pershan: Phys. Rev. **127**, 1918 (1962)
3.4 Y.R. Shen: *The Principles of Nonlinear Optics* (Wiley, New York 1984)

Chapter 4

4.1 P.D. Maker, R.W. Terhune, M. Niesenoff, C.M. Savage: Phys. Rev. **8**, 21 (1962)
4.2 D.M. Bloom, G.W. Bekkers, J.F. Young, S.E. Harris: Appl. Phys. Lett. **26**, 687 (1975)
4.3 D.M. Bloom, J.F. Young, S.E. Harris: Appl. Phys. Lett. **27**, 390 (1975)
4.4 N. Bloembergen: *Nonlinear Optics* (Benjamin, New York 1965) p. 11
4.5 M.D. Levenson: Phys. Today **30**, 44 (1977)
4.6 W.E. Bron, J. Kuhl, B.K. Rhee: Phys. Rev. B **34**, 6961 (1986)
4.7 C.R. Guiliano: Phys. Today **34**, 27 (April 1980)

Chapter 5

5.1 H. Goldstein: *Classical Mechanics* (Addison Wesley, Reading, MA 1950) Chap. 10
5.2 K. Huang, M. Born: *Dynamical Theory of Crystal Lattices* (Clarendon, Oxford 1954) p. 57
5.3 A.D. Boardman, D.E. O'Connor, A.P.Young: *Symmetry and Its Applications in Science* (Wiley, New York 1973) Chap. 7
5.4 J.D. Jackson: *Classical Electrodynamics* 1st edn. (Wiley, New York 1973) p. 272
5.5 R.W. Woods: Philos. Mag. **6**, 729 (1928)
5.6 G. Baym: *Lectures on Quantum Mechanics* (Benjamin, New York 1969) p. 271
5.7 [Ref. 5.4, p. 123ff.]
5.8 [Ref. 5.6, p. 276ff.]
5.9 Y.R. Shen: *The Principles of Nonlinear Optics* (Wiley, New York 1984) Chap. 10
5.10 M.D. Levenson, N. Bloembergen: J. Chem. Phys. **60**, 1323 (1974)
5.11 L.D. Landau, E.M. Lifshitz: *Theory of Elasticity* (Pergamon, Oxford 1959) Chap. 1
5.12 D.F. Nelson, M. Lax: Phys. Res. Lett. **34**, 379 (1970)
5.13 [Ref. 5.11, Chap. 3]
5.14 C. Kittel: *Quantum Theory of Solids* (Wiley, New York 1963)
5.15 G.B. Wright (ed.): *Light Scattering Spectra of Solids* (Springer, Berlin, Heidelberg 1969)

Chapter 6

6.1 G. Baym: *Lectures on Quantum Mechanics* (Benjamin, New York 1969)
6.2 A.A. Abragam: *Principles of Nuclear Magnetic Resonance* (Clarendon, Oxford 1962)
6.3 R. Loudon: *Quantum Theory of Light* (Oxford Univ. Press, London 1973)
6.4 S.L. McCall, E.L.Hahn: Phys. Rev.Lett. **18**, 908 (1967); Phys. Rev.Lett. **183**, 457 (1969)
6.5 J.D. Jackson: *Classical Electrodynamics*, 2nd edn. (Wiley, New York 1975) p. 299
6.6 K. Leung, D. Hone, P. Riseborough, S.E. Trullinger, D.L. Mills: Phys. Rev. B **21**, 4017 (1980)
6.7 [Ref. 6.1, Chap. 22]
6.8 G.L.Lamb: *Elements of Soliton Theory* (Wiley, New York 1980)

Chapter 7

7.1 A. Yariv: *Quantum Electronics*, 2nd edn. (Wiley, New York 1975) p. 498ff.
7.2 M. Abramowitz, I.A. Stegun: *Handbook of Mathematical Functions*, NBS, Applied Mathematics Series No. 55 (National Bureau of Standards, U.S. Government Printing Office, Washington, DC 1965) p. 374
7.3 [Ref. 7.2, p. 355ff.]
7.4 J.D. Jackson: *Classical Electrodynamics*, 2nd edn. (Wiley, New York 1975) p. 232
7.5 L.F. Mollenauer: Philos. Trans. R. Soc. London A **315**, 437 (1985)
7.6 V.E. Zakharov, A.V. Mikailov: Soc. Phys.–JETP **47**, 1017 (1978)
7.7 J. Satsuma, N. Yajima: Prog. Theor. Phys., Suppl. **55**, 284 (1974)
7.8 L.F. Mollenauer, R.H. Stolen, J.P. Gordon: Phys. Rev. Lett. **45**, 1095 (1988)

7.9 B. Nikolaus, D. Grischkowski: Appl. Phys. Lett. **43**, 228 (1983)
7.10 Wei Chen, D.L. Mills: Phys. Rev. Lett. **58**, 160 (1987)
7.11 U. Mohideen et al.: Opt. Lett. **20**, 1674 (1995)
7.12 C. Kittel: *Quantum Theory of Solids* (Wiley, New York 1963) Chap. 10
7.13 D.L. Mills, S.E. Trullinger: Phys. Rev. Lett. B **36**, 947 (1987)
7.14 D.K. Campbell, M. Peyrard, P. Sodano: Physica **190**, 165 (1986)
7.15 Wei Chen, D.L. Mills: Phys. Rev. B **36**, 6269 (1987)
7.16 C.M. de Sterke, J.E. Sipe: Phys. Rev. A **38**, 5149 (1988)

Chapter 8

8.1 J.D. Jackson: *Classical Electrodynamics*, 2nd edn. (Wiley, New York 1975) p. 278ff.
8.2 J.T. Cushing, *Applied Analytical Mathematics for Physical Scientists* (Wiley, New York 1975) Sect. 8.5
8.3 For a discussion of the construction of the relevant Green's functions, and application to a multilayer geometry, see D.L. Mills, A.A. Maradudin: Phys. Rev. B **12**, 2943 (1975)
8.4 T.F. Heinz: In *Nonlinear Surface Electromagnetic Phenomena*, ed. by H.E. Ponath, G.I. Stegeman, Modern Problems in Condensed Matter Sciences, Vol. 29 (North-Holland, Amsterdam 1991) p. 353
8.5 H.M. van Driel: Appl. Phys. A **59**, 545 (1994)
8.6 J. Rudnick, E. Stern: Phys. Rev. B **4**, 4274 (1971)
8.7 J.C. Quail, H.C. Simon: Phys. Rev. B **31**, 4900 (1985)
8.8 M.G. Weber, A.Liebsch: Phys. Rev. B **37**, 6187 (1988); A. Liebsch, W.L. Schaich: Phys. Rev. B **40**, 5401 (1989)
8.9 R. Murphy, M. Yeganch, K.J. Song, E.W. Plummer: Phys. Rev. Lett. **63**, 318 (1989)
8.10 T.F. Heinz, M.M.T. Loy, W.A. Thompson: Phys. Rev. Lett. **54**, 63 (1985)
8.11 H.W.K. Tom, C.M. Mate, X.D. Zhu, J.E. Crowell, T.F. Heinz, G.A. Somorjai, Y.R. Shen: Phys. Rev. Lett. **52**, 348 (1984)
8.12 X. Zhuang, D. Wilk, L. Marruci, Y.R. Shen: Phys. Rev. Lett. **75**, 2144 (1995)
8.13 J.D. Byers, H.I. Yee, J.M. Hicks: J. Chem. Phys. **101**, 6233 (1994)
8.14 W. Daum, K.A. Friedrich, C. Klünker, D. Kuabhen, U. Stimmung, H. Ibach: Appl. Phys. A **59**, 553 (1994)
8.15 [Ref. 8.1, Sect. 3.4]
8.16 [Ref. 8.1, p. 149ff.]
8.17 J.D. Eversole, H.P. Broida: Phys. Rev. B **15**, 1644 (1977)
8.18 L.D. Landau, E.M. Lifshitz: *Electrodynamics of Continuous Media* (Pergamon, Oxford 1960), Sect. 8, p. 42
8.19 [Ref. 8.1, Eq. (9.24), p. 396]
8.20 *Surface Polaritons; Electromagnetic Waves at Surfaces and Interfaces*, ed. by V.M. Agranovich, D.L. Mills (North-Holland, Amsterdam 1982)
8.21 A.A. Maradudin, D.L. Mills: Phys. Rev. B **11**, 1392 (1975). This paper contains a small technical error corrected in a most interesting argument by G.S. Agarwal: Phys. Rev. B **14**, 846 (1976). A study of the influence of surface polaritons on the reflectivity of oxidized aluminum surfaces has been given by D.L. Mills, A.A. Maradudin: Phys. Rev. B **12**, 2943 (1975). A detailed and careful summary of the theory has been given by A.A. Maradudin [Ref. 8.20, p. 405]
8.22 D.L. Mills: Phys. Rev. B **12**, 4036 (1975)
8.23 M. Weber, D.L. Mills: Phys. Rev. B **26**, 1075 (1982); ibid. B **27**, 2698 (1983)
8.24 M. Fleischmann, P.J. Hendra, A.H. McQuillan: Chem. Phys. Lett. **26**, 163 (1974)
8.25 *Surface Enhanced Raman Scattering*, ed. by R.K. Chang and T.E. Furtak (Plenum, New York 1982)
8.26 A. Otto: J. Raman Spectrosc. **22**, 743 (1991)
8.27 J.C. Tsang, J.R. Kirtley, T.N. Theis: Solid State Commun. **35**, 667 (1980)
8.28 H. Sano, S. Ushioda: Phys. Rev. B **53**, 1958 (1996)

8.29 C.K. Chen, A.R.B. de Castro, Y.R. Shen: Phys. Rev. Lett. **46**, 145 (1981)

8.30 G.A. Farias, A.A. Maradudin: Phys. Rev. B **30**, 3002 (1984)

Chapter 9

9.1 C. Kittel: *Introduction to Solid State Physics*, 2nd edn. (Wiley, New York 1956) Chap. 15

9.2 E. Burstein, D.L. Mills: Reports on Progress in Physics **37**, 317 (1974) Sect. IV

9.3 L.D. Landau, E.M. Lifshitz: *Statistical Physics*, 3rd edn., Part 1 (Pergamon, Oxford 1980) Sect. 120, p. 365

9.4 M. Tinkham: *Group Theory and Quantum Mechanics* (McGraw-Hill, New York 1964) p. 142

9.5 C. Lin, S.D. Bader: J. Appl. Phys. **67**, 5758 (1990)

9.6 S.D. Bader: J. Magn. Magn. Mater. **100**, 440 (1991)

9.7 R.P. Pan, H.D. Wei, Y.R. Shen: Phys. Rev. B **39**, 1229 (1989)

9.8 J. Reif, J.C. Zink, C.M. Schneider, J. Kirschner: Phys. Rev. Lett. **67**, 2878 (1991)

9.9 R. Stolle, K.J. Veenstra, F. Mandeco, Th. Rasing, H. van den Berg, N. Persat: Phys. Rev. B **55**, R4925 (1997)

9.10 D.L. Mills: "Surface Spin Waves on Magnetic Crystals," Chap. 3 of *Surface Excitations*, ed. by V.M. Agranovich and R. Loudon (Elsevier, Amsterdam 1984) Sect. 3.2

9.11 R. Damon, J. Eshbach: J. Chem. Phys. Solids **19**, 308 (1961)

9.12 R.Q. Scott, D.L. Mills: Phys. Rev. **15**, 3545 (1977)

9.13 L.D. Landau, E.M. Lifshitz: *Theory of Elasticity*, 2nd edn. (Pergamon Press, Oxford 1970) p. 101ff.

9.14 J.J. Brion, R.F. Wallis, A. Hartstein, E. Burstein: Phys. Rev. Lett. **28**, 1455 (1972)

9.15 R.E. Camley, D.L. Mills: Phys. Rev. B **26**, 1280 (1982)

9.16 M.R.F. Jensen, S.A. Feiven, T.J. Parker, R.E. Camley: Phys. Rev. B **55**, 2745 (1997)

9.17 B. Laks, D.L. Mills: Phys. Rev. B **22**, 4445 (1980)

9.18 For a review, see E.F. Schloemann: Proc. IEEE **76**, 188 (1988)

9.19 C. Kittel: *Quantum Theory of Solids* (Wiley, San Francisco 1973) Chap. 4

9.20 J. Sandercock: J. Appl. Phys. **50**, 7784 (1979)

9.21 R.E. Camley, D.L. Mills: Phys. Rev. B **18**, 4821 (1978)

9.22 R.E. Camley, T.S. Rahman, D.L. Mills: Phys. Rev. B **23**, 1226 (1981)

9.23 Chen S. Tsai, D. Young: IEEE Trans. MTT **38**, 560 (1990)

Chapter 10

10.1 E.N. Lorenz: J. Atmos. Sci. **20**, 130 (1963)

10.2 B.A. Huberman, J.P. Crutchfield: Phys. Rev. Lett. **43**, 1743 (1979)

10.3 J.R. Ackerhalt, P.W. Milonni, M.L. Shih: Phys. Rep. **128**, 207 (1985)

10.4 J.P. Eckerman: Rev. Mod. Phys. **53**, 463 (1981)

10.5 M. Feigenbaum: J. Stat. Phys. **19**, 25 (1978); ibid. **21**, 665 (1979); Commun. Math. Phys. **77**, 65 (1980)

10.6 Paul S. Linsay: Phys. Rev. Lett. **47**, 1349 (1981)

10.7 D. Ruelle, F. Takens: Commun. Math. Phys. **20**, 167 (1971)

10.8 Y. Pomeau, P. Manneville: Commun. Math. Phys. **77**, 189 (1980)

10.9 R. Loudon: *Quantum Theory of Light* (Clarendon, Oxford 1973)

10.10 L.W. Casperson: J. Opt. Soc. Am. B **2**, 62, 73 (1985)

10.12 C.O. Weiss, A. Godone, A. Olafsson: Phys. Rev. A **28**, 892 (1983)

10.13 Y.R. Shen: *The Principles of Nonlinear Optics* (Wiley, New York 1984)

10.14 Wei Chen, D.L. Mills: Phys. Rev. **58**, 160 (1987)

10.15 H. Nakatsuka, S. Asaka, H. Itoh, K. Ikeda, M. Matsuoka: Phys. Rev. **50**, 109 (1983)

10.16 H.M. Gibbs, F.A. Hopf, D.L. Kaplan, R.L. Shoemaker: Phys. Rev. **46**, 474 (1981)

Appendices

A.1 J.A. Armstrong, N. Bloembergen, J. Ducuing, P.S. Pershan: Phys. Rev. **127**, 1918 (1962)
B.1 R. Hirota: J. Phys. Soc. Jpn. **33**, 1459 (1972)
B.2 A.C. Scott, F.Y.F. Chen, D. McLaughlin: Proc. IEEE **61**, 1443 (1973)
B.3 J.F. Currie: Phys. Rev. A **16**, 1692 (1977)
B.4 R. Rajaraman: *Solitons and Instantons* (North-Holland, Amsterdam 1982)
C.1 L.D. Landau, E.M. Lifshitz: *Statistical Physics* (Addison Wesley, London 1958) p. 350ff.
C.2 A.A. Abrikosov, L.P. Gor'kov, I.E. Dzyaloshinskii: *Methods of Quantum Field Theory in Statistical Physics* (Prentice Hall Inc., Englewood Cliffs, New Jersey 1963) Chap. 6
C.3 D.L. Mills, Y.J. Chen, E. Burstein: Phys. Rev. B **13**, 4419 (1976)
C.4 R.E. Camley, D.L. Mills: Phys. Rev. B **18**, 4821 (1978)
C.5 R.E. Camley, T.S. Rahman, D.L. Mills: Phys. Rev. B **23**, 1226 (1981)

Subject Index

Production: Druckhaus Beltz, Hemsbach

Springer and the environment

At Springer we firmly believe that an international science publisher has a special obligation to the environment, and our corporate policies consistently reflect this conviction.
We also expect our business partners – paper mills, printers, packaging manufacturers, etc. – to commit themselves to using materials and production processes that do not harm the environment. The paper in this book is made from low- or no-chlorine pulp and is acid free, in conformance with international standards for paper permanency.

Springer